D1671849

Microwave Mixers
Second Edition

For a complete listing of the *Artech House Microwave Library*,
turn to the back of this book

Microwave Mixers

Second Edition

Stephen A. Maas

Artech House
Boston • London

Library of Congress Cataloging-in-Publication Data

Maas, Stephen A.
Microwave mixers / Stephen A. Maas. — 2nd ed.
p. cm.
Includes bibliographical references and index.
ISBN 0-89006-605-1
1. Microwave mixers. I. Title.

TK7872.M5M33 1993 92-21762
621.381'33—dc20 CIP

© 1993 ARTECH HOUSE, INC.
685 Canton Street
Norwood, MA 02062

All rights reserved. Printed and bound in the United States of America. No part of this book may be reproduced or utilized in any form or by any means, electronic or mechanical, including photocopying, recording, or by any information storage and retrieval system, without permission in writing from the publisher.

International Standard Book Number: 0-89006-605-1
Library of Congress Catalog Card Number: 92-21762

10 9 8 7 6 5 4 3 2 1

To the memory of my father,
Joseph Henry Maas

Contents

Preface

I began writing the first edition of *Microwave Mixers* in 1985. I had been out of graduate school for a year at that point, and, as a confirmed workaholic, I was still having difficulty adjusting to the negative-going discontinuity in my workload. Writing a book solved the problem admirably. Not only was the task therapeutic, but it saved me from other therapies that my wife and employer were all too ready to prescribe. Anyway (how easily we digress!) I approached the task with a clear objective: to concentrate on the fundamentals and to avoid material that would quickly be outdated. This way, I expected the book to have a much greater lifetime than other books we all have seen, which concentrate on the state of the art in darling technologies, and seem to be little more than extended survey papers.

Well, it never occurred to me that the fundamentals could also be dated. People just don't do things today the way they did in 1985. Although it is a banal observation that hardware technology has advanced in the past seven or eight years, it is less obvious that the ways to analyze mixer circuits have also been streamlined. This is especially true of FET mixers; the theory presented in Chapter 9 of the first edition is now so dated that I chose simply to eliminate it from the second. In fact, Chapter 9 has been completely rewritten, with a much stronger emphasis on the intuitive aspects of the operation and design of such mixers, and a weaker emphasis on analysis. Indeed, with the availability of general-purpose harmonic-balance simulators, software products that simply did not exist in 1985, there is little need to concentrate on specific analytical techniques. Other chapters were similarly in need of updating.

One of the main reasons for preparing a second edition of a book is to do the things one wishes he had done in the first edition. Writing a book is easily the world's second strangest form of on-the-job training (teaching in a university is the first), and like other apprentices, an author often fails to get things the way he wants them first time. A second edition is, in effect, a second chance. This edition is considerably different from the first: it is completely rewritten, and virtually all the figures have been redrawn. It is, I believe, cleaner, more understandable, and easier to read and use.

Most important, of course, is the change in technical content. I made major changes to almost all of the chapters; only a couple of chapters escaped heavy revision. The question of what to retain from the original has occasionally been a tough one. Chapters 5 and 6 were modified only minimally; I have always liked them. I retained much of the material in Chapter 2 on dot-matrix diodes that could be considered obsolete; however, much of this represents the pinnacle of mixer-diode technology, and I was reluctant to abandon it. I did, however, include new material, especially a section on monolithic diodes. Similarly, Chapter 3, on MESFETs, has quite a bit of new material and concentrates on understanding and modeling these important devices.

Chapter 4 is far more comprehensive than its earlier incarnation; it includes much more specific information on the analysis of diode mixers. Chapter 7 has been criticized for its lack of specific design information for baluns; this deficiency has been rectified. Likewise, Chapter 8 contains more design information on baluns and modern balanced-mixer structures. Finally, Chapter 10 has been added to examine the most important technology in microwave engineering, monolithic circuits. I am optimistic that readers will find a good helping of new and valuable material.

Finally, some acknowledgments. I am indebted to my family for their patience while I pursued this project, especially to my wife, Julie, who is becoming accustomed to competing with a computer for my attention, and to my younger son David, who isn't. I must also thank my older son, Ben, for typing large amounts of text into the computer and serenading me with Shostakovich preludes as I wrote. I would also like to thank my students at UCLA, who read parts of the book and gleefully pointed out all the errors they could find.

S. A. Maas
Long Beach, CA
April, 1992

Chapter 1
Introduction

1.1 HISTORY AND FUNDAMENTALS

Credit for the invention of the mixer is generally given to Major Edwin Armstrong, one of the most creative inventors in the history of radio communication. Crude mixing processes had been employed previously, mostly in attempts to down-convert received signals directly to baseband. Because the stability of the local oscillator was usually inadequate, these techniques worked poorly. Armstrong, however, used vacuum-tube mixers to shift the received signal to an "intermediate frequency," where it could be amplified with good selectivity, high gain, and low noise, and finally demodulated. Armstrong is also credited with the invention of *frequency modulation* (FM). FM would be impossible without such receivers, because tunable FM demodulators were–and still are–very difficult to design. His *superheterodyne receiver* (he coined the name) is still the model for communication receivers in use today. It is found in receivers ranging in sophistication from cheap transistor radios to communication satellites.

The subsequent history of mixers is essentially the history of low-noise receivers. The need for high-quality microwave radar receivers became urgent during World War II, and was addressed in part by the establishment of the MIT Radiation Laboratory*. Receiver sensitivity in a radar system is critical in establishing its maximum range, and radars must operate at high frequencies, where good sensitivity is relatively difficult to achieve. Early radars operated in the UHF

---

* The author asks the reader's forbearance as he indulges in a little editorializing: it is unfortunate that virtually all of the interest surrounding the recent fiftieth anniversary of the MIT Radiation Laboratories centered almost exclusively on the development of microwave transmitters. The development of receiver technology, especially point-contact diode mixers, was an important part of the work of the "Rad Labs," and was at least as important to the successful development of radar as the transmitters were. Virtually nothing has been said about this part of the Rad Labs' efforts.

region, where vacuum-tube mixers were available, and had reasonably good sensitivity. The range resolution of UHF radars was thoroughly inadequate, however, and with the development of the magnetron tube, which could generate high levels of microwave power, microwave radar became a possibility. Because there were no low-noise receivers, the mixer was the first stage in early radar receivers, and as such its sensitivity was critical. Thus, the scientists at the "Rad Labs" devoted much work to the improvement of mixer sensitivity.

Very little theoretical work had been performed on mixers before 1940, and the quality of available point-contact diodes was poor. In less than 10 years, the theoretical underpinnings of mixer design were in place, and the quality of diodes (and especially diode packages) had improved dramatically. Conversion losses of microwave mixers dropped from around 20 dB in 1940 to 10 dB in 1945, eventually bottoming out at 6 dB around 1950. Today the theory of diode mixers is well established, and high-quality Schottky-barrier diodes are readily available, as is effective mixer-design software. As a result, mixers having conversion losses below 4 dB at a frequency of 50 GHz are regularly produced.

Diode mixers are the highest-frequency low-noise microwave components in existence: at high frequencies a diode mixer is often the only device that can be used at all. At the time of this writing (1992), *low-noise transistor amplifiers* (LNAs) are unavailable at frequencies above 120 GHz, and receivers having mixer front ends exhibit performance superior to those using LNAs at frequencies above approximately 100 GHz. As better amplifiers are produced, they may supplant mixers at progressively higher frequencies, but interest in plasma diagnostics, radio astronomy, and radar imaging are already creating a need for receivers well into the terahertz region. Virtually all such receivers employ Schottky-barrier diode mixers.

Mixers also have important, if prosaic, applications in more ordinary low-frequency and microwave receivers, where they are used to shift signals to frequencies where they can be amplified and demodulated most effectively. Mixers can also be used as phase detectors and in demodulators, and must perform these functions while adding minimal noise and distortion.

Figure 1.1 shows, for example, the block diagram of a VHF or UHF communication receiver. The receiver has a single-stage input amplifier; this "preamp" increases the strength of the received signal so that it exceeds the noise level of the following stages. The preamp has no more gain than necessary to overcome the noise of the later stages; if its gain were too high, the strongest received signals might generate distortion that would interfere with weaker signals. The first IF frequency is relatively high (in a VHF or UHF receiver, the widely accepted standard is 10.7 MHz); this high IF moves the image frequency well away from the RF, thus allowing the image to be rejected effectively by the input filter.

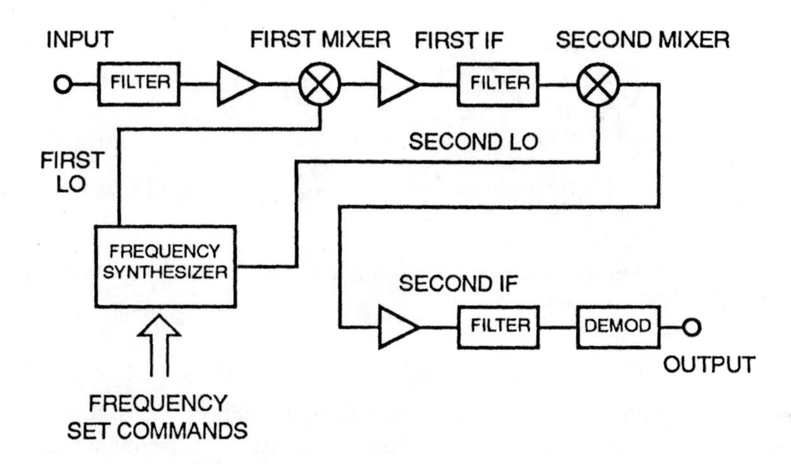

Figure 1.1 Dual-conversion VHF/UHF communication receiver.

The second conversion occurs after considerable amplification (mostly by the first IF stages), and is used to select some particular signal within the input band and to shift it to the second IF frequency (a frequency at which the demodulator works well). Because narrow bandwidths are generally easier to achieve at this lower frequency, the selectivity of the filter used before the detector is much better than that of the first IF. The frequency synthesizer generates the fixed-frequency *local oscillator* (LO) signal for the first mixer, and the variable-frequency LO for the second mixer.

Many of the same receiver-design considerations apply at microwave frequencies. At higher frequencies, however, there may be greater concern for minimizing receiver noise, and consequently a greater number of input amplifier stages may be needed. In the millimeter-wave region, it may be impossible to design a low-noise amplifier, so considerable effort may be expended to minimize the conversion loss (hence the noise figure) of the mixer.

A mixer is fundamentally a multiplier. This point is illustrated in Figure 1.2, which shows an ideal analog multiplier with two sinusoids applied to it. The signal applied to the RF port has a carrier frequency ω_s and a modulation waveform $A(t)$. (This represents a double-sideband, suppressed-carrier signal; any other type of modulation could be used in this example as well.) The other, the LO, is a pure, unmodulated sinusoid at frequency ω_p. The LO is sometimes called the *pump waveform*, and the mixer is often said to be *pumped* when the LO is applied.

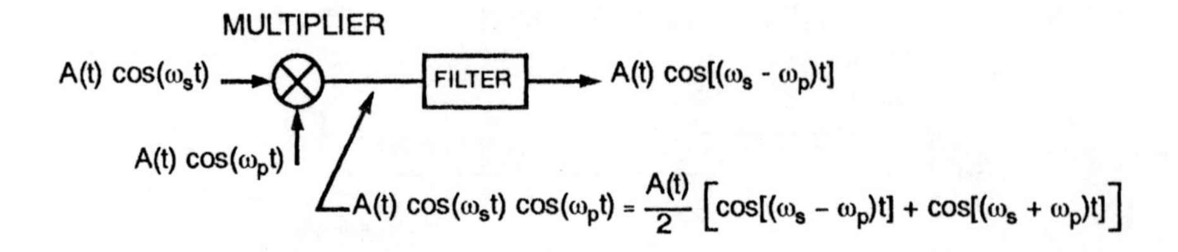

Figure 1.2 A mixer is fundamentally a multiplier; the difference frequency in the IF results from the product of sinusoids.

Through some trigonometry too simple to be worth repeating, the output is found to consist of modulated components at the sum and difference frequencies. The sum frequency is rejected by the IF filter, leaving only the difference.

Fortunately, an ideal multiplier is not the only thing that can realize a mixer. Any nonlinear device can perform the multiplying function. The use of a nonideal multiplier results in the generation of LO harmonics and in mixing products other than the desired one. The desired output frequency component must be filtered from the resulting mess.

The use of a nonideal multiplier can be illustrated by describing the *I*/*V* characteristic of the nonlinear device via a power series,

$$I = a_0 + a_1 V + a_2 V^2 + a_3 V^3 + \dots \tag{1.1}$$

and letting V equal the sum of the two inputs in Figure 1.2. After an appreciable amount of algebraic and trigonometric manipulation, the output is found to be a signal having the original modulation, but shifted to the difference frequency. If it is assumed that the voltage of the modulated input signal is much smaller than that of the LO, the current contains small-signal components at the frequencies

$$\omega_n = \omega_0 + n\omega_p \tag{1.2}$$

where ω_0 is the difference frequency $\omega_s - \omega_p$. The current also includes harmonics of the LO. Again, it is generally easy to filter out the desired difference frequency. Occasionally, the sum frequency is desired (an up converter), or even one of the mixing frequencies other than the simple difference frequency. These can also be selected if desired.

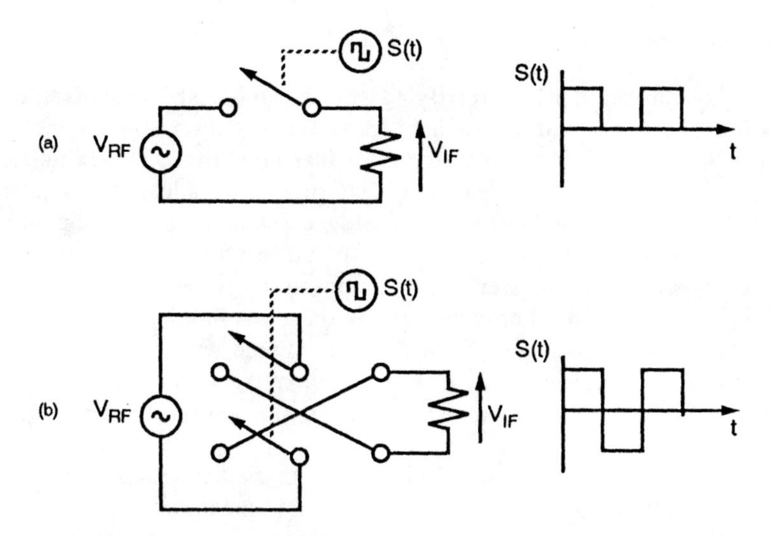

Figure 1.3 Two switching mixers. The IF is the product of the switching waveform $s(t)$ and the RF input, making these mixers a type of multiplier: (a) a simple switching mixer; (b) a polarity-switching mixer.

Another way to view the operation of a mixer is as a switch. Indeed, in the past, diodes used in mixers have been idealized as switches operated at the LO frequency. Figure 1.3(a) shows a mixer modeled as a switch; the switch interrupts the RF voltage waveform periodically at the LO frequency. The IF voltage is the product of the RF voltage and the switching waveform. In some cases the switching waveform might not have a 50% duty cycle, so, in general, it includes all harmonics of its fundamental frequency, plus a dc component; thus the IF includes a large number of mixing products. The desired output, again, can be separated from the others by filtering.

Another switching mixer is shown in Figure 1.3(b). Instead of simply interrupting the current between the RF and IF ports, the switch changes the polarity of the RF voltage periodically. The advantage of this mixer over the one in Figure 1.3(a) is that the LO waveform has no dc component, so the product of the RF voltage and switching waveform does not include any voltage at the RF frequency. Thus, even though no filters are used, the RF and LO ports of this mixer are inherently isolated. Doubly balanced mixers, examined in Chapters 7 and 8, are realizations of the polarity-switching mixer.

1.2 DEVICES FOR MIXERS

Although, in theory, any nonlinear or rectifying device can be used as a mixer, only a few devices satisfy the practical requirements of mixer operation. Any device used as a mixer must have a strong nonlinearity, electrical properties that are uniform between individual devices, low noise, low distortion, and adequate frequency response. The nonlinear device most often employed for mixing is the Schottky-barrier diode, a diode consisting of a rectifying metal-to-semiconductor junction. (Because of frequency-response limitations imposed by recombination in the junction, *pn*-junction diodes are thoroughly unsuited for use in microwave mixers.) In the past, point-contact diodes, which consist of a pointed-wire contact to a bulk semiconductor, were used extensively in microwave mixers. Indeed, since they rectify by means of a metal-to-semiconductor junction, the point contact is a type of primitive Schottky barrier.

Schottky-barrier diodes are superior to point-contact diodes because the former are fabricated photolithographically on an epitaxial substrate. This type of fabrication results in lower junction capacitance and series resistance (two important parasitics), and a better *I*/*V* characteristic. Both are majority-carrier devices and thus are not subject to the recombination-time limitations that affect *pn* diodes. The successful development of small, high-quality Schottky junctions, sometimes with diameters below 1 μm, has led to the development of practical mixers at frequencies above 1,000 GHz. Schottky-barrier diodes are also used in more mundane applications, in virtually all microwave mixers, and are available in a wide variety of packages, as well as unmounted chips. They are now inexpensive enough to be used in balanced mixers costing less than $20 each.

The successful development of good-quality Schottky barriers on *gallium arsenide* (GaAs) semiconductors has been responsible for the dramatic improvement in millimeter-wave mixers throughout the last decade. GaAs is decidedly superior to silicon for high-frequency mixers because of its higher electron mobility and saturation velocity. These allow lower series resistance, for a specified junction capacitance, to be achieved in GaAs devices than in silicon. The performance of GaAs diodes regularly exceeds that of silicon, and reliability and damage resistance (because of GaAs diodes' higher breakdown voltages) are also superior. GaAs diodes are, however, significantly more expensive than silicon, and thus are normally reserved for more critical applications.

Bipolar-transistor mixers are used occasionally below 1 GHz, but are uncommon at microwave frequencies. Progress in the development of *heterojunction bipolar transistors* (HBTs) may bring about a resurgence in the use of bipolar devices as mixers. HBTs are often used as analog multipliers operating at frequencies approaching the microwave range; the most common form is a Gilbert multiplier circuit, similar to the Gilbert multipliers used in analog integrated circuits (for a further description of Gilbert multipliers, see any text on analog integrated circuits).

Field-effect transistor (FET) mixers have gained considerable popularity in recent years. The primary reason for their increased use seems to be the concomitant progress in GaAs *monolithic microwave integrated circuit* (MMIC) technology. FETs are generally better suited to MMICs than diodes: FETs do not require the nonplanar balun structures commonly required by balanced diode mixers, and they are compatible with the MMIC technology of FET amplifiers and other components (mixing diode and FET technology in a single chip often presents fabrication problems). Of course, a major advantage of FET mixers over diode mixers is the FET's ability to provide several decibels of conversion gain, while most diode mixers exhibit at least 5 to 6 dB of loss. FET mixers are regularly operated at frequencies extending well into the millimeter range; conversion gain has been achieved at frequencies above 90 GHz.

FET mixers are usually operated as active circuits, with the LO and RF signals applied to the gate, and with the IF filtered from the drain; the LO signal modulates the FET's transconductance to produce mixing. This type of operation generally results in the best conversion gain and lowest noise temperature. FET mixers have also been operated with the LO applied to the drain, although without clear advantages over "gate mixing." A FET can also be operated as a resistive mixer. In this mode of operation, the LO is applied to the gate and the RF is applied to the drain; the IF is filtered from the drain. No dc drain bias is used, and the channel operates as a time-varying resistance. Because the linearity channel's resistance is very good, such mixers have very low levels of intermodulation distortion.

Dual-gate FETs have advantages over single-gate devices in mixer applications. The most significant advantage is that the LO signal and RF can be applied to separate gates, thus achieving approximately 20 dB of LO-to-RF isolation without the use of filters or balanced structures. Dual-gate FET mixers have exhibited low distortion, probably because their transconductances are very linear. However, their noise figures and conversion gains are not, on the whole, as good as those of mixers using single-gate FETs.

1.3 BALANCED AND SINGLE-DEVICE MIXERS

Single FETs or diodes can be used as mixers. However, designs sometimes use combinations of two, four, or even eight diodes in a balanced structure. Balanced mixers have important performance advantages compared to single-device mixers. One of those advantages, the inherent RF-to-IF isolation of the polarity-switch mixer (which must always be realized as a balanced circuit), has already been described. Another advantage is that the RF and LO are inherently isolated. A balanced mixer also rejects the AM noise from the LO source, and rejects certain spurious responses. The amount of rejection depends upon practical aspects of the circuit design, but in most cases is at least 10 to 20 dB. This rejection occurs because of the phase relationships of the voltages in the circuit, and does not require any filtering.

Because the input power is divided between multiple devices, the power-handling capability of a balanced mixer is invariably better than that of a single-device mixer. (Unfortunately, the LO power is divided between the diodes as well, so the LO power requirements are commensurately greater.) For this reason, as well as its spurious-response rejection, a balanced mixer is usually chosen for an application where strong signals are anticipated.

Balanced mixers are divided loosely into two classes, called *singly balanced mixers* and *doubly balanced mixers*. Singly balanced mixers usually use two devices, and are usually realized as two single-device mixers connected by a 180- or 90-deg hybrid. Doubly balanced mixers usually consist of four untuned devices interconnected by multiple hybrids, transformers, or baluns. They are usually too complicated to allow for individual tuning of the devices, so they may have higher conversion loss (or lower gain) than single-device or singly balanced mixers. The lack of tuning elements, combined with broadband hybrids, often gives such mixers very wide bandwidths. A great variety of balanced diode mixer circuits are in common use, but virtually all of them are simply different realizations of the same basic circuits and make use of the same principles.

Figure 1.4 shows the most commonly encountered realizations of the three mixer types. Figure 1.4(a) shows a single-diode mixer with the RF, LO, and IF separated by filters. Figure 1.4(b) shows a singly balanced mixer, realized with a transformer hybrid (Chapter 7), although many types of 180-deg hybrids could be used. The ring mixer in Figure 1.4(c) is a classical realization of the doubly balanced mixer, which will be described further in Chapter 8.

For reasons given in Chapter 9, balanced mixers—especially doubly balanced mixers—are more difficult to realize with FETs than with diodes. This situation is especially true of hybrid integrated circuits, where the parasitics associated with interconnections of several FETs limit the mixer's performance. Balanced FET mixers are somewhat more practical in MMICs, but their high-frequency performance is often unspectacular. It is likely that better high-frequency performance will be obtained in the future as device and circuit technologies mature.

1.4 MIXER DESIGN

In the past, designing a mixer involved either a great deal of trial and error or the use of a program designed specifically for the purpose of analyzing a diode or FET mixer. Because of recent improvements in the technology of nonlinear circuit analysis by computer, there now exist general-purpose computer programs for the analysis (and sometimes optimization) of nonlinear circuits. These programs are

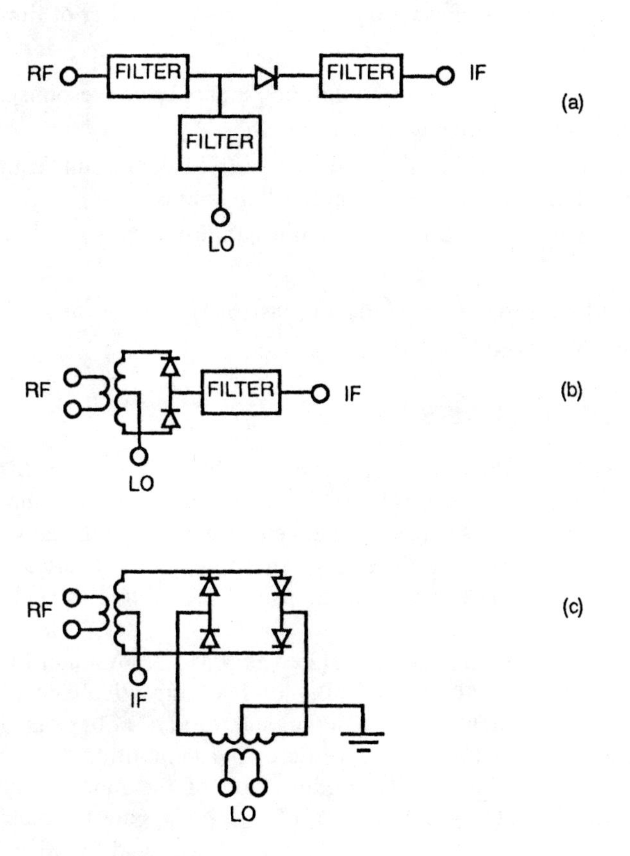

Figure 1.4 The most common forms of the three mixer types: (a) single-device; (b) singly balanced; (c) doubly balanced.

based on harmonic-balance analysis, and are used in a manner analogous to that of ubiquitous linear-circuit analysis programs. Many are (or can be) integrated in design "frameworks" that include the graphical capabilities necessary for mask layout.

Such programs have measurably improved the reliability of the mixer-design process, especially for diode mixers. They can be used to analyze single- or multiple-device mixers using either diodes or FETs and mixers having very complex structures.

At this writing, however, these programs still have a number of limitations:

- The lack of noise analysis;
- Unreliable computation of intermodulation and spurious responses;
- Limited ability to optimize nonlinear circuits;
- Limited accuracy in modeling the linear parts of the circuit (especially transmission-line discontinuities and coupled-line baluns);
- The high cost of both the programs and computers powerful enough to run them.

Because of these limitations (especially the last one), the use of such programs is not as widespread as it might be.

1.5 MONOLITHIC CIRCUITS

Microwave mixers are often realized as monolithic circuits (or MMICs). Such circuits offer the advantages of small size, low-cost fabrication in large quantities, and uniform performance. The costs of development are high, however, and the technology is appropriate in a limited range of applications. Fabrication costs will undoubtedly decrease in the future, and the use of monolithic mixers will be more widespread.

Either diode or FET mixers can be realized as MMICs, although FET mixers are usually preferred. Because of the difficulty of realizing wideband planar baluns, classical diode designs such as the ring or star mixers (Chapters 7 and 8) are sometimes impractical. Another problem in creating monolithic diode mixers is the difficulty of fabricating high-quality diodes. One of the most common ways to create a monolithic diode is by means of a FET's gate-to-channel junction. Unfortunately, because such diodes are lateral structures and the MMIC's epitaxial layer is optimized for FETs instead of diodes, they often have relatively high series resistance. The high series resistance causes the conversion loss and noise figure to be higher than those of mixers using discrete diodes.

Chapter 2
Schottky-Barrier Diodes

Although point-contact diodes have been the dominant devices for microwave mixers in the past, Schottky-barrier diodes are now almost universal. The primary reason for the preeminence of the Schottky diode is that it is a majority-carrier device, and, as such, it does not suffer from the charge-storage effects that limit the switching speed of *pn* junctions. Furthermore, Schottky diodes can be fabricated more precisely than either *pn* junctions or point-contact diodes, and it is possible to achieve very good and very uniform electrical characteristics. For these reasons this chapter is concerned almost exclusively with Schottky-barrier devices; we mention point-contact diodes only briefly for comparison.

2.1 SCHOTTKY-DIODE OPERATION

2.1.1 Ideal Junction Characteristics

The Schottky-barrier diode is formed by a metal contact (the anode) to a semiconductor (the cathode), instead of the more common junction between *p*- and *n*-type semiconductors. Schottky diodes differ from *pn*-junction devices in that rectification occurs because of differences in work function between the metal contact and the semiconductor, rather than a nonuniform doping profile. Conduction is not controlled by minority carrier recombination in the semiconductor, but by thermodynamic emission of majority carriers over the barrier created by the unequal work functions. The Schottky diode is, therefore, a majority carrier device whose switching speed is not limited by minority carrier effects.

In practical mixer diodes, the diode is usually realized on a silicon or GaAs semiconductor. Many metals can create a Schottky barrier on either silicon or GaAs; the most common barrier metals for GaAs are platinum and titanium, and occasionally gold is used. Carrier mobility is greater in *n*-type material than in *p*-

type, so diodes using *n*-type semiconductors have lower series resistances and higher cutoff frequencies (Section 2.2.2). Hence, *n*-type semiconductors are used almost exclusively for mixer diodes.

In order to understand the use of Schottky diodes in mixers, it is necessary to understand their *I*/*V* characteristics, junction capacitance, and series resistance. These will be found first for the ideal diode, and will be modified to include nonideal behavior. To determine these parameters, we must begin by examining the energy band structure, depletion width, depletion charge, and, finally, junction capacitance. We first examine the band structure and use it to derive the *I*/*V* characteristic.

Figure 2.1 shows the energy-band diagrams of a metal and an *n*-type semiconductor. The difference between the Fermi level E_f for each material and the free-space energy level E_0 is the work function, $q\phi_m$ or $q\phi_s$, where q is the electron charge. The work function is therefore the average energy required to remove an electron from the material. The electron affinity qX is the energy required to remove an electron from the conduction band to free space. X is a constant for each material and must remain constant throughout it. However, the Fermi level of the semiconductor, and hence its work function, depends on doping density.

When the metal and semiconductor are in equilibrium and are not in contact, the energy levels are constant throughout the materials. The Fermi levels are unequal, indicating that the electrons in one material (in this case the metal) have less energy, on the average, than those in the semiconductor. Therefore, when the materials are joined, some of the electrons in the semiconductor move spontaneously into the metal and collect on its surface. These leave behind ionized donor locations, which

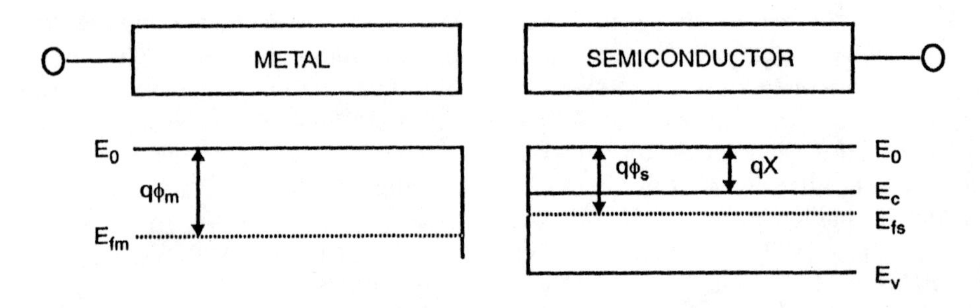

Figure 2.1 Band structure of the metal and semiconductor before contact. E_0 is the free-space energy level, E_c is the bottom of the conduction band, and E_v is the top of the valence band. E_{fm} and E_{fs} are the Fermi levels in the metal and semiconductor, respectively.

are positively charged, and create a negative surface charge where they collect on the surface of the metal. The positively charged region is called a *depletion region,* since it is almost completely depleted of mobile electrons.

The shape of the energy diagram of the metal-to-semiconductor junction is governed by three rules:

1. In equilibrium, the Fermi levels of the semiconductor and metal must be constant throughout the system.
2. The electron affinity must be constant.
3. The free-space energy level must be continuous.

Figure 2.2(a) shows the resulting band structure when the metal and semiconductor are joined. In order to satisfy all three rules simultaneously, the valence and conduction bands of the semiconductor are forced to bend at the junction; the upward bend of the conduction band of the *n*-type semiconductor indicates a depletion region. The resulting potential difference across this region, as shown in the figure, is simply the difference between the work functions, $\phi_{bi} = \phi_m - \phi_s$. This is called the *built-in potential* of the junction.

The positively charged depletion region in the semiconductor, like a capacitor, can be considered an area of stored charge. Indeed, charge has been moved onto the metal contact, one "plate," by the application of the built-in potential difference. Before it is possible to determine the capacitance, it is necessary to find the quantity of charge that has been moved; this quantity is equal to the total depletion-region charge. The charge density in the depletion region is known: because the depletion region's charge comes from donor atoms, all of which are ionized, the charge density is equal to the doping density. The junction area is, of course, known, but the width of the depletion region still must be found in order to determine the total charge.

The electric field in the depletion region is found by applying Gauss' law to the region. It should be obvious that the electric field is in the negative x direction (Figure 2.2), and that it is maximum at the junction. The electric field must also be zero at the edge of the depletion region, because $E = -d\phi/dx = 0$, as evidenced by the flat band at this point. Secondly, the voltage across the junction, found by integrating the electric field, must equal ϕ_{bi}. Applying Gauss' law in one dimension, we have

$$\frac{d}{dx}E(x) = \frac{\rho(x)}{\varepsilon_s} = \frac{qN_d}{\varepsilon_s} \tag{2.1}$$

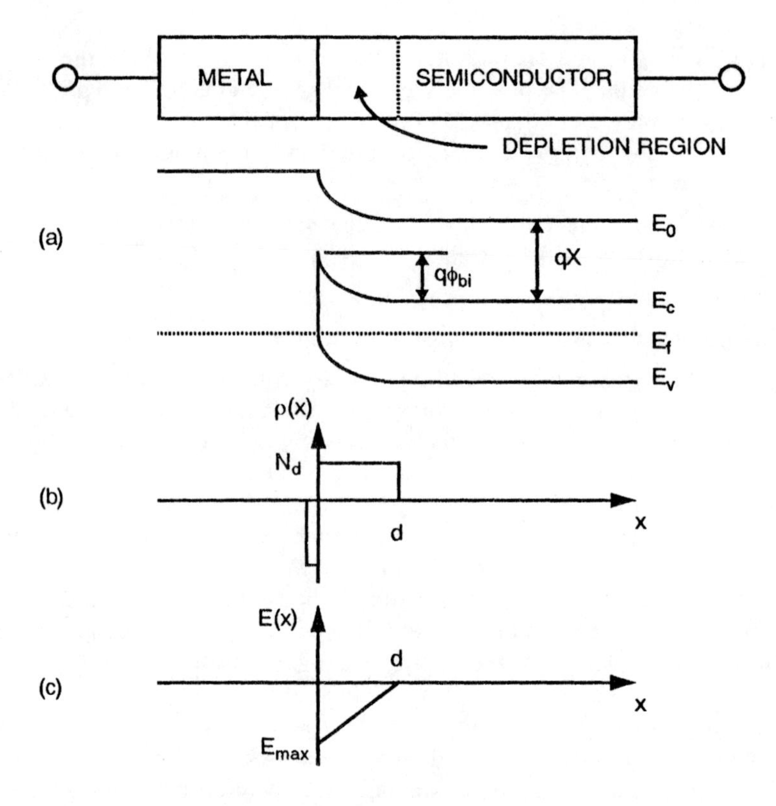

Figure 2.2 (a) Band structure of the Schottky junction; (b) charge densities at the junction (the negative component is the electron component on the surface of the metal); (c) electric field in the depletion region.

$$E(x) = E_{max}(1 - x/d) \tag{2.2}$$

where

$$E_{max} = \frac{-qN_d d}{\varepsilon_s} \tag{2.3}$$

E_{max} is the maximum electric field, d is the depletion width, N_d is the doping density (assumed to be uniform), and ε_s is the dielectric permittivity of the semiconductor. The dielectric permittivity ε_s is equal to 13.1 ε_0 for GaAs and 11.9 ε_0 for silicon, where ε_0, the permittivity of free space, equals 8.854×10^{-14} F cm^{-1}. An assumption used in deriving Equations (2.1) to (2.3) is that the edge of the depletion region is abrupt; that is, there is no gradual variation in charge density between the depletion region and undepleted semiconductor. This assumption is called the *depletion approximation.* In fact, a narrow transition region does exist, but its effect is negligible for most purposes.

Since $E(x)$ is a simple triangle function, it is easily integrated to give

$$\phi_{bi} = \frac{E_{\max} d}{2} = \frac{q d^2 N_d}{2\varepsilon_s} \tag{2.4}$$

The resulting depletion width d is

$$d = \sqrt{\frac{2\phi_{bi}\varepsilon_s}{qN_d}} \tag{2.5}$$

The charge in the depletion region is found from the donor density and the dimensions of the region, which are now known. The depletion charge Q_j is

$$Q_j = qWdN_d = W\sqrt{2q\varepsilon_s\phi_{bi}N_d} \tag{2.6}$$

where W is the area of the junction. The quantity relates directly to the junction capacitance.

2.1.2 Ideal *I/V* Characteristic and Junction Capacitance

Figure 2.3 shows a biased Schottky junction. Since bias is applied, the junction is no longer in equilibrium, and the requirement that the Fermi levels be constant throughout the diode no longer applies. Instead, the Fermi levels (which should rightly be called *quasi-Fermi levels* for the nonequilibrium case), move with applied voltage. The offset from their equilibrium position is simply equal to qV, where V is the applied voltage. The voltage across the junction then is $\phi_{bi} - V$, where V is defined as positive with polarity that forward-biases the junction. The expressions for electric field $E(x)$, maximum electric field $E_{\max}$, depletion region d, and charge Q_j are still valid for the biased diode as long as the potential ϕ_{bi} is replaced by

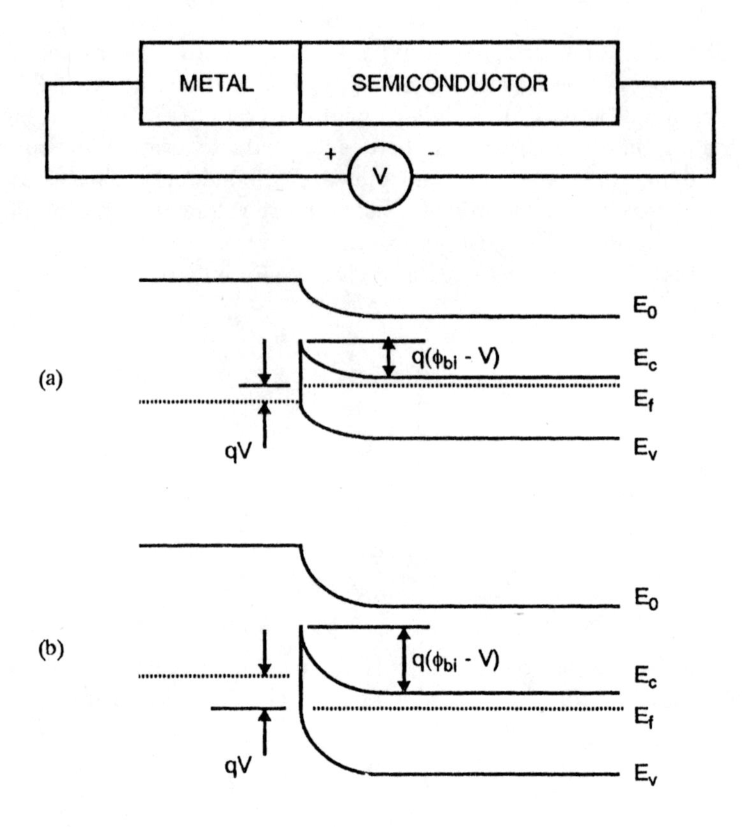

Figure 2.3 Biased Schottky junction: (a) forward bias ($V > 0$); (b) reverse bias.

$\phi_{bi} - V$. The resulting expressions for charge and depletion width are as follows:

$$Q_j = W\sqrt{2q\varepsilon_s N_d(\phi_{bi} - V)} \tag{2.7}$$

$$d = \sqrt{\frac{2\varepsilon_s(\phi_{bi} - V)}{qN_d}} \tag{2.8}$$

The capacitance of a nonlinear capacitor is defined as the derivative of charge with respect to the junction voltage:

$$\frac{dQ_j}{dV} = C(V) = W\sqrt{\frac{q\varepsilon_s N_d}{2(\phi_{bi} - V)}} = \frac{W\varepsilon_s}{d} \tag{2.9}$$

This can be put in the form

$$C(V) = \frac{C_{j0}}{(1 - V/\phi_{bi})^{1/2}} \tag{2.10}$$

which is most useful for circuit analysis. C_{j0} is the junction capacitance at zero bias voltage.

The exponent 1/2 in the denominator of (2.10) comes from the assumption that the doping density N_d is uniform throughout the semiconductor. In practice, N_d may not be uniform, thus changing the exponent. One of the most dramatic examples of this is the Mott diode, whose capacitance has relatively weak dependence on voltage. Mott diodes are discussed in Section 2.3.

The junction current can be found by several methods. The following derivation is simple and intuitively satisfying. Other approaches are those of Schottky [1] and Bethe [2]; these are described by Sze [3, Chapter 5].

Electron conduction occurs primarily by thermionic emission over the barrier. This emission occurs equally in both directions in equilibrium (i.e., at zero bias), giving no net current. When forward bias is applied, electron energy is increased relative to the barrier height, allowing increased electron emission from the semiconductor into the metal. The current component in the opposite direction stays constant because the barrier height (to a first approximation, at least) stays constant.

The electron density at the junction n_1 can be found from the Maxwell-Bolzmann distribution. It is given by

$$n_1 = N_d \exp\left(\frac{-q\phi_{bi}}{KT}\right) \tag{2.11}$$

under zero bias conditions. The current in each direction is equal, and must be proportional to this electron density. Under bias, the potential barrier becomes $\phi_{bi} - V$, and therefore the density of forward-conducted electrons is

$$n_2 = N_d \exp\left(\frac{-q(\phi_{bi} - V)}{KT}\right) \tag{2.12}$$

where K is Bolzmann's constant (1.37×10^{-23} J/K) and T is absolute temperature. The current is proportional to the difference between these densities:

$$I(V) = I_0\left(\exp\left(\frac{qV}{KT}\right) - 1\right) \tag{2.13}$$

Equation (2.13) is called *the ideal diode equation.* In order to compensate for nonideal behavior, it is usually modified by the addition of an extra parameter, η, to form

$$I(V) = I_0\left(\exp\left(\frac{qV}{\eta KT}\right) - 1\right) \tag{2.14}$$

The parameter η is a number close to 1.0, usually between 1.05 and 1.25. It is called the *slope parameter* or the *ideality factor.* As η increases, the strength of the diode's nonlinearity decreases; and virtually all aspects of a mixer's performance–noise, conversion loss, and LO power requirements–become worse. We shall discuss this further in later chapters.

Calculation of the current parameter I_0 is a much more complicated task and probably futile, since I_0 can be dominated by second-order effects such as leakage, charge generation, and tunneling. Nevertheless, an ideal expression for I_0 can be found by assuming that all current conduction is by thermionic emission. It is given by [4]

$$I_0 = A^{**} T^2 W \exp\left(\frac{-q\phi_b}{KT}\right) \tag{2.15}$$

where A^{**} is the modified Richardson constant [4], W is the junction area, and ϕ_b is the barrier height (the difference between the Fermi level and the peak of the conduction band). A^{**} is approximately 96 A cm^{-2} K^{-2} for silicon and 4.4 A cm^{-2} K^{-2} for GaAs. The low value of the Richardson constant for GaAs implies that the knee of the I/V characteristic occurs at higher junction voltages in GaAs diodes.

2.1.3 Deviations From Ideality

Real Schottky diodes do not always follow the expressions derived in Section 2.1.2. Deviations from ideal behavior arise from imperfections in fabrication or factors that are not included in this relatively simple theory. A few of the major limitations are given below.

Schottky Barrier Lowering

In the previous section we assumed that the barrier height remained constant under all conditions of applied voltage. In fact, the barrier height varies with applied voltage because conduction electrons experience a force from their image charges in the metal. This force attracts the electrons toward the metal surface, effectively lowering the barrier and allowing voltage-dependent deviations form ideal behavior. In theory, this "image force" should give the reverse current I_0 a fourth-power dependence upon voltage, rather than the constant value implied by (2.14). However, this dependence is usually not observed, because carrier generation in the depletion region at high reverse bias and tunneling effects dominate reverse leakage.

At forward biases above approximately 0.1V, the effect of barrier lowering is to increase the ideality factor η slightly and to make it a weak function of junction voltage. A diode that is ideal except for barrier lowering has the ideality factor

$$\eta = \frac{1}{1 - \frac{d\phi_b}{dV}} \tag{2.16}$$

where $d\phi_b/dV$ is the variation in barrier height with applied voltage. An expression for this quantity is [4]

$$\frac{d\phi_b}{dV} = \frac{1}{4}\frac{q^3 N_d}{8\pi^2 \varepsilon_s}\left(\phi_b - V - \phi_{fc} - \frac{KT}{q}\right)^{-3/4} \tag{2.17}$$

where ϕ_{fc} is the potential difference between the Fermi level and the bottom of the conduction band. As with the reverse case, this quantity rarely dominates the ideality factor; for $N_d = 10^{17}$ cm^{-3}, η is only 1.02.

Surface Imperfections

The semiconductor surface on which the junction is fabricated must be extremely clean in order to obtain good *I/V* characteristics. However, in spite of scrupulous care in fabrication, a small amount of junction contamination cannot be avoided. The deposition of the junction metal may also damage the crystal structure of the surface, especially if sputtering techniques are used. Formation of undesired chemical compounds between the junction metal and the semiconductor may also occur, especially if the diode is subjected to high temperatures. Although diodes are rarely exposed to high temperatures in use, they are frequently exposed to high temperatures as part of the fabrication process: annealing to repair sputtering damage, attaching to a circuit, or when they are innocent bystanders while other components are soldered into the mixer. This damage increases both the ideality factor and, in some cases, reverse leakage. Surface imperfections are probably the major cause of nonideal behavior in Schottky diodes.

Tunneling

Thermal emission is not the only mechanism by which electrons can cross the potential barrier at the junction. Quantum mechanical tunneling through the barrier is also possible, and may have a significant effect on the *I/V* characteristic at low temperatures and high doping densities. Tunneling is often responsible for "soft" *I/V* characteristics (i.e., high η) at low currents. It is particularly significant in devices designed for cryogenic operation because, as temperatures are lowered, the current component due to tunneling does not decrease as rapidly as the thermionic component. Tunneling also increases the noise temperature of the diode. Tunneling will be discussed further in Section 2.2.4.

Series Resistance

Schottky junctions generally require lightly doped semiconductors having relatively high bulk resistivities. A lightly doped substrate is not practical for diode fabrication because it results in high series resistance and a poor ohmic contact to the cathode. Practical diodes are fabricated on a lightly doped, thin epitaxial layer that is grown on a heavily doped, low resistance substrate (Section 2.4.1); the lightly doped layer is used for the junction and the heavily doped substrate minimizes series resistance. A high-quality ohmic contact to this heavily doped substrate can be made relatively easily.

In such diodes, the greatest part of the series resistance is the resistance of the undepleted epitaxial layer under the junction. The epilayer is normally made thick enough to contain the depletion region, even at high reverse junction voltages, so at intermediate voltages the layer of undepleted semiconductor material may be fairly thick. For example, a diode having an epitaxial doping density of 2×10^{17} has a zero-voltage depletion depth of approximately 750 Å; however, it must have an epilayer thickness of 1,000 to 2,000 Å to contain the depletion layer at a reverse voltage of 5V to 6V. Thus, at normal voltages there may be 500 to 1,500 Å of undepleted, high-resistance expitaxial material under the junction.

The remaining bulk resistance of the substrate and its ohmic contact, as well as the undepleted epitaxial area, may leave several ohms of resistance in series with the junction. The resistance creates power losses, which are often substantial, especially in millimeter-wave mixers, where the junction area is very small. Series resistance often creates a lower limit to the useful diode size; at present, submicron diameter diodes can be fabricated, but series resistance usually limits practical sizes to 1.5 to 2.0 μm.

Because it is strongly dependent on diode structure, it is difficult to describe a general procedure for estimating series resistance. Estimation is further complicated by the fact that the skin effect causes the diode's current density to be greatest in the surface of the substrate, rather than in its bulk, at frequencies above approximately 50 GHz. Similarly, at high frequencies, the whisker or other connecting wire may have several ohms of series resistance resulting from skin effect.

We shall illustrate the calculation of series resistance by examining a dot-matrix diode (Section 2.4) at high frequencies. First we must consider the current path in the chip, shown in Figure 2.4. The current path is from the anode through the expitaxial layer, spreading out because of skin effect along the top surface of the chip to its edge. Because the substrate resistivity is much lower than that of the epitaxial layer, the current in the top of the chip is primarily in the substrate. The final part of the current path is along the sides of the chip. Therefore, the series resistance consists of three components: the undepleted epitaxial layer under the junction, the spreading resistance of the top surface of the diode between the anode and the sides of the chip, and the resistance of the vertical sides. Because the epitaxial layer is thin compared to the diode diameter, there is little current spreading in the epitaxial layer. The resistance of the epitaxial layer is that of a cylinder of material:

$$R_{d1} = \frac{4(t-d)}{\pi a^2 q \mu N_d} \tag{2.18}$$

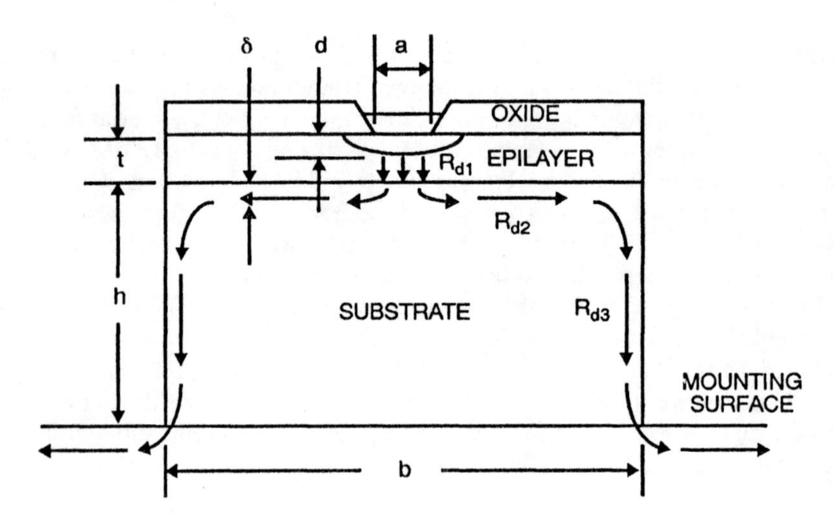

Figure 2.4 Current distribution in the dot-matrix diode.

where t is the epitaxial layer thickness, d is the depletion width and μ is the electron mobility. The spreading resistance component is found by first approximating the chip as a cylinder with the anode in the center, and is approximately

$$R_{d2} = \frac{\ln(b/a)}{2\pi\delta q\mu N_d} \tag{2.19}$$

where a is the anode diameter, b is the diameter of the chip (which can be approximateu as a side length for the more common square chip), and δ is the skin depth in the substrate material. The sidewall resistance is given by

$$R_{d3} = \frac{h}{\pi b\delta q\mu N_d} \tag{2.20}$$

where h is the chip height. This estimate for R_{d3} may be low because mechanical damage or roughness of the side of the chip increases R_{d3} substantially. Finally, the high-frequency series resistance is the sum of the three components:

$$R_s = R_{d1} + R_{d2} + R_{d3} \tag{2.21}$$

Of course, the series resistance measured at dc includes only R_{d1}; without skin effect, R_{d2} and R_{d3} become the bulk dc resistance of the substrate, which is negligible compared to R_{d1}. Diode series resistance specified by manufacturers is invariably the dc value.

Edge Effects

The expressions in Section 2.1.2 are based on an assumption that the electric field is perpendicular to the junction over its entire area. However, practical diodes consist of a small anode on a large semiconductor surface; consequently, the fringing electric field near the edge of the metal anode is greater than the field in the center. The current density is therefore greatest at the edge of the junction and may be relatively low at the center. As a result, the series resistance depends more strongly on the junction's periphery than its area. For this reason, diodes occasionally have anode geometries, such as a cross shape, that increase the periphery relative to the junction area. Edge effects are the main reason why series resistance does not "scale" with area.

Edge effects are also evident in the junction capacitance: the same fringing field contributes a component to the junction capacitance that is proportional to the anode's periphery. Because of this component, the capacitance is often not proportional to the diode's area and, like series resistance, does not scale.

2.2 DIODE CIRCUIT MODEL

2.2.1 Equivalent Circuit

In order to analyze mixers, it is necessary to have a circuit model of the Schottky-barrier diode that is valid for both large-signal and small-signal analysis. Because the Schottky diode is largely immune to minority carrier effects, the junction capacitance and current described by (2.10) and (2.14) change almost instantaneously with junction voltage, and the dc expressions for these quantities are valid to several hundred gigahertz. Therefore, we can assume that the large-signal diode model is *quasistatic*; that is, the capacitance and current are functions of the junction voltage alone and change instantaneously with that voltage. In Schottky diodes, the quasistatic assumption has been shown experimentally to be valid in mixers to at least 250 GHz.

Figure 2.5 shows a circuit model of the Schottky diode. It consists of a nonlinear resistance and capacitance representing the junction, and a series resistance (this circuit represents the intrinsic diode junction; other elements may be added to

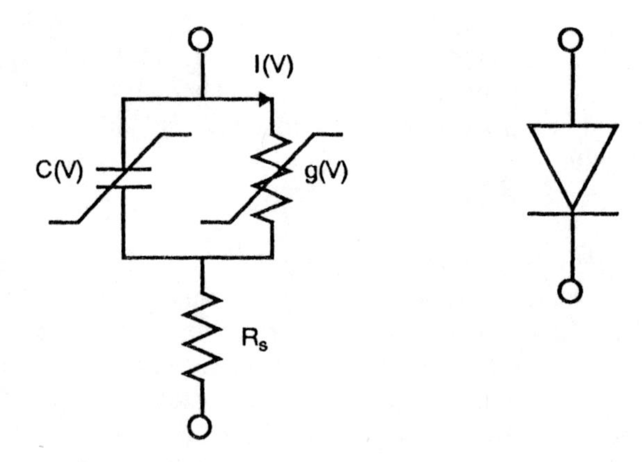

Figure 2.5 Equivalent circuit of a chip diode. This circuit represents only the intrinsic diode; package parasitics are not shown.

describe package capacitance or bond-wire inductance). Although the series resistance varies slightly with junction voltage, its nonlinearity is usually negligible, and for virtually all purposes, it is treated as a linear element. The control voltage for the capacitance and the junction current is the *junction* voltage, not the *terminal* voltage; that is, it must not include the voltage drop across the series resistance.

Although the equivalent circuit of Figure 2.5 applies to both large- and small-signal analysis, it is used in different ways. In large-signal analysis (such as calculating the LO waveforms in a mixer), the nonlinear model is used as shown. The capacitance and junction current are given by (2.10) and (2.14), where V represents the instantaneous value of that time-varying voltage. In small-signal analysis (which is used to determine the conversion loss), we linearize the I/V and C/V characteristics around the instantaneous large-signal voltage and treat the junction conductance and capacitance as linear, time-varying elements. This is justified by the fact that the RF and IF voltage components are small perturbations of the large-signal (LO) junction voltage. The small-signal junction conductance $g(V)$ is the derivative of junction current with voltage:

$$g(V) = \frac{d}{dV}I(V) = \frac{q}{\eta KT}I_0 \exp(\frac{qV}{\eta KT}) \tag{2.22}$$

or

$$g(V) = \frac{q}{\eta KT}(I(V) + I_0) \tag{2.23}$$

which gives the result that the junction conductance is proportional to its current. Because it is small compared to $I(V)$, the I_0 term in (2.23) is often ignored.

It is important to recognize that Equation (2.10) describes only the depletion capacitance, but practical diodes may have parasitic capacitances from the diode's metallizations or leads (e.g., in a beam-lead diode). Also, in certain types of diodes, where the doping density is not uniform throughout the depletion region, Equation (2.10) may not describe the capacitance adequately.

2.2.2 Diode Measurements

If the semiconductor doping profile is uniform, the Schottky diode can be characterized by its built-in potential ϕ_{bi}, the zero-voltage junction capacitance C_{j0}, the ideality parameter *n*, the current parameter I_0, and the series resistance R_s. These can be measured in a straightforward manner.

DC I/V Characteristic

The *I/V* characteristic can be measured directly with a semiconductor curve tracer or by point-by-point measurements. It is helpful to have a measuring device that can indicate currents below 1.0 μA in order to detect reverse leakage (which should be all but immeasurable in undamaged GaAs millimeter-wave devices) and to examine the *I/V* characteristic at low currents. Several decades of current should be recorded, from the lowest measurable value to a level where at least a few millivolts are developed across R_s. The data are plotted on semilog graph paper. The low-current points should lie on a straight line; however, at the high-current end of the *I/V* curve, the curve will deviate noticeably from the straight line. The deviation represents the voltage drop across the series resistance.

The ideality factor η, current parameter I_0, and series resistance R_s can be found from the *I/V* curve. A typical Schottky diode *I/V* curve is shown in Figure 2.6, with a summary of *I/V* calculations. At low currents, little voltage drop occurs across the series resistance, and the change in voltage per decade of current is constant. Simple manipulation of (2.14) gives an expression for η:

$$\eta = \frac{q}{KT}\Delta V \log(e) \tag{2.24}$$

where ΔV is the change in junction voltage per decade of current, and *e* is

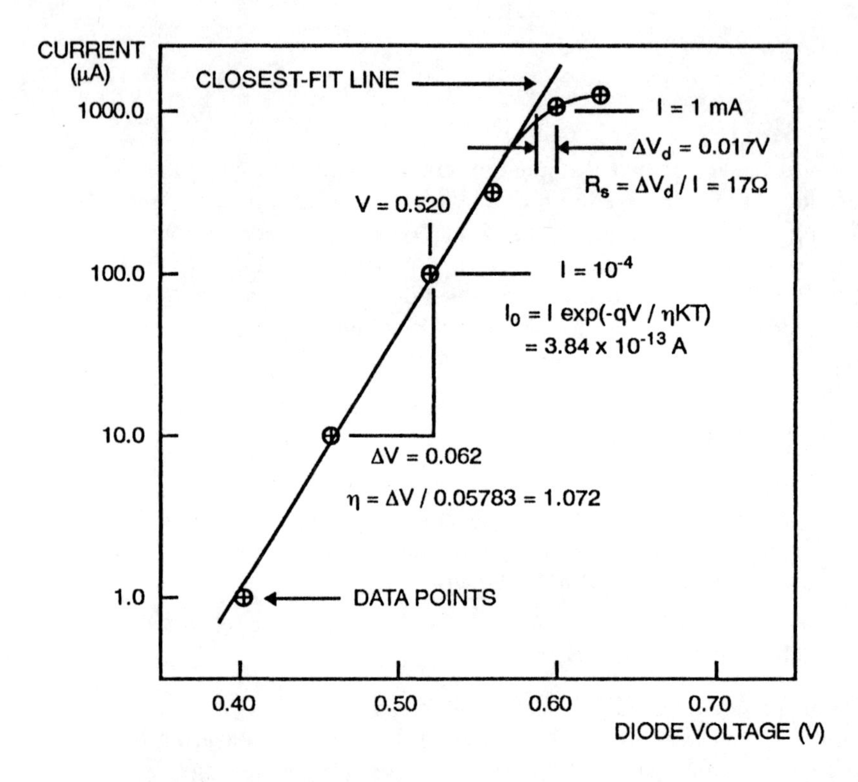

Figure 2.6 Schottky diode *I*/*V* curve and parameter calculations. Parameters η and I_0 are found from the straight line that fits the measured data most closely, and R_s is found from the deviations at the high-current end of the curve.

2.7182818. At 72°F, this becomes

$$\eta = \frac{\Delta V}{0.05783} \tag{2.25}$$

Knowing η, we can find I_0 from any point along the closest fitting straight line by solving (2.14). This gives

$$I_0 = I(V)\exp\left(\frac{-qV}{\eta KT}\right) \tag{2.26}$$

The additive −1.0 in (2.14) has been ignored in (2.24) through (2.26) because it is insignificant compared to the term $\exp(qV/\eta KT)$ at any measurable current.

The deviation in the *I/V* characteristic from a straight line at high currents is caused by the voltage drop across R_s. If, at current I_d, the deviation in voltage between the straight line and the measured *I/V* curve is ΔV_d, then R_s is simply

$$R_s = \frac{\Delta V_d}{I_d} \tag{2.27}$$

Heating of the junction at high current densities may cause dc measurements of R_s to be erroneously low [5]. The error occurs because the diode's *I/V* curve shifts to the left as temperature increases at high currents, causing ΔV_d in Figure 2.6 to be too low. This effect is most often observed in very small GaAs diodes; junction heating effects are usually insignificant in GaAs or silicon diodes having anode diameters greater than 3 to 4 μm.

Capacitance-Voltage (C/V) Characteristic

Equation (2.10) shows that the intrinsic junction capacitance has a square-root dependence on junction voltage. Thus, if N_d is uniform throughout the semiconductor, a plot of $1/C^2$ under reverse bias is a straight line, as shown in Figure 2.7. The curve intersects the $1/C^2$ axis at $1/C_{j0}^2$ and, if extrapolated, intersects the V axis at $V = \phi_{bi}$. (Because of junction conductance, it is invariably impossible to measure capacitance near ϕ_{bi}.)

In principle, it is a simple matter to find the parameters C_{j0} and ϕ_{bi}: measure $C(V)$ at a few points, plot $1/C^2$, and fit a straight line to them. In practice, it is a bit more difficult. One problem is that $C(V)$ often includes a constant component, which must be subtracted before a straight line can be obtained. This component can be very difficult to measure; it may be necessary to estimate it or simply to subtract from $C(V)$ whatever constant quantity is necessary to obtain the expected straight-line characteristic.

A greater problem is the difficulty of measuring $C(V)$ with a capacitance bridge, since C_{j0} may be as small as a few femtofarads. To circumvent this problem, a few oversized test diodes may be fabricated on the same wafer as the mixer diodes. The capacitance of a test diode can be measured easily, and C_{j0} of the mixer diodes is found by scaling according to area. If the diode diameters are greater than approximately 10 times the epitaxial layer thickness, edge effects are not too severe, and this scaling is probably valid. The major disadvantage of scaling C_{j0} from test

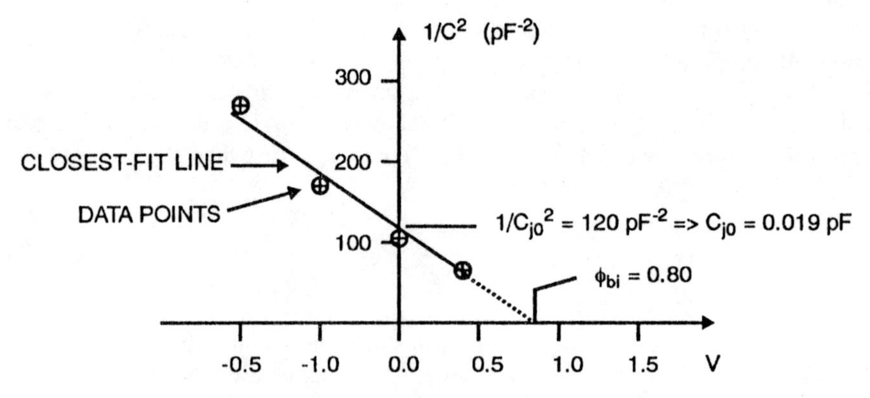

Figure 2.7 C/V characteristic of a Schottky junction formed on a uniformly-doped semiconductor. The data points may deviate from a straight line at high reverse voltages because of nonuniform doping near the epilayer-to-substrate interface.

diodes is in small devices, where edge effects dominate. Also, they are available only to the users who make their own diodes.

Test diodes can also be used to determine the *C/V* characteristic and the doping profile of the wafer. If the doping density of the epitaxial layer is not uniform (and it is not uniform near the bottom of the epitaxial layer), the $1/C^2$ function is not a straight line at voltages that deplete the epitaxial layer completely. The doping density can be found from the relation:

$$N_d = -\frac{2}{q\varepsilon_s \frac{d}{dV}(1/C^2)} \tag{2.28}$$

which may be used with (2.9) to determine the profile. Wafer profiling is particularly important for diodes such as the Mott structure, which have intentionally nonuniform doping.

Cutoff Frequency

A figure of merit often used for mixer diodes is the cutoff frequency f_c, where

$$f_c = \frac{1}{2\pi R_s C_{j0}} \tag{2.29}$$

R_s is generally determined by means of dc measurements, and C_{j0} by a capacitance bridge operating at a few megahertz. Measured this way, f_c may be considerably higher than it would be if measured at the mixer's operating frequency, where the skin effect would increase R_s. Because of the inaccuracy introduced by skin-effect resistance and the difficulty in measuring the junction capacitances of small diodes, methods for measuring f_c directly at microwave frequencies have been developed. Two classical techniques are those of Houlding [6] and DeLoach [7]. A third method is to use a vector network analyzer to measure the S parameters of the diode mounted in a transmission-line fixture; a model can then be fitted to the device in a manner similar to that used for transistors (Chapter 3).

2.2.3 Microwave-Frequency Diode Measurements

Both the Houlding and DeLoach methods are useful for determining diode Q (defined as the ratio of cutoff frequency to operating frequency), or simply the cutoff frequency. Houlding's technique is based on reflection measurements and is most accurate close to the diode's cutoff frequency. It cannot be used to determine R_s and C_{j0} separately. DeLoach's technique requires a transmission measurement, and determines the diode's series resistance, junction capacitance, and contacting whisker or bond-wire inductance independently. It is best used at frequencies well below cutoff, and must be made at the series resonant frequency of the junction capacitance-inductance combination. If the diode Q is not high, it may require measurements over a very broad bandwidth, with concomitant inaccuracies due to calibration problems. Both methods require only scalar measurements.

Since the cutoff frequencies of diodes used for high-performance mixers are on the order of 2,000 GHz, and it is usually necessary to know both the capacitance and resistance, not just the cutoff frequency, DeLoach measurements are generally preferred. Houlding measurements may be valuable, however, where only cutoff frequencies are of interest. This is often the case in large-area, low-frequency mixer diodes or varactors, where junction capacitance can be measured independently with a capacitance bridge. Both methods are subject to error, particularly in R_s, caused by circuit losses.

In both methods, it is important that the incident power be kept low enough that it does not "pump" the diode; that is, it does not cause the diode to rectify. Pumping the diode changes the measured values of junction capacitance.

Houlding Technique

The Houlding technique is based on the measurement of a change of the Q of the diode, which is the ratio of junction reactance to series resistance. The diode, mounted in a fixture, is first connected to a reflectometer (Figure 2.8) and matched via an appropriate tuner. The Q is then changed by varying the diode bias voltage slightly, thereby mistuning the load and raising the *voltage standing wave ratio* (VSWR). The change in Q can be determined from the VSWR as

$$\text{VSWR} = \frac{1 + \left(\dfrac{\Delta Q^2}{4 + \Delta Q^2}\right)^{1/2}}{1 - \left(\dfrac{\Delta Q^2}{4 + \Delta Q^2}\right)^{1/2}} \tag{2.30}$$

which can be approximated for $\Delta Q^2 << 2$ as

$$\text{VSWR} \approx 1 + \Delta Q^2 \tag{2.31}$$

If the C/V characteristic of the diode is known, it is possible to deduce the original Q of the diode. The cutoff frequency is found from the relation

$$Q = f_c/f_m \tag{2.32}$$

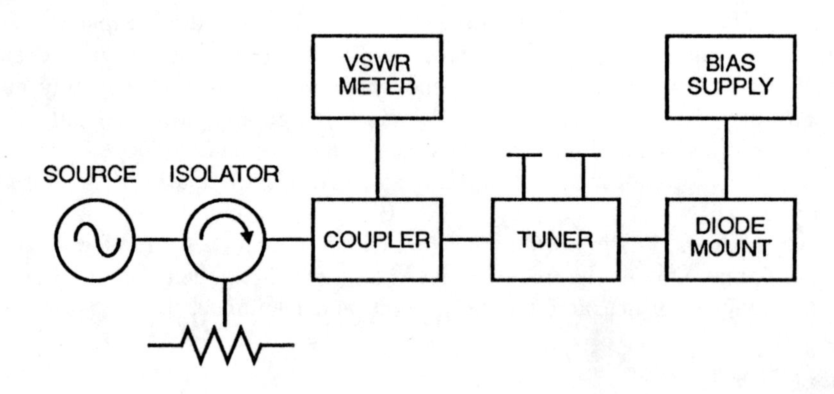

Figure 2.8 Test setup for Houlding measurements.

where f_m is the measurement frequency.

The measurement procedure is as follows:

1. The diode must be mounted in a coaxial or reduced-height waveguide mount similar to that which would be used in a mixer. The choke structure used to bias the diode must have minimal reactance at the measurement frequency.
2. If the diode is waveguide-mounted, an adjustable waveguide short must be used behind the diode and adjusted so that its open-circuit plane is at the diode; that is, it adds no reactance of its own to the mount. This can be done by first adjusting the short so its shorting plane is at the diode (i.e., adjusted to the point where no change in VSWR occurs when diode's bias voltage is varied). The waveguide short is then moved precisely one-quarter wavelength away from this position.
3. The diode mount is then connected to the reflectometer shown in Figure 2.8. The diode's bias is set to 0V and the tuner is adjusted until a VSWR below 1.01 is observed.
4. The diode's bias voltage is then varied slightly (approximately 0.5V to 1.0 V, reverse) and the VSWR is again determined.
5. The cutoff frequency is

$$f_c = \frac{f_m}{\dfrac{C_{j0}}{C_{j1}} - 1} \frac{2}{\sqrt{\Gamma^{-2} - 1}} \tag{2.33}$$

where C_{j1} is the capacitance corresponding to the bias voltage in Step 4, and Γ is the reflection coefficient corresponding to the VSWR measured in Step 4. Note that the ratio of capacitances in (2.33) can be found directly from (2.10), without knowledge of C_{j0} or C_{j1}, as long as ϕ_{bi} is known and the functional form of the C/V characteristic is known.

DeLoach Technique

The DeLoach technique is preferred for measurements of high-performance mixer diodes because it can be used to determine R_s and C_{j0}, not only f_c. Since it is based on transmission measurements, it is not necessary to have an accurately calibrated waveguide short, and the mount can be extremely simple. For millimeter-wave measurements a waveguide mount is needed, although at lower frequencies a coaxial mount may suffice. Since DeLoach measurements involve series resonances, the technique can be applied with good accuracy to packaged diodes.

The principle is illustrated by Figure 2.9, where the diode is described by its equivalent circuit. If the bias is zero or reverse, the junction resistance R_j is large enough to be negligible. Hence, at the series resonant frequency, only R_s is in parallel with the transmission line, and R_s can be found from the increased insertion loss. If the frequency is changed slightly, the transmission through the network depends on the diode Q. When Q is known, junction capacitance can be determined because R_s has been found from the transmission loss at resonance.

The technique is as follows:

1. The diode is mounted in a transmission-type coaxial or reduced-height waveguide test fixture.
2. The test system shown in Figure 2.10 is used to find the frequency of maximum loss (which is the series resonant frequency f_0), the maximum loss L, and the frequencies where the loss is exactly 3 dB less than that at resonance, f_1 and f_2. The diode should be biased at zero volts.
3. The circuit elements of a diode mounted in a coaxial or other *transverse electromagnetic* (TEM) structure are found from the following relations:

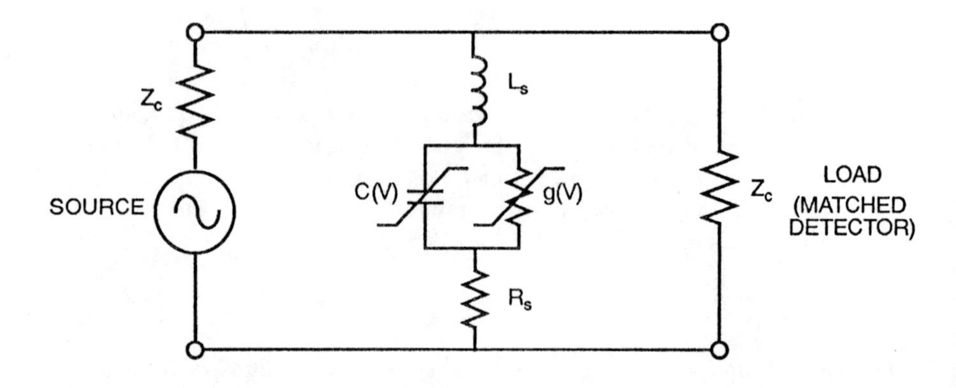

Figure 2.9 Equivalent circuit for DeLoach measurements. L_s is the inductance of the external lead or contacting whisker.

$$R_s = \frac{Z_c}{2(\sqrt{L} - 1)} \tag{2.34}$$

$$C_{j0} = \frac{1}{\pi Z_c}\left(\frac{f_1 - f_2}{f_1 f_2}\right)(\sqrt{L} - 1)(1 - 2/L)^{1/2} \tag{2.35}$$

$$L_s = \frac{1}{4\pi^2 f_1 f_2 C_{j0}} \tag{2.36}$$

4. For a waveguide-mounted diode, the relations are

$$C_{j0} = \frac{1}{2\pi f_1 f_2}\left(\frac{f_1^2 - f_2^2}{f_2 \delta_1 + f_1 \delta_2}\right) \tag{2.37}$$

$$L_s = \frac{1}{2\pi}\left(\frac{f_1 \delta_1 + f_2 \delta_2}{f_1^2 - f_2^2}\right) \tag{2.38}$$

$$R_s = \frac{Z_{g0}}{2(\sqrt{L} - 1)} \tag{2.39}$$

where

$$\delta_{1,2} = \left(\frac{2Z_{g(1,2)}(Z_{g0} + Z_{g(1,2)}(\sqrt{L} - 1))^2 - Z_{g0}^3 L}{4(Z_{g0}L - 2Z_{g(1,2)})(\sqrt{L} - 1)^2}\right)^{1/2} \tag{2.40}$$

Z_c is the characteristic impedance and L is the loss at resonance ($L > 1$). L_s is the series inductance of the whisker or other connection wire. Z_g is the waveguide impedance via the power-voltage definition; that is,

$$Z_{g(0,1,2)} = 754\frac{b}{a}\frac{1}{\sqrt{1-\dfrac{f_{gc}^2}{f_{(0,1,2)}^2}}} \tag{2.41}$$

where b and a are the waveguide height and width, respectively, and f_{gc} is the waveguide's cutoff frequency. Z_{gn} and δ_n are calculated at f_n, $n = 0, 1, 2$. The waveguide dimensions must be chosen so that only the TE_{01} mode propagates; additionally, it is necessary that $R_s << 1.2\,Z_c$ or $1.2\,Z_{gn}$.

One should recognize that, even in the absence of the skin effect, the value of f_c determined by Houlding or DeLoach measurements is generally not the same as that found from dc measurements. In Houlding or DeLoach measurements, R_s is determined at zero bias; however, in dc measurements, R_s is measured at

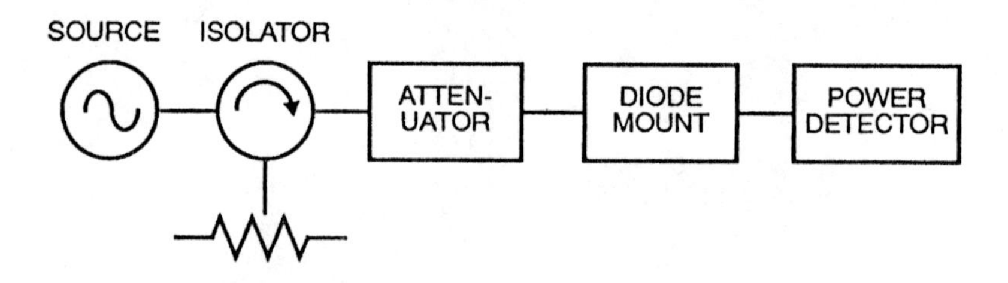

Figure 2.10 Test setup for DeLoach measurements.

0.4V to 0.6V forward bias, depending on material and diode size. In the latter case R_s represents a larger amount of undepleted epilayer, and thus is slightly greater.

Measuring Diode Parameters With a Network Analyzer

It is possible to measure a diode's resistance and capacitance by mounting the diode in a calibrated transmission-line test fixture and measuring its S parameters. The techniques for de-embedding the S parameters of a device from a fixture are well established. When the S parameters are known, it is a relatively simple matter to fit the diode's equivalent circuit to the measured S parameters, or to calculate the diode's parameters analytically.

This method entails the use of a large amount of very expensive test equipment, and therefore may not be a reasonable option for many mixer designers. Aside from this practical problem, there is a more significant limitation: this method can be used only at frequencies where vector network analyzers are available (i.e., below approximately 60 GHz). Unfortunately, the most critical need for accurate diode parameters exists at frequencies above 100 GHz, where those parameters have the most profound effect on mixer performance. This method is best suited to low-frequency measurements of monolithic or beam-lead diodes in microstrip fixtures or on semiconductor wafers; it is not well suited for use with many kinds of diodes that are used in millimeter-wave mixers, especially whiskered dot-matrix diodes.

2.2.4 Effect of Temperature on Diode Characteristics

Temperature effects are important for two reasons. First, mixer diodes exhibit enough temperature sensitivity to affect mixer performance at temperature extremes normally encountered in the environment, and it may be necessary to compensate in some way for those changes. Second, the noise temperature of a mixer can be dramatically reduced by cooling it to cryogenic temperatures. Mixers for very low-noise applications, such as radio astronomy, are often cooled to 12K to 15K using closed-cycle (Gifford-McMahon cycle) helium refrigerators, or even to 4.2K with liquid helium or Joule-Thompson refrigerators. Because of the higher low-temperature mobility of GaAs and absence of carrier freeze-out (which occurs in silicon around 40K), cooled mixers invariably use GaAs diodes instead of silicon.

Equation (2.14) indicates that the diode current becomes more sensitive to voltage at low temperatures, and (2.15) indicates a square law dependence of I_0 on temperature. Therefore, as temperature is reduced, the slope of the I/V characteristic becomes steeper and its knee moves to higher voltages. In GaAs diodes, the ideality factor η stays reasonably constant at temperatures as low as 100K, and then rises as temperature drops further. At 30K or below, the increase in η almost completely compensates for further temperature reduction, keeping the slope of the I/V curve nearly constant. Examples of diode I/V characteristics at low temperatures are shown in Figure 2.11.

One reason for the disappointing rise of the ideality factor at low temperatures is that conduction ceases to be dominated by thermionic emission, and quantum-mechanical tunneling becomes more significant. When thermionic emission dominates, the tunneling component of current varies exponentially with the square root of doping density. Therefore, tunneling can be reduced in diodes designed for cryogenic use by doping them very lightly. For example, GaAs Schottky diodes used at room temperature usually have doping densities around 2×10^{17} cm^{-3}. For cryogenic operation, doping may be as low as 10^{16} cm^{-3}.

Tunneling alone is not adequate to explain the very high ideality factors ($\eta = 5$ to 10) experienced with most diodes at very low temperatures (<20K). The causes in this case are not entirely known, although it may be that the basic thermionic emission theory is not valid at these temperatures [8].

Throughout the cryogenic temperature range, however, and at temperatures up to those that physically damage the diode, the *I*/*V* characteristic remains exponential. Because of the exponential characteristic, the diode's conductance waveform can be kept constant over extremely wide temperature ranges, as long as diode bias and LO power level are modified to compensate. Although the diode's junction capacitance does not behave as conveniently when temperature is reduced, the mixer's conversion loss (although not noise figure) remains essentially constant as well. Therefore, the key to optimizing mixer performance over environmental temperature variations is to change the bias and LO power level appropriately with temperature. It is rarely necessary to retune the mixer. It is a simple matter to make bias circuits that provide this temperature compensation (Chapter 5).

Series resistance is affected both by changes in semiconductor mobility and

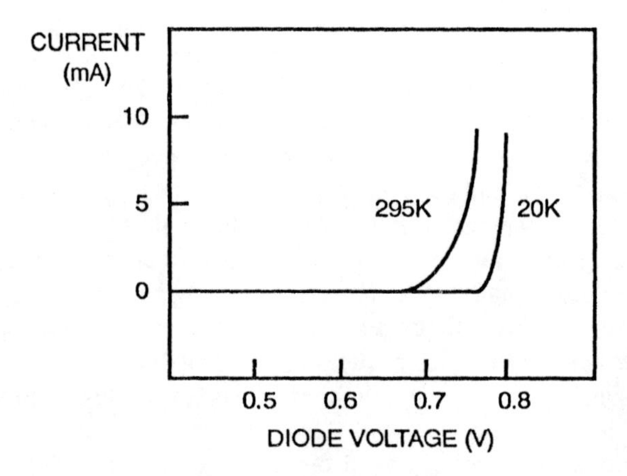

Figure 2.11 Comparison of *I*/*V* characteristics of a Schottky diode at 295K and 20K. At the lower temperature, the knee rises 0.1V to 0.2V and the slope is steeper. However, the *I*/*V* characteristic remains exponential.

quality of the ohmic contact, although at very low temperatures the latter usually dominates. Because it is more difficult to find metals with the appropriate work functions and to dope the semiconductor heavily, ohmic contacts are more difficult to produce on GaAs than on silicon. At low temperatures, the ohmic contact may revert to a rectifying one, which is unfortunately reverse-biased when the diode is forward-biased. The result may be anything from a slight rise in R_s to complete failure of the diode to conduct.

The junction capacitance characteristic generally remains relatively constant with temperature change. For most materials, ϕ_{bi} drops slightly as temperature is reduced, but its temperature coefficient is only about -1.5 mV/K at room temperature. Much more significant is the effect of the change in bias on the junction-capacitance waveform. The knee of the *I/V* curve rises as temperature decreases, so the diode must be biased at a higher voltage at low temperatures than at room temperature, resulting in increased junction capacitance. This capacitance increase degrades the mixer's conversion loss on cooling, and increases the LO power requirements.

2.3 MOTT DIODES

In very-low-noise applications, mixer diodes are sometimes cooled to cryogenic temperatures. However, we noted in Section 2.2.4 that the ideality factor of a Schottky diode increases significantly below 100K, and cooling a mixer to temperatures below about 30K gives very little improvement in noise temperature. If a conventional Schottky diode is used, the reduction in mixer noise temperature on cooling from room temperature to 15K is at most a factor of 2.0 to 2.5, although the mixer's physical temperature may be reduced by a factor of 20 or more [8], [9].

At least three phenomena are responsible for this disappointing situation (there may be more that are not yet understood). One is the effect of tunneling, described in Section 2.2.4, which becomes worse at low temperatures and high doping densities. Reducing the doping density in the epilayer may help reduce tunneling at the expense of increased series resistance in the undepleted part of the epilayer. A second phenomenon is that the nonlinear junction capacitance causes minimum conversion loss and minimum noise temperature to occur under different conditions of tuning, bias, and LO level. The result is that mixers with very low conversion loss often have disappointingly high noise temperatures. A third is that the rise in the diode's "knee" voltage at low temperatures must be compensated by an increase in diode bias, causing the average junction capacitance to increase. Thus, if the nonlinearity of the junction capacitance is reduced, lower noise temperature results.

Although a constant diode capacitance does not absorb power, in conjunction with R_s it creates an RC filter between the LO port and the diode, and thus contributes to LO power loss. The capacitive nonlinearity also causes the reactive part of the diode's input impedance, at the LO frequency, to be a function of LO level, making LO matching more difficult. Hence, if the junction-capacitance nonlinearity is decreased, LO-power requirements can be reduced. Performance may be improved as well, because in many mixers, especially those above 100 GHz, performance is often limited by the availability of LO power, not by the properties of the mixer itself.

Mott diodes have been used to overcome some of these problems. The Mott diode consists of a metal contact to a semiconductor having a thin, lightly doped expitaxial layer and a very heavily doped buffer layer. The epitaxial layer is only a few hundred angstroms thick (compared to 1,000Å to 2,500Å for a conventional Schottky diode). Because it is thin and typically doped at one-tenth the density of a conventional Schottky diode, the depletion region extends through the epitaxial layer and into the heavily doped buffer. Figure 2.12 shows the resulting band structure, charge density, and electric field in an ideal Mott diode at zero junction voltage; one should compare this to Figure 2.2, which shows the same characteristics of the conventional Schottky junction.

The doping density in the semiconductor of a Mott diode must change by a factor of at least 10^3 at the transition between the epitaxial and buffer layers, in a distance of a few tens of angstroms. This abrupt change in doping density cannot be achieved with conventional liquid- or vapor-phase epitaxial growth. Semiconductors for Mott diodes must therefore be fabricated by means of molecular-beam epitaxy.

The light doping in the epitaxial later helps to prevent tunneling at low temperatures; because of the diode's heavily doped buffer layer, minimizing tunneling is particularly important. The Mott diode's ideality factor at very low temperatures is often not significantly different from that of the Schottky diode. Even so, the Mott diode exhibits substantially better low-temperature performance than the conventional Schottky diode.

The junction capacitance of the Mott diode consists of two components that are electrically in series: the capacitance of the fully depleted epitaxial layer, which is small and is not significantly nonlinear, and the much larger nonlinear capacitance of the depleted part of the buffer layer. The total capacitance is dominated by the smaller epilayer capacitance, which is linear, so the nonlinearity of the combination is relatively weak. However, its very thin epitaxial layer causes the Mott diode's zero-voltage junction capacitance to be greater than that of an equal-size Schottky diode.

Keen [8] compares mixer performance of a Schottky diode to that of an equivalent Mott diode. The results are shown in Table 2.1. The Mott diode exhibits

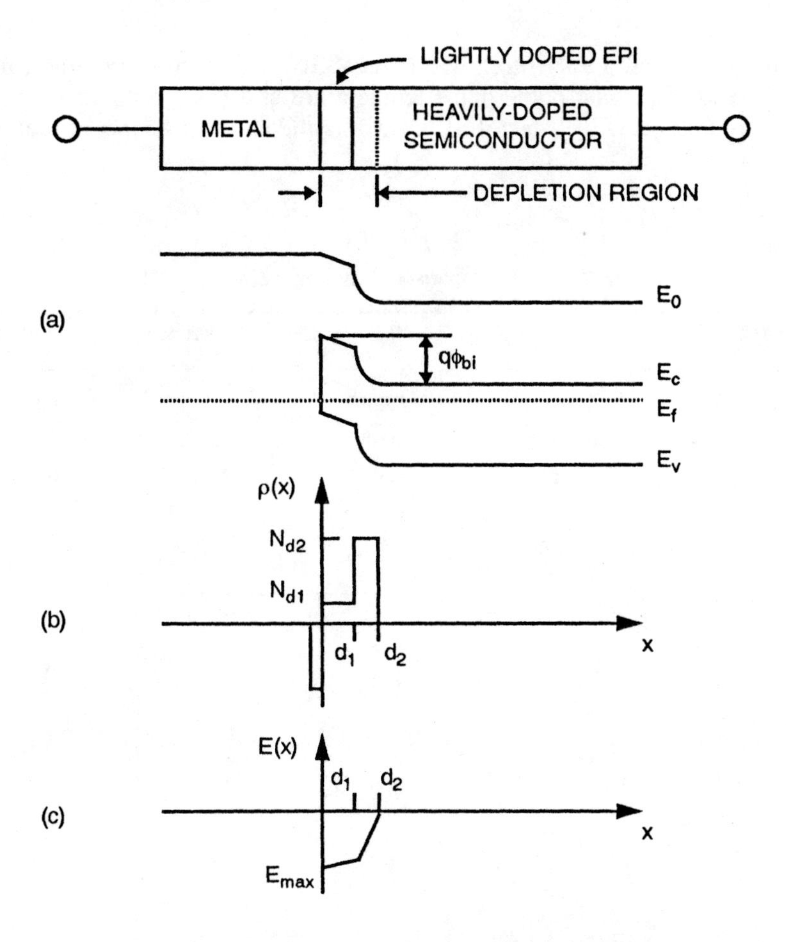

Figure 2.12 (a) Band structure, (b) charge density, and (c) electric field of a Mott diode. Most of the junction voltage drop is across the lightly doped epitaxial layer, which is fully depleted.

poorer room-temperature performance than the Schottky, possibly because of its higher junction capacitance, but is much better at low temperatures and requires 8 dB less LO power. Even at high temperatures the Mott diode requires less LO power. Table 2.2 compares the diodes, and Figure 2.13 shows their *C/V* characteristics. Note that the series resistance of the Mott diode increased, for unknown reasons, from 6Ω at room temperature to 20Ω at 20K. Another 115-GHz mixer reported by Keen, Kelly, and Wrixon [10], using a Mott diode doped at

2.5×10^{16} cm^{-3}, had a 98K *single-sideband* (SSB) noise temperature at a physical temperature of 42K, and 600K noise temperature at 295K. In this case, the noise temperature decreased by a factor of 6.7, while the physical temperature decreased by a factor of 7.0.

Table 2.1

Mixer Performance Comparison: Schottky vs. Mott, $f = 115$ GHz

	300K Schottky	*20K Schottky*	*300K Mott*	*20K Mott*
Conversion loss	6.2 dB	6.2 dB	6.6 dB	6.6 dB
SSB noise temp.	600K	300K	800K	200K
LO power	2.5 mW	1.0 mW	1.0 mW	0.15 mW

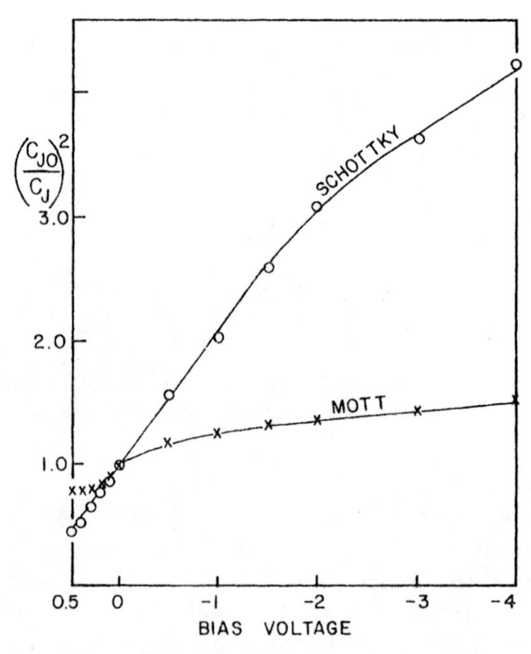

Figure 2.13 *C/V* characteristics of a Mott diode and a Schottky diode having the same C_{j0}. (*Source*: N. Keen et al. [10])

Table 2.2
Diode Comparison

	Schottky	*Mott*
Diode size	2.5 μm (round)	5 × 1.3 μm (oval)
Epitaxial thickness	3000Å	700Å
Epitaxial doping density	2×10^{17} cm^{-3}	3×10^{16} cm^{-3}
Series resistance (300K)	8Ω	6Ω
C_{j0}	8 fF	16 fF

2.4 PRACTICAL DIODE STRUCTURES

2.4.1 Dot-Matrix Diode

Because its series resistance and junction capacitance are small and other parasitics are virtually nonexistent, the dot-matrix chip diode has consistently achieved the best performance of any mixer diode; thus, it is preferred for high-performance mixers at very high (millimeter-wave) frequencies. Dot-matrix diodes frequently have anode diameters as small as 1.5 μm and have junction capacitances as low as 5 fF (5×10^{-15} F). Cutoff frequencies are typically 1,500 to 2,500 GHz, although 3,500 GHz or even 4,000 GHz is not unheard of. Such high cutoff frequencies allow dot-matrix diodes to be used at several hundred or even 1,000 GHz. Either Mott or conventional Schottky diodes can be realized in this structure; the only differences are in the epilayer dimensions and doping densities.

Figure 2.14 shows a cross section of a dot-matrix diode, and Figure 2.15 shows a photograph of a diode. The most striking aspect of the appearance of the dot-matrix diode is the literally thousands of closely spaced diode “dots” (each of which is the anode of an individual diode) on the surface of the chip. A single anode “dot” is contacted by a sharply pointed wire, called a *whisker*; the cathode connection is to the underside of the chip, usually by soldering the chip to an appropriate mounting surface. The large number of anodes allows an anode to be contacted easily by the whisker, even if the individual dots cannot be seen, a situation which is more often the case than not. The contacting whisker need only be touched to the surface of the chip, and the probability is good that it will contact an anode (for more reliable and controlled contacting, a scanning electron microscope with a micromanipulator is

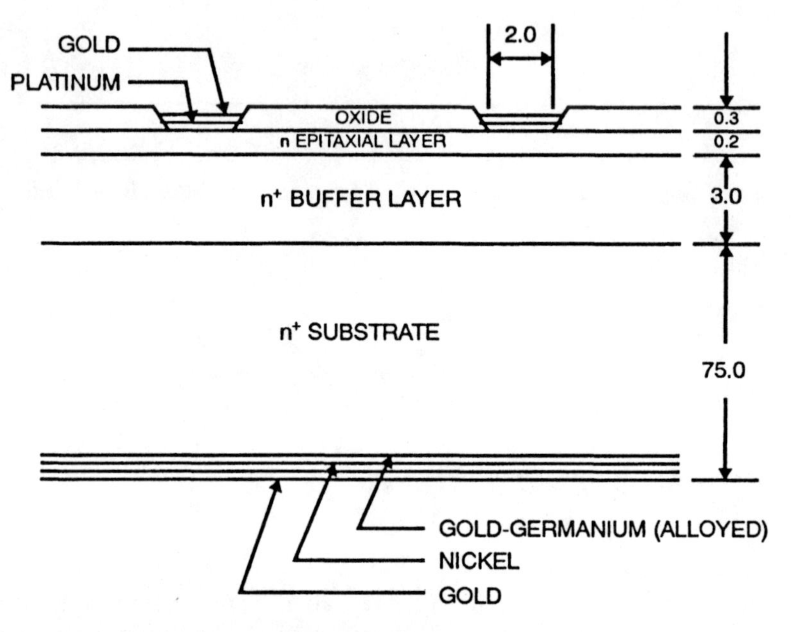

Figure 2.14 Cross section of a dot-matrix diode. All dimensions are in microns, and may vary substantially for different frequencies.

Figure 2.15 Scanning electron microscope photograph of the surface of a dot-matrix diode. This 3.0-μm device has C_{j0} = 0.020 pF, R_s = 3.5Ω, and η = 1.18. (Manufacturer is Farran Technology, Ltd., Cork, Ireland.)

often used to position the whisker [11]). An advantage of this structure is that the chip need not be discarded if a few of the diodes on its surface are defective. If a diode is destroyed by electrical overstress, the whisker need only be moved and recontacted to another dot, and the chip may be used again. Because the dots are closely spaced, it is possible occasionally to contact two anodes simultaneously. In order to prevent this occurrence, some manufacturers have recessed the anode metal in a fairly deep (0.5 to 1.0 μm) oxide hole. This hole may also help to keep the whisker in place when the mixer is subjected to shock or vibration.

GaAs dot-matrix diodes have completely supplanted silicon. Silicon dot-matrix diodes, although used in the past, have been rendered obsolete by the rapidly falling costs of GaAs dot-matrix diodes and high-performance GaAs beam-lead devices. Because the electron mobility of silicon is much lower than that of GaAs, silicon diodes require extremely thin epitaxial layers to achieve low series resistance. Unfortunately, these thin epilayers give silicon diodes low reverse breakdown voltages (approximately 2 V, compared to 6V to 8V for GaAs). Hence, they are more easily damaged by electrostatic discharge or excessive LO power, and may generate noise due to avalanche breakdown. Unlike GaAs, silicon diodes often have a distinct minimum in noise temperature as a function of LO power: as LO power is increased beyond the conversion-loss minimum, noise temperature rises sharply. This rise may be caused by avalanche breakdown.

The GaAs dot-matrix diode is formed on a low-resistivity substrate having a thin (1,000Å to 2,500Å), moderately doped (0.5 to 2×10^{17} cm^{-3}) epitaxial layer. Diodes designed for cryogenic operation may have doping densities as low as 1×10^{16} cm^{-3}. A very pure, heavily doped buffer layer is often used between the epitaxial layer and substrate in order to minimize series resistance and to prevent diffusion of impurities from the substrate into the epilayer during processing. (This is a common practice in the fabrication of many types of diodes.) An oxide layer is grown on top of the epitaxial layer, holes are etched after conventional optical photolithography, and metal is deposited by sputtering, plating, or electron-beam evaporation. To prevent corrosion and to reduce whisker-contact resistance, a gold layer is applied on top of the junction metal by electroplating. An ohmic contact, most frequently alloyed gold-germanium, is applied to the reverse side. Usually, the anode dots are all the same diameter, although diodes having dots of different diameters have been produced. The diameter is a tradeoff between series resistance and junction capacitance, and depends primarily upon the frequency at which the diode will be used.

The choice of a junction metal can affect the performance, electrical characteristics, and reliability of the diode, and, thus, the mixer in which it is used. The most commonly used junction metal for low-noise, millimeter-wave GaAs dot-matrix diodes is platinum. Platinum affords a moderate barrier height, allowing the diode to be operated without dc bias at reasonable LO levels. The main disadvantage of platinum is that it forms $PtAs_2$ at the semiconductor interface at temperatures above approximately 150°C. Exposure to temperatures of 270°C or greater (such high temperatures are frequently used for eutectic soldering) for more than a few seconds can measurably degrade the diode's ideality factor. Indium alloy solders, which have melting points as low as 98°C, are frequently used for attaching platinum-anode dicdes to a circuit.

Gold is rarely used as a junction metal in practical diodes, although it can be used to produce high-quality junctions in research devices. Gold forms an excellent Schottky barrier on GaAs and can be applied very easily by electroplating. However, it creates a relatively large barrier potential, 0.9V, making the use of dc bias necessary. Gold junctions can be damaged by even very modest heating; exposure to temperatures above 100°C for more than 20 or 30 seconds can degrade the junction. Consequently, gold diodes must be attached to the circuit by low-temperature fluxed solders or silver-filled epoxies, both of which may have high resistivity (especially at millimeter wavelengths), may contaminate the diode surface, or may be chemically incompatible with a gold-metallized substrate. Furthermore, a gold anode is so soft that the whisker can be pushed through it easily, forming a point contact with the semiconductor. This point contact, which has very poor *I*/*V* characteristics, is effectively in parallel with the Schottky junction.

Titanium forms an excellent Schottky barrier on GaAs. Titanium junctions can survive temperatures of 350°C for more than one hour, and are hard enough to resist whisker penetration. However, titanium may react with the gold overlayer. This problem can be circumvented by using a thin platinum or tungsten layer between the gold and titanium.

Gold-germanium is used almost exclusively for ohmic contacts. The ohmic contact is often plated with nickel and gold to allow either fluxed solder or eutectic attachment.

One of the main disadvantages of the dot-matrix diode is the difficulty of achieving a mechanically reliable whisker contact. It is very difficult and expensive to create a packaged, whisker-contacted diode, and probably futile, since package parasitics may dominate the mixer's performance, obviating the advantage of the diode's low junction capacitance. An acceptable compromise between performance and practicality involves the use of a dot-matrix diode having a relatively large(>12 μm) diameter. The diode is mounted in a miniature, low-parasitic "pill" package, and instead of a whisker, a small gold wire or ribbon is thermocompression bonded to the anode. These packaged diodes are capable of good performance at frequencies below approximately 50 GHz; at higher frequencies, the package parasitics and excessive junction capacitance limit the bandwidth and conversion loss. Because they are often very lightly doped (to keep the junction capacitance low in spite of their large anode areas), these diodes usually have lower cutoff frequencies than small diodes. High-performance beam-lead diodes, or other whiskerless structures, can also be used to produce good mixers without the need for whiskers. We shall examine these in Section 2.4.4.

2.4.2 Dot-Matrix Diode Variants

Many unusual structures have been proposed to circumvent some of the problems of dot-matrix diodes or to improve their performance. The need for easier handling, applicability to hybrid integrated circuits, and the desire for reduced skin-effect resistance have been the driving forces in their design. Unfortunately, these diodes are not unequivocally superior to conventional dot-matrix devices, and, thus, have never been widely accepted.

The notch-front [12], [13] diode represents an attempt to reduce skin-effect resistance by replacing the relatively lossy current path through the semiconductor with low-resistance metal. In a notch-front diode, the semiconductor wafer is processed much like any other dot-matrix diode. However, the wafer is not sawed into individual chips; instead, saw cuts are made most of the way through the wafer from the epilayer side (Figure 2.16). This leaves a plaid pattern of individual chips connected by a thin piece of substrate at the ohmic side of the wafer. The sides of

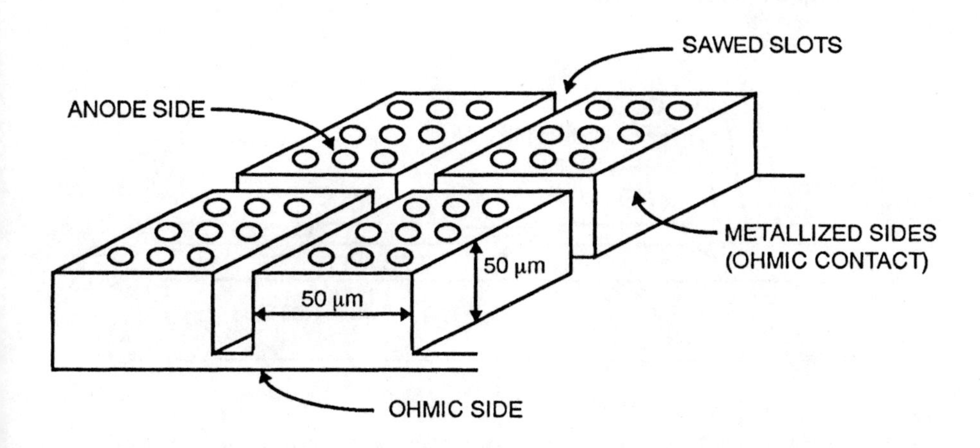

Figure 2.16 Notch-front diode wafer (before separating the chips).

the chips are then metallized and the wafer is broken or sawed into individual diodes. Because the chip is metalized on its side, it can be mounted on its side in microstrip or suspended-substrate stripline circuits, creating a very short path between the epilayer and the mounting surface.

Another dot-matrix variant is the moat-etched or "bathtub" diode [11] (Figure 2.17). In this structure, the epilayer is rounded under the anode by a series of etching and anodizing steps performed before the junction metal is deposited. Because the anode is round instead of flat, the current over its surface is relatively uniform, and edge effects that would otherwise increase the series resistance are minimized.

Another way to reduce edge effects is to use a diode anode having a large periphery relative to its area (i.e., having any shape other than a circle). According to this reasoning, cross-shaped anodes have occasionally been created. These have not exhibited significantly better RF performance than conventional diodes, however, and are probably difficult to fabricate and whisker.

2.4.3 Point-Contact Diodes

It is worth examining point-contact diodes, if only to make the neophyte appreciate modern Schottky diodes. Packaged point-contact diodes are still in occasional use in test equipment such as test mixers and detectors, where high performance is not needed and ease of diode replacement is important. It is worth remembering, however, that point contact diodes were the high-performance devices of their day, and were supplanted only when precision photolithography and high-quality

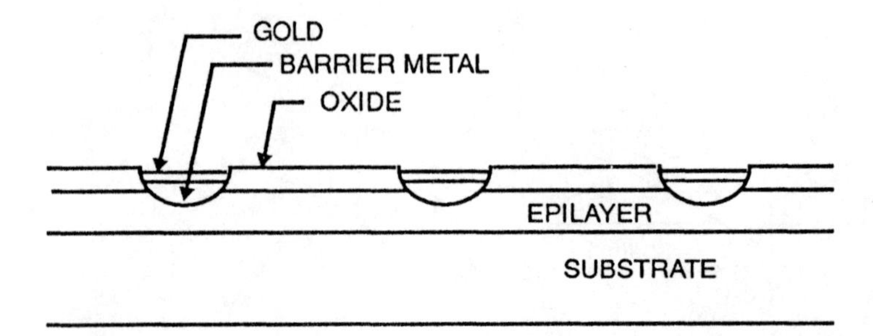

Figure 2.17 Moat-etched diode cross section. Except for the rounded junction, dimensions and doping densities are conventional.

epitaxy made very small Schottky diodes possible.*

The point contact diode consists of a whisker contact to a bulk silicon crystal (i.e., one that does not have an epitaxial layer). The crystal is usually mounted in a large cartridge package, or, for high frequencies, in a Sharpless mount (Section 2.5). The point-contact diode is therefore a type of Schottky diode, although the haphazard nature of its construction does not allow the consistently good performance achieved by modern Schottky diodes.

To minimize series resistance, the silicon crystal is heavily doped. However, the Schottky junction requires a lightly doped contact layer. To create this layer, an oxide layer is first grown on the silicon. The dopant near the silicon-to-oxide interface is then diffused into the oxide by heating the substrate to 900° to 1,100°C, and the oxide is etched away. The whisker, usually tungsten (a very hard metal), is electrochemically etched to a sharp point and contacts the surface of the diode.

The crystal surface inevitably develops an oxide coating, and perhaps other contaminants, upon exposure to air. Therefore, when the whisker contact is made, a poor *I*/*V* characteristic results. To achieve a good metal-semiconductor contact, the whisker must be forced through the contaminant layer by mechanical shock. This process is euphemistically called "tuning" the diode mount, and involves hitting it with a small hammer while observing the *I*/*V* curve on a curve tracer. When the process is complete, the shape of the whisker tip, the area of the junction, the degree of damage to the crystal near the contact, and their effect on mixer performance are all random variables.

The point-contact diode is a relatively expensive device to produce compared to the Schottky diode, because it requires more human labor. Therefore, replacement point-contact diodes are often fabricated as Schottky diodes and mounted in packages having the same external dimensions.

2.4.4 Beam-Lead Diodes

The use of hybrid microwave integrated circuit technology for high-frequency components has created a need for a small, low-parasitic diode having ribbon leads. Beam-lead diodes meet this need extremely well.

There are many different designs for beam-lead diodes. A common type is shown in Figure 2.18. The diode is formed on a substrate similar to that used for a

---

* One could make the case that nothing is ever completely obsolete. In 1987, the author had the indescribable experience of working on a point-contact diode mixer that was part of an Atlas launch vehicle pressed into service when the Space Shuttle was grounded. Markings on the unit indicated that it was originally built in the late 1950s as part of the Mercury program.

conventional dot-matrix diode; however, the anode is located near the edge of the chip and the ohmic contact is made from the top side. The anode and ohmic contacts have integral gold ribbons for attaching the device to a circuit.

The major limitations of the beam-lead diode are related to its junction capacitance. In order to create an adequate anode connection, the anode diameter must be at least 10 to 15 μm, and the resulting junction capacitance is at least 0.08 to 0.10 pF. Furthermore, the diode has additional parasitic in the form of an overlay capacitance between the anode ribbon and the semiconductor, and between the anode and cathode metallizations. In small diodes, this overlay capacitance may be even greater than the junction capacitance; the resulting large total capacitance restricts the use of conventional beam-lead diodes to frequencies below approximately 30 GHz.

The parasitic capacitance can be minimized only by increasing the oxide thickness (which is limited by the difficulty of etching accurate holes through a thick oxide layer), by increasing the spacing between the cathode and anode (which increases series resistance), or by moving the anode close to the edge of the chip (which makes the structure extremely fragile). The latter option is the one usually employed. In spite of these difficulties, beam-lead diodes can be fabricated at low cost and often have parasitic capacitances lower than those of other types of packaged diodes. Creative ways to reduce the parasitics of beam-lead diodes are the subject of the next section.

A second limitation that the beam-lead diode shares with all planar diodes is that

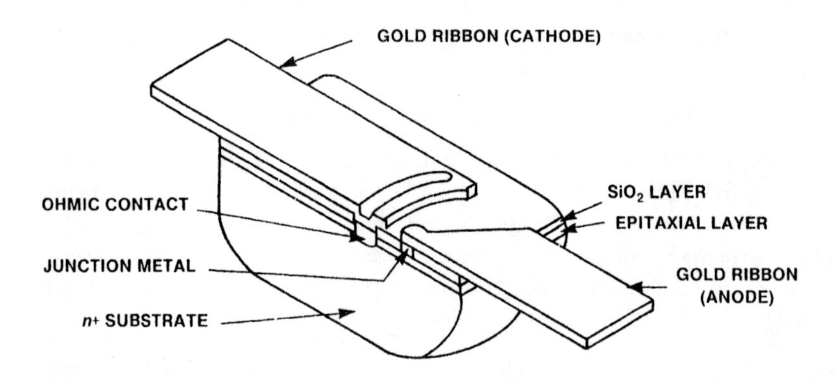

Figure 2.18 Cutaway view of a beam-lead diode.

the cathode connection must be made from the top of the structure. This requirement usually results in a relatively high series resistance, which is caused by two factors: the first is that electrons must travel through a relatively long semiconductor path to the cathode, and the second (and usually more significant) is that the current injection is nonuniform over the area of the anode.

2.4.5 Low-Parasitic Diodes for Millimeter-Wave Applications

Designs for low-parasitic beam-lead diodes have been published periodically [14]–[17], but few have been sufficiently well received to result in commercial products. One exception is that of [15], shown in Figure 2.19(a).

This device achieves low parasitic capacitance by using a very thick oxide layer on top of the semiconductor. The oxide layer is 2 to 3 μm thick, and extends well beyond the edge of the chip. The support provided by the oxide layer allows the anode ribbon to be very narrow; the combination of the narrow ribbon and the thick oxide results in a low overlay capacitance of approximately 10 fF. The manufacturers of this diode have reportedly succeeded in making very small holes in this thick oxide layer, and achieving junction capacitances of 20 fF. The series resistance is 6Ω, resulting in a cutoff frequency of 2,650 GHz. This is adequate for use at 100 GHz.

Unfortunately, this structure is at least as fragile as a conventional beam-lead device. The main shortcoming in its mechanical design is that it replaces the wide yet flexible anode lead of the conventional device with a thin, brittle oxide layer. This oxide layer may crack under any but the most careful handling.

Another approach to the creation of a low-parasitic diode is the "surface-channel" device of [17]. The low overlay capacitance of this device is achieved through the use of an air gap (the "surface channel") between the anode contact pad (it does not use ribbons) and the cathode. A cross section of this device is shown in Figure 2.19(c); because the open space in the surface channel has a very low dielectric constant, the only significant parasitic capacitance arises in the long path through the semi-insulating layer under the channel. This capacitance is obviously very small. Furthermore, the diode is mechanically very rugged, and can be wire bonded or flip-chip mounted.

The surface-channel device, with a 2.5-μm anode, has a junction capacitance of 6 fF and a parasitic capacitance of 15 fF. The series resistance is 6Ω. It has been used to achieve a 950K single-sideband noise temperature and 6.4-dB conversion loss at 110 GHz. This is comparable to a good whiskered dot-matrix diode.

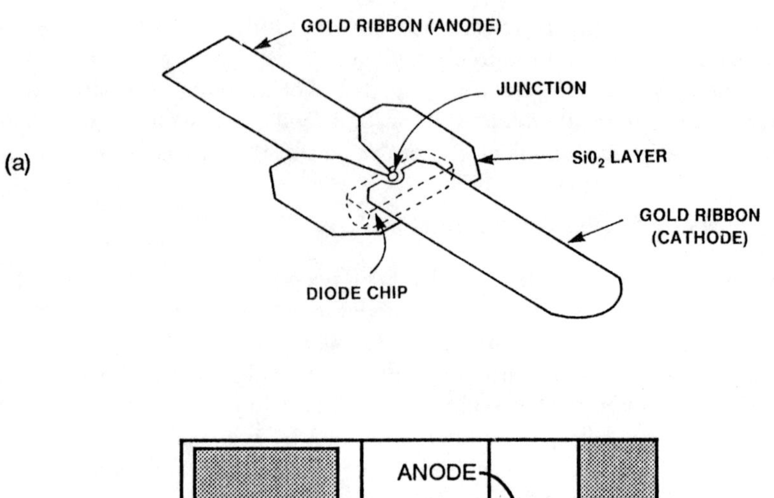

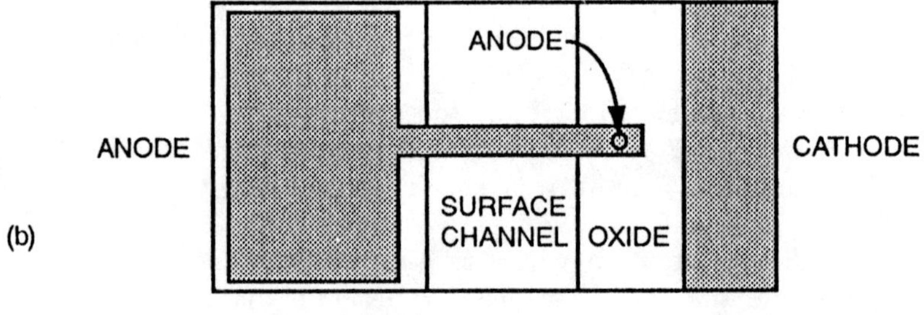

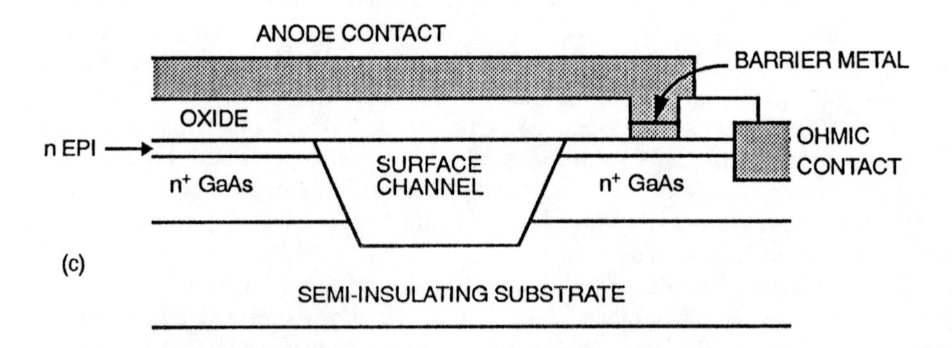

Figure 2.19 Millimeter-wave low-parasitic diodes: (a) oxide-insulated beam lead; (b) surface-channel diode, top view; (c) surface-channel diode, cross section. (Parts (b) and (c) after [17], © 1987 IEEE.)

2.4.6 Diodes for Monolithic Circuits

Diode mixers are commonly realized as GaAs monolithic (MMIC) circuits. Because FETs and diodes are often fabricated on the same wafer, the fabrication technologies for diodes must often be compatible with FET technologies. To simplify fabrication processes, this requirement for compatibility is often invoked even when FETs and diodes are not mixed.

There are two approaches to the design of MMIC diodes: the first is to use a FET structure to realize a diode; the second is to develop a unique yet compatible diode process. The first of these alternatives can be accomplished most easily by using a FET gate metallization as an anode and the drain and source regions as the cathode. It is clearly undesirable to use the standard FET geometry for such "FET diodes," because the FET's gate resistance is intolerable. Instead, it is customary to use a structure having the same vertical structure as the FET (i.e., the same epilayer and ohmic-contact region thicknesses and doping densities, but a circular or square anode). To minimize series resistance, the anode is surrounded by the cathode region.

The disadvantage of this approach is that the FET structure is simply not a very good one for use as a diode. Although thicker than that of a dot-matrix diode, the epilayer is usually too thin for a device that has considerable lateral current. Furthermore, the current injection from the anode is nonuniform, causing the series resistance to be high. It is not unusual for such diodes to have series resistances on the order of 50Ω, and cutoff frequencies of only 100 GHz.

FET diodes are usually acceptable for low-frequency, moderate-performance applications. However, at high frequencies or when low conversion loss is necessary, they are usually inadequate, and an uncompromised diode design must be employed. One of the most successful is a mesa structure [18]-[22] using an air-bridge anode. Such a device is shown in Figure 2.20.

The mesa diode is fabricated on a semi-insulating substrate having a high-conductivity buffer layer and an ordinary epilayer. The high-conductivity buffer is used as the cathode. Because it extends under the epilayer, current uniformity is good and series resistance is low. However, the rest of the circuit must be insulated from this layer, so the epilayer and buffer must be etched away everywhere except at the diode. As a result, the diode is formed on a mesa. The anode is formed near the edge of the mesa, and the ohmic contact encircles it on three sides. To minimize junction area and overlay capacitance, and to insulate the anode lead from the mesa, the anode connection consists of an air bridge.

Mesa diodes have been realized as both conventional Schottky [19] and Mott [18] structures. The diode in [18] had 7 fF overlay capacitance, 25 fF junction

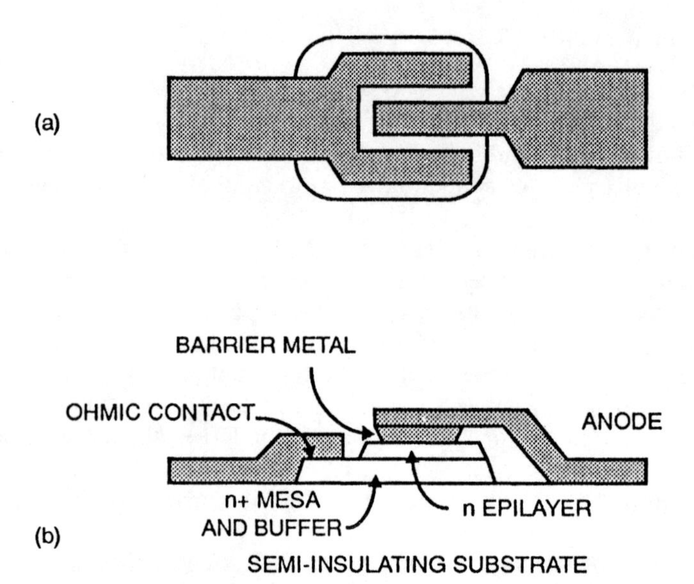

Figure 2.20 Mesa diode: (a) top view; (b) cross section. The ohmic contact surrounds the anode to minimize the series resistance.

capacitance, and 10Ω series resistance. Additionally, the air-bridge anode connection had 0.01 nH inductance. Conversion loss of a 30-GHz mixer using this diode was 6 dB. The diode in [19] had 30 fF junction capacitance and 6Ω series resistance, and achieved approximately 6 dB conversion loss in the 30- to 40-GHz range.

It is possible also to use a proton-bombarded region, instead of a surface channel, as an insulator. This creates a planar diode and obviates the need for an air bridge; however, the proton-bombarded region probably has a greater parasitic capacitance than an open channel. Such a diode is described in [23]. Although this diode was designed for detector applications, the same technology is applicable to mixers. The anode, which is relatively long and narrow to minimize series resistance, was defined by a self-alignment process. The series resistance is 16Ω, the junction capacitance is 5.5 fF, and the parasitic capacitance is 3.5 fF. This results in a cutoff frequency of 1.1 THz, good enough for many millimeter-wave mixers.

2.5 CONSIDERATIONS IN USING DOT-MATRIX DIODES

2.5.1 Mount Mechanical Design

At high frequencies, diodes are often mounted in a waveguide. One popular mount design, which was originally developed for millimeter-wave point-contact diodes, is that of Sharpless [24]. *Sharpless mounts* are still in common use, and most other waveguide mounting schemes for dot-matrix diodes are variations on this theme.

A Sharpless mount is shown in Figure 2.21. The diode is mounted on the end of the IF pin, which protrudes into the top of the waveguide slot. Another pin, with the contacting whisker attached, is pressed into a hole on the opposite side. Because the wafer is quite thin (approximately 0.05 inches), the diode surface is easy to see under an optical microscope, and whiskering the diode is relatively easy. If the lengths and characteristic impedances of the coaxial sections are appropriately chosen, the IF pin can be used as a low-pass filter. The diode can also be biased through the IF pin.

The main disadvantage of the Sharpless mount is its fabrication cost. Such mounts usually must be handmade, because the mechanical tolerances for their

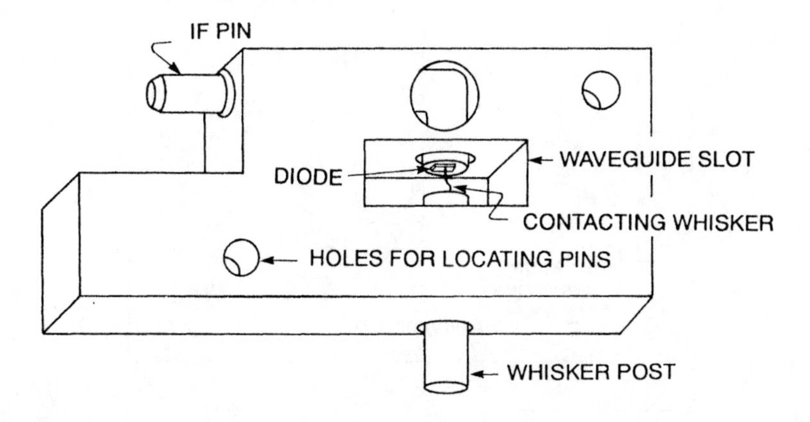

Figure 2.21 The classical Sharpless wafer. The post on which the diode is mounted and the IF pin are a single piece, which serves as a rudimentary low-pass filter. The whisker pin is press-fit into a counterbored hole in the wafer, and it contacts the wafer over approximately 0.050 to 0.100 inch of its length.

alignment with the waveguide are often very tight: at frequencies above 100 GHz, alignment tolerances of ±0.0005 inch are often necessary. Other designs for diode mounts have varying degrees of reliability or ease of assembly [25] – [27]; however, these have not solved the problem of the high fabrication cost.

A modern variant of the Sharpless mount, having excellent mechanical features (but relatively high cost), is shown in Figure 2.22. In this mount, the IF pin is replaced by a coaxial low-pass filter, supported by a dielectric insulator. The design of this filter (which we shall discuss in Chapter 6) is a critical factor in achieving good performance.

Attention to the mechanical design of a Sharpless wafer or similar structure can do much to guarantee that the diode whisker will not lose contact with the anode when the structure is subjected to mechanical shock or vibration. Because the whisker has very little mass, shock or vibration sensitivity of the wafer is rarely caused by movement of the whisker. Instead, the movement of the IF pin, on which the diode is mounted, causes loss of contact. Similarly, temperature sensitivity is usually caused by thermal instability of some part of the mount, not simply expansion or contraction of the whisker. (One of the most common causes is the

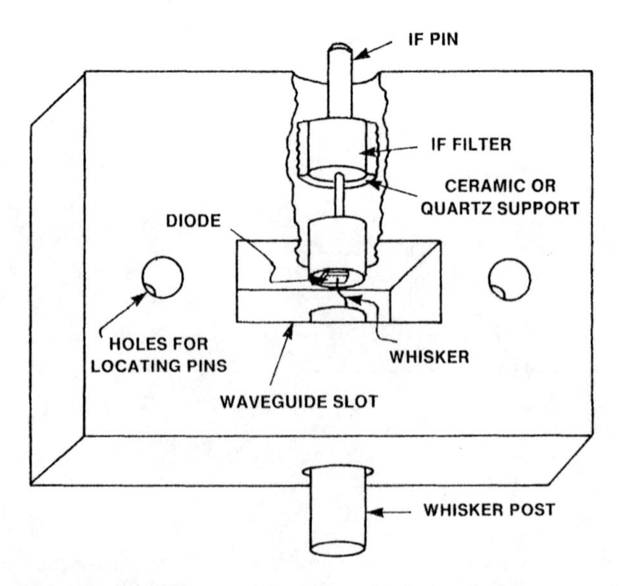

Figure 2.22 A modern variant of the Sharpless wafer. The ceramic support is soldered in place for thermal stability. The structure allows a more careful and controlled design for the IF filter, which is critical to the performance of the mixer.

practice of embedding the IF pin in a dielectric, especially plastic or epoxy; dielectrics have relatively high thermal expansion coefficients.) Loss of contact as the mixer is cooled to very low temperatures is often caused by the whisker slipping sideways in response to increased vertical (axial) forces. It is possible to design the shape of the whisker so that axial forces are not translated to lateral forces [11].

2.5.2 Diode Whiskering

The problem of creating a reliable whiskered diode contact often obviates the use of a dot-matrix diode, where it would otherwise be ideal. It may be surprising to some to learn that such diodes have been qualified to spacecraft reliability standards and are frequently cooled to cryogenic temperatures. A well-designed diode mount is, in fact, quite rugged.

The material used for the whisker wire is critical to the mixer's mechanical reliability. Unfortunately, the most commonly used metals are tungsten and phosphor bronze, chosen more for historical reasons (their earlier use in point-contact diodes) than for technical ones. The use of one of these hard metals for the whisker often creates problems, especially in cooled mixers: when the mixer is cooled, the mount contracts and increases the force that the whisker tip applies to the anode. After several temperature cycles, the point penetrates the anode metal and contacts the epitaxial layer, causing a poor I/V characteristic. A better whisker material is an alloy called *nioro*, which is 82% gold and 18% nickel. Nioro has far better mechanical properties and lower RF loss than either tungsten or phosphor bronze. Nioro whisker wire should be fully annealed.

It is important in fabricating or selecting diodes for millimeter-wave applications that parasitic capacitance be minimized by preventing metal buildup on the oxide surface around the edge of the dot. It is also clearly advantageous for the whisker point to be sharp and well centered on the dot. Minor increases in capacitance caused by a poor whiskering job (in which the whisker tip deforms or slides part way off the diode's anode) can increase conversion loss by as much as 1 dB at 115 GHz.

The fit of the whisker pin into the diode wafer and the material selected for both the pin and the wafer are important. Soft metals (such as aluminum) and metals having high expansion coefficients should be avoided; steel or one of the harder brass alloys are best. A good choice of material for the whisker post is a hardened steel pin. These are available as "gage pins," used by machinists to measure hole diameters. Their diameters are commonly available in 0.0001-inch increments, so a pin of the appropriate size can be obtained easily. The gage pin can be ground to the correct length. It is a good idea to bevel its edges slightly to prevent gouging as it is pressed into the wafer.

The whisker pin should be gold-plated to a thickness of 50 to 100 microinches; a copper flash before the gold plating will ensure good gold adhesion and will allow the soldering of the whisker wire to the pin if desired. The gold plating serves as a lubricant, allowing the pin to move smoothly, and prevents the pin from "creeping" after the diode has been whiskered. The hole in the wafer should be reamed with a new precision reamer, and its diameter measured. The whisker pin diameter should be 0.0001 to 0.00015 inch greater than the hole diameter.

The whisker wire can be soldered or welded to a plated pin. It must have a bend to absorb stress after it contacts the diode; straight whiskers are rarely reliable. The bend can be formed by hand under a microscope or with a specially designed micromanipulator. Bending the wire around a blunt knife edge or a cleanly cut piece of stainless-steel shim stock works well.

The whisker point should be etched as the last step before inserting the whisker pin into the diode wafer. If nioro wire is used, it can be etched by dipping the end into a potassium-cyanide solution and passing a dc current through it. To prevent corrosion of the electrode and contamination of the etchant, the electrode that is immersed in the solution should be titanium. To obtain a symmetrical point, the electrode should be oriented so that the current is uniform over the end of the wire. The shape of the point after it is etched depends strongly upon the chemical concentration; the higher the concentration, the sharper the point will be. Some type of fixture is necessary to hold the whisker pin and to lower it into the solution precisely. It is also necessary to be careful with the solution because it is highly toxic; a few drops, if swallowed, could be fatal. The fumes from it (which are hydrogen cyanide) are also highly toxic. Above all, cyanides must be kept far away from acids.

Whiskering the diode can be performed under a medium-power optical microscope that looks down the axis of the waveguide slot in the wafer and views the diode edge-on. It is not necessary to see individual diode dots or even to tilt the wafer for a view of the diode's surface. It is helpful, however, to have a diffuse light source behind the wafer. As the whisker tip is brought close to the surface of the diode, its reflection can be seen in the oxide layer; the whisker is advanced until the tip and the reflection meet. If a semiconductor curve tracer indicates that a diode has been contacted, the whisker should be advanced another 0.0001 to 0.0005 inch, depending on its spring constant, to provide additional mechanical loading to keep the whisker in place. If a diode has not been contacted, the whisker can be withdrawn, the pin rotated slightly to move the whisker to a new location, and the contacting procedure repeated. It is necessary to move the whisker pin with a micrometer adjustment.

2.5.3 Packaged Diodes

Any diode that can be ribbon-bonded can be mounted in a metal-ceramic package. In order to be ribbon-bonded reliably, the anode diameter must be at least 8 μm, and preferably greater than 12 μm. The use of a relatively large diode limits packaged diodes to prosaic applications and to frequencies below approximately 50 GHz. Package parasitics may also make wideband operation considerably less successful than would be the case with chip devices. However, packaged devices are sometimes necessary, for example, in severe environments or for high-reliability applications. Although packaged devices cost more than chips, they may reduce overall manufacturing costs by reducing hand labor and obviating the need for specialized whiskering equipment. Packaged diodes are also useful for balanced mixer circuits such as the crossbar type (Chapter 8), where two diodes having different polarities are needed, and whiskered chips are therefore not practical.

Two of the most commonly used diode packages are shown in Figure 2.23. Both packages are made of alumina ceramic ($\varepsilon_r = 9.8$) and have gold-plated kovar end-caps matched to the thermal expansion coefficient of the ceramic. If properly

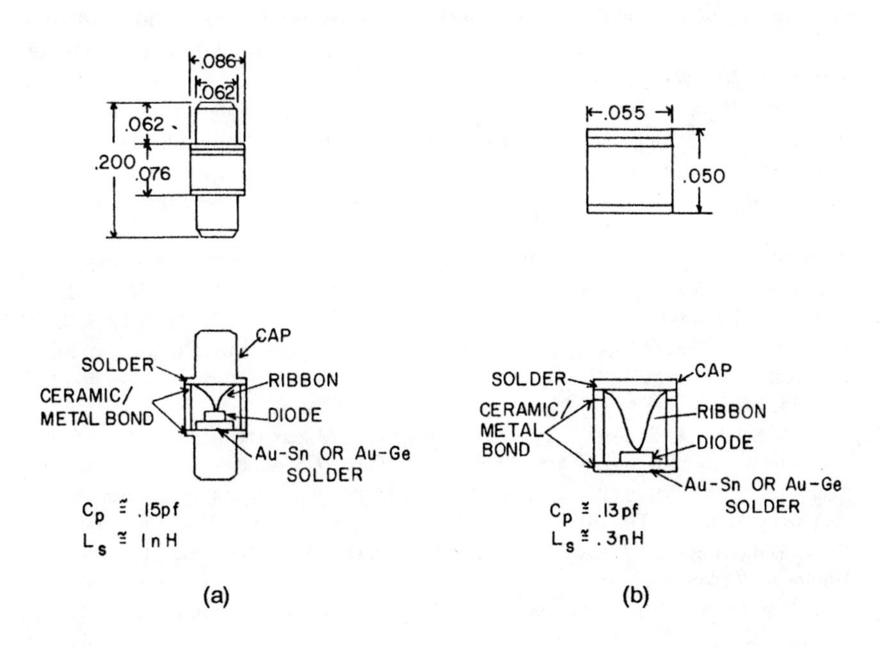

Figure 2.23 Ceramic-metal packages for chip diodes: (a) "pill" package; (b) "micro-pill" package.

assembled, the packages are hermetic. The package in Figure 2.23(a) is commonly called a "pill" package; its pins are useful for mounting the diode. It is, however, somewhat larger than the "micro-pill" package of Figure 2.23(b), and has larger parasitics.

The micro-pill package is somewhat more difficult to mount, but has substantially smaller parasitics. It also fits better into reduced-height waveguide, which is usually necessary for optimal matching (Chapter 6). A common method of attaching this package is to solder it into its circuit with low temperature solder (to avoid high temperatures that might damage the diode (Section 2.4.1) or melt the solder that holds the package together). It is also possible to bond a ribbon to either terminal.

It is important in mounting these packages to avoid stress at thermal extremes, as might be caused by soldering both terminals into a rigid structure. One way to avoid such stress is to connect one terminal by soldering and another by a spring finger, a miniature bellows, or a wire or ribbon bond.

2.6 REFERENCES

[1] Schottky, W., "Halbertheorie der Sperrschicht," *Naturwissenschaften*, Vol. 26, 1938, p. 843.

[2] Bethe, H. A., "Theory of the Boundary Layer of Crystal Rectifiers," MIT Radiation Laboratory Report 43-12, 1942.

[3] Sze, S. M., *Physics of Semiconductor Devices,* 2nd ed., New York: John Wiley and Sons, 1981.

[4] Rhoderick, E. H., "Metal-Semiconductor Contacts," *IEE Proc.*, Part I, Vol. 129, 1982, p. 1.

[5] Held D. N., and A. R. Kerr, "Conversion Loss and Noise of Microwave and Millimeter-Wave Mixers: Part I—Theory; Part 2—Experiment," *IEEE Trans. Microwave Theory Tech.*, MTT-26, 1978, p. 49.

[6] Houlding, N., "Measurement of Varactor Quality," *Microwave J.*, Vol. 3, 1960, p. 40.

[7] DeLoach, "A New Technique to Characterize Diodes and an 800-GHz Cutoff Frequency Varactor at Zero Volts Bias," *IEEE Trans. Microwave Theory Tech.*, MTT-12, 1964, p. 15.

[8] Keen, N. J., "The Mottky Diode: A New Element for Low-Noise Mixers at Millimeter Wavelengths," *Proc. AGARD Conference on MMW and Sub-MMW Propagation and Circuits*, No. 245, 1978, p. 16-1.

[9] Weinreb, S., and A. R. Kerr, "Cryogenic Operation of Mixers for Millimeter and Centimeter Wavelengths," *IEEE J. Solid-State Circuits*, Vol. SC-8, 1973, p. 58.

[10] Keen, N. J., W. M. Kelly, and G. T. Wrixon, "Pumped Schottky Diodes with Noise Temperatures of Less Than 100 K at 115 GHz," *Electron. Lett.*, Vol. 15, 1979, p. 689.

[11] "New Dimensions in Microwave and MMW Power Measurements," (Product Feature) *Microwave J.*, Vol. 27, Sept. 1984, p. 229.

[12] Schneider, M. V., "Low-Noise MM-Wave Schottky Mixers," *Microwave J.*, Vol. 21, Aug. 1978, p. 78.

[13] Schneider, M. V., and E. R. Carlson, "Notch-Front Diodes for MMW ICs," *Electron. Lett.*, Vol. 13, 1977, p. 745.

[14] Clifton, B. J., "Schottky-Diode Receivers for Operation in the 100-1000 GHz Region," *Radio and Electronic Engineer*, Vol. 49, 1979, p. 333.

[15] Cardiasmenos, A. G., "New Diodes Cut the Cost of MMW Mixers," *Microwaves*, Sept. 1978, p. 78.

[16] Ballamy, W. C., and A. Y. Cho, "Planar Isolated GaAs Devices Produced by Molecular Beam Epitaxy," *IEEE Trans. Electron Devices*, Vol. ED-23, 1976, p. 78.

[17] Bishop, W.L., et al., "A Novel Whiskerless Schottky Diode for Millimeter and Submillimeter Wave Applications," *IEEE MTT-S International Microwave Symposium Digest*, 1987, p. 607.

[18] Nightingale, S. J., et al., "A 30 GHz Monolithic Single Balanced Mixer with Integrated Dipole Receiving Element," *GaAs Monolithic Circuits Symp. Digest*, 1985, p. 74.

[19] Meier, P. J., et al., "IC Techniques Slice Cost of Integrated Receiver," *Microwaves and RF*, June, 1985, p. 90.

[20] Archer, J. W., R. A. Batchelor, and C. J. Smith, "Low-Parasitic Schottky Planar Diodes for Millimeter-Wave Integrated Circuits," *IEEE Trans. Microwave Theory Tech.*, Vol. MTT-38, 1990, p. 15.

[21] Jacomb-Hood, A. W., et al., "30 GHz and 60 GHz GaAs MMIC Microstrip Mixers," *IEEE GaAs IC Symposium Digest*, 1986, p. 195.

[22] Goldwasser, R., et al., "Millimeter-Wave Monolithic IC in Production," *Microwave Systems News* (special issue), July 1988, p. 16.

[23] Zah, C., et al., "Millimeter-Wave Monolithic Schottky Diode Imaging Arrays," *Int. J. Infrared and Millimeter Waves*, Vol. 6, 1985, p. 981.

[24] Sharpless, W. M., "Gallium Arsenide Point-Contact Diodes," *IRE Trans. Microwave Theory Tech.*, Vol. 9, 1961, p. 6.

[25] Kerr, A. R., R. J. Mattauch, and J. A. Grange, "A New Mixer Design for 140-220 GHz," *IEEE Trans. Microwave Theory Tech.*, MTT-25, 1977, p. 399.

[26] Archer, J. W., "All Solid-State Low-Noise Receivers for 210–240 GHz," *IEEE Trans. Microwave Theory Tech.*, MTT-30, 1982, p. 124.

[27] Archer, J. W., "An Efficient 200–290 GHz Frequency Tripler Incorporating a Novel Stripline Structure," *IEEE Trans. Microwave Theory Tech.*, MTT-32, 1984, p. 421.

Chapter 3
MESFETs and HEMTs

The primary device for microwave mixers above 1 GHz is the Schottky-barrier diode. Nevertheless, the successful development of low-noise GaAs *metal-semiconductor field-effect transistors* (MESFET) has made it possible to design microwave and millimeter-wave FET mixers having performance equal or superior to that of diode mixers. Dual-gate MESFETs are particularly well suited to mixer applications: with one gate used for the LO and the other for the RF input, dual-gate devices have inherently good LO-to-RF isolation. They can also be used as self-oscillating mixers for low-cost commercial microwave applications.

FET mixers can achieve lower noise and intermodulation than diode mixers, and require less LO power. However, the most significant difference between a FET mixer and a diode is that the FET is capable of achieving conversion gain instead of loss. Therefore, a receiver using a FET mixer needs fewer stages and requires fewer low-noise amplifier stages ahead of the mixer.

Because it is often difficult to fabricate high-quality diodes in a monolithic circuit technology that is compatible with FETs, FET mixers are often better suited than diodes for use in MMICs. Recent progress in the development of microwave FET mixers has been motivated largely by the growth of MMIC technology. Furthermore, the successful development of *high-electron-mobility transistors* (HEMT) (MESFETs fabricated on heterojunction materials) and HEMT MMICs has opened the millimeter-wave region to active FET mixers.

3.1 MESFET STRUCTURE AND OPERATION

3.1.1 Structure

The GaAs MESFET is a *junction field-effect transistor* (JFET) realized on GaAs instead of silicon and having a Schottky-barrier gate instead of a *pn* junction. Its length (i.e., its dimension in the direction of electron flow) is usually less than 0.5 μm, and may be as short as 0.15 μm for millimeter-wave devices. Because of the high mobility and saturation velocity of electrons in GaAs and the FET's short gate length, electron transit times under the gate are on the order of a few picoseconds. MESFETs are thus capable of operation at frequencies in the microwave and millimeter-wave ranges.

Figure 3.1 shows a cross section of a primitive MESFET. The active region, the channel, is a moderately doped epitaxial layer 1,000Å to 2,500Å thick, grown on an undoped substrate. All conventional GaAs MESFETs use *n*-type material for the channel because it has higher mobility than *p*-type. A thick, undoped buffer layer forms the lower boundary of the channel. This layer prevents substrate impurities from diffusing into the epitaxial layer during processing. The drain and source metallizations form ohmic contacts at each end of the channel; a dc voltage is applied to these, creating a longitudinal electric field. In normal operation, the field is strong enough to accelerate the electrons to their saturated drift velocity, creating an electron current from the source to drain. The gate, usually of aluminum or titanium, forms a Schottky barrier with the channel. The gate creates a depletion region that extends part way into the channel at zero gate bias (as with the Schottky diode, this depletion region represents a capacitance). Varying the gate-to-source

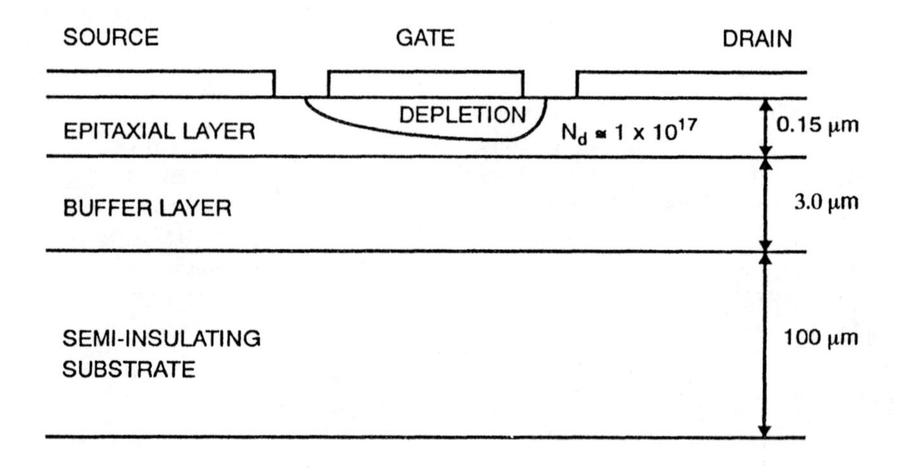

Figure 3.1 Cross section of a primitive MESFET, showing its structure and typical dimensions.

voltage modulates the depletion depth, and, hence, the thickness of the conductive channel; the channel current varies accordingly. As in the diode, the gate voltage also varies the depletion capacitance. The doping density in the channel is typically $10^{17}cm^{-3}$.

Many structural modifications to the FET of Figure 3.1 are made to reduce the FET's parasitic resistances and capacitances, thus improving its gain in both mixer and amplifier applications. The most important of these parasitics are the capacitance of the gate-to-channel junction, the resistance of the gate, and the parasitic resistance in series with the source.

One of the most effective ways to reduce these resistances is to create a relatively thick epitaxial layer and etch a deep recess for the gate. If the bottom of the recess is rounded, the length of the depletion region is less than it would be with a flat surface, and the FET operates as if its gate length were less than its actual length. The thick piece of epitaxial material between the source and the channel reduces the source resistance. Today, virtually all discrete FETs use some type of recessed-gate structure. Ion implantation can be used to minimize source resistance further by heavily doping the regions where the ohmic contacts are formed.

Reducing the length of the gate reduces the FET's gate-to-channel capacitance and increases its transconductance; however, it also increases the gate resistance. Very short gates often have a T-shaped cross section; this shape minimizes the gate's resistance by giving it a large cross-sectional area (these are often called *tee* or *mushroom* gates, according to the gastronomical preferences of the fabricator). Separating the gate into many short pieces also reduces gate resistance, although sometimes at the expense of increased parasitic capacitance from the increased amount of interconnecting metal. A modern FET structure incorporating these improvements is shown in Figure 3.2.

Other properties of the MESFET's structure involve trade-offs between cost and intended application. In low-noise devices, it is always desirable to minimize gate length. Short-gate devices have low fabrication yields, however, and are therefore relatively expensive. High-frequency, low-noise devices may have gates as short as 0.15 μm; to minimize the cost of devices designed for frequencies below approximately 12 GHz, gate lengths of 0.5 μm may be used. The gate must be wide enough so that the device can conduct sufficient channel current to deliver the required power and to achieve adequate transconductance. The cost of increasing the gate width is reduced fabrication yield and increased gate-to-source and gate-to-drain capacitance.

Figure 3.3 shows one of the most common FET geometries. It is commonly used for low-noise amplifiers at frequencies up to 18 GHz, although it is also suitable for FET mixers. The gate length is typically 0.5 μm or less, and its width is

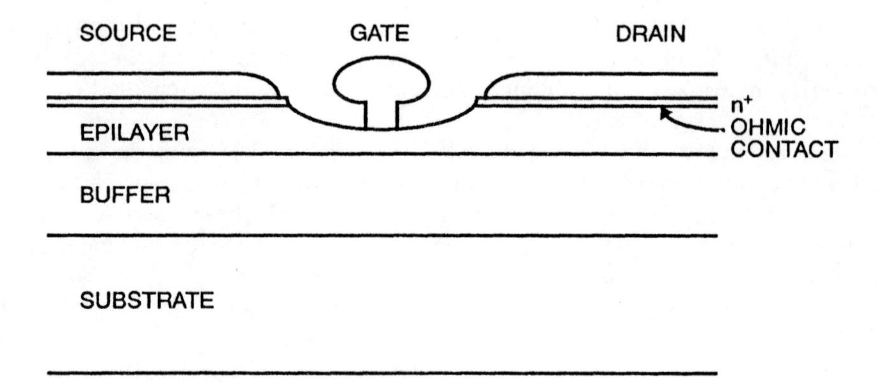

Figure 3.2 Cross section of a modern MESFET with a recessed *tee* gate.

approximately 300 μm. To minimize gate resistance, connections are made in two places along the gate, and each "tap" is connected to a contact pad. The large source metallization that encircles the gate pads allows multiple connections to the source, minimizing parasitic inductance (source inductance has a surprisingly strong, deleterious effect on the FET's gain). The large metallized area allows the source connection to be made with ribbon or gold mesh instead of the more commonly used 0.7-mil gold wire. Two drain connection points are also provided, although the inductance of the drain connection is not as critical to performance as those of the gate and source.

A detail of the channel cross section is also shown in Figure 3.3. Ideally, a recessed gate is used and the gate metal effectively spans the recess, minimizing the parasitic resistance between the channel and the source and drain contacts. Sometimes, to reduce drain-to-gate capacitance, the gate and its recess are slightly offset toward the source.

3.1.2 Operation of GaAs MESFETs

Although the structure of the MESFET is superficially similar to that of the silicon JFET, and their drain *I*/*V* characteristics are qualitatively similar, the operation of the two devices is very different. The most important difference is in the velocity-field characteristics of the materials: in GaAS, electrons reach a high saturation velocity at a relatively low electric field strength; in silicon, the saturation velocity

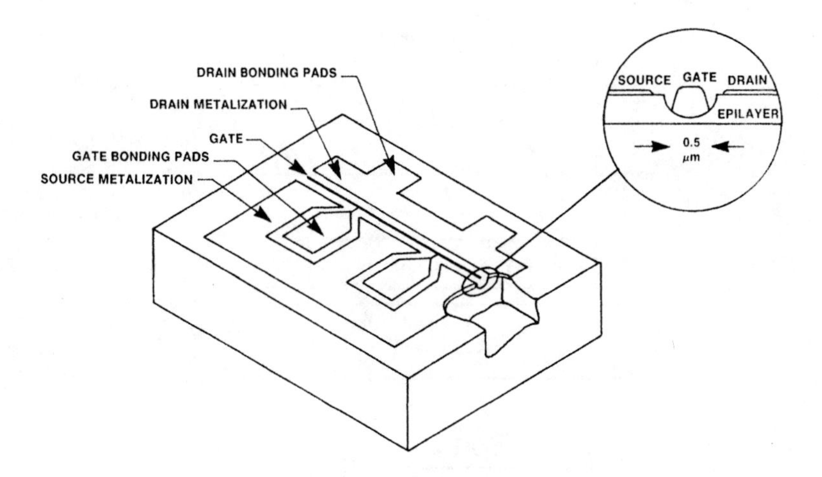

Figure 3.3 Small-signal MESFET suitable for mixer applications, with a detail of its channel.

is lower and the mobility is relatively constant up to a much greater field strength. Thus the electrons in GaAs devices move at saturated drift velocity through most of the channel, rather than at a velocity proportional to the electric field strength, and the resulting short transit times provide higher-frequency operation. Velocity saturation in GaAs also contributes to the formation of a large dipole layer at the drain end of the channel. Current saturation occurs in GaAs devices not because the channel is "pinched off," as in silicon JFETs, but because increases in drain voltage are absorbed by this dipole layer.

Figure 3.4 and Figure 3.5 illustrate the formation of the dipole layer (sometimes called a *charge domain*) under the gate. At very low drain-to-source voltages (Figure 3.4(a) and (b)), below 0.2V to 0.3 V, channel electrons move at a velocity proportional to electric field strength, which is in turn proportional to drain-to-source voltage. The gate depletion region is nearly symmetrical and, at a prescribed drain-to-source voltage, the channel current is proportional to the thickness of the undepleted channel. The channel operates as a gate-voltage-controlled resistor. Because of the similarity of the drain *I/V* characteristic to that of a triode vacuum tube, the FET is said to be operated in its *linear*, or *triode*, region. As drain voltage is increased, the potential between the gate and channel becomes greater at the drain end than at the source end, and the depletion region therefore becomes wider near the drain than near the gate (Figure 3.4(c)). The current must be the same throughout the channel, however, so the electron velocity must be greater near the drain, where

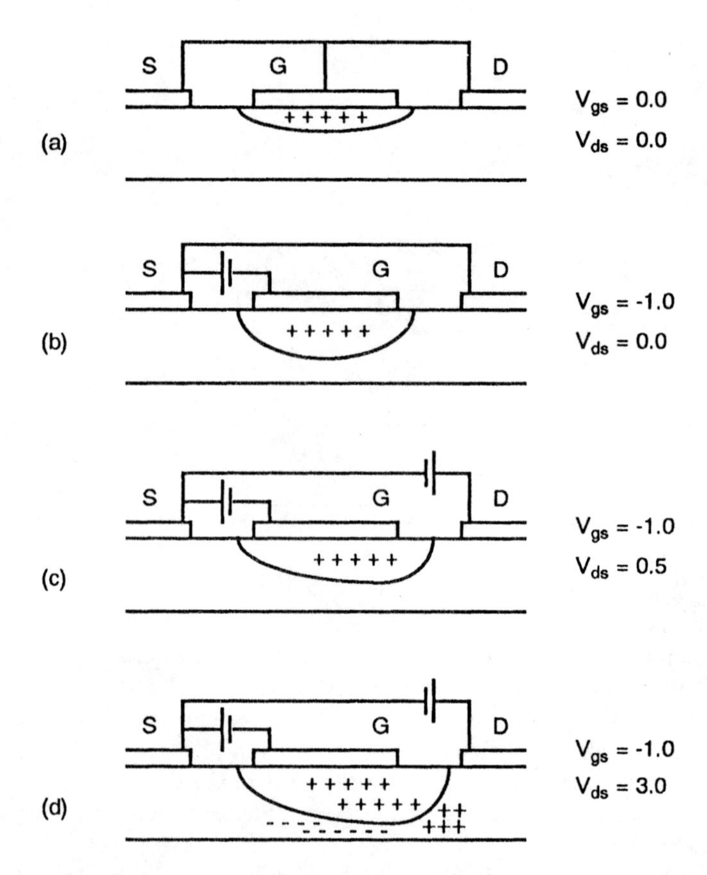

Figure 3.4 Gate depletion region under four different bias conditions: (a) zero bias; (b) a negative gate bias increases the depletion depth without changing its symmetrical shape; (c) drain bias below saturation causes the depletion region to "pull" toward the drain; (d) full drain bias causes negative charge to accumulate in the channel.

it is narrower, than near the source. The electric field, which at low field strengths is proportional to electron velocity, must also be greater near the drain. At a drain voltage of a few tenths of a volt, the electrons reach saturation velocity at the drain end of the channel.

At high reverse gate voltage, the depletion region is relatively wide and the channel thickness is narrow, so the drain-to-source voltage at which velocity saturation occurs is lower than at low gate voltages. As drain voltage is increased further, the electric field becomes stronger throughout the channel, especially near the drain, the depletion region becomes more distorted, and the point where electrons reach saturated drift velocity moves toward the source. At practical operating voltages, the electrons travel at saturated drift velocity throughout most of the channel.

The distortion of the depletion region causes the channel to become narrow between the saturation point and the drain. In order to satisfy the requirements of current continuity in the channel, the electrons must move faster through the narrow part of the channel, or the charge density must increase beyond the doping density. Because the electrons cannot exceed saturation velocity (except in some special cases, which we will consider later), the latter must occur. A region of electron accumulation, with negative net charge density, forms under the depletion region between the saturation point and the drain end of the gate. Because the channel becomes progressively narrower near the drain, the negative charge density increases toward the drain.

Between the edge of the depletion region and the drain contact, the electrons remain at saturated velocity. The electrons enter the region between the edge of the gate and the drain contact, where the channel opens abruptly. Here precisely the opposite situation exists: the electrons are moving very fast, but the channel thickness is also very great. The current-continuity requirement dictates that an electron deficit must exist in this region, resulting in electron depletion and a net positive channel charge (from the ionized donor atoms).

Beyond saturation voltage, virtually all of any further increases in drain-to-source voltage are absorbed by this dipole layer, increasing the stored charge and the magnitude of the electric field in that region, but causing very little increase in channel current. In this case, the FET is said to be in *current saturation,* or simply in *saturation*. Figure 3.5 shows the channel electric field, electron velocity, and charge density in a MESFET operating in its saturation region.

The terms *linear region* and *saturation region* are admittedly poorly chosen. However, they are widely used, and there are no commonly accepted substitutes. Remember that they refer to conditions in the channel, not the mode of operation of a circuit using these devices. If the rationale for choosing the terms were the latter, it would be more appropriate to interchange them.

Other phenomena may complicate this neat description of FET operation. One is sub-threshold conduction. Figure 3.4 and the above discussion imply that the FET can be turned off sharply by increasing the reverse gate-to-source voltage until the depletion region extends completely across the channel. The FET can indeed be turned off this way, but for two reasons the turn-off is not nearly as sharp as might

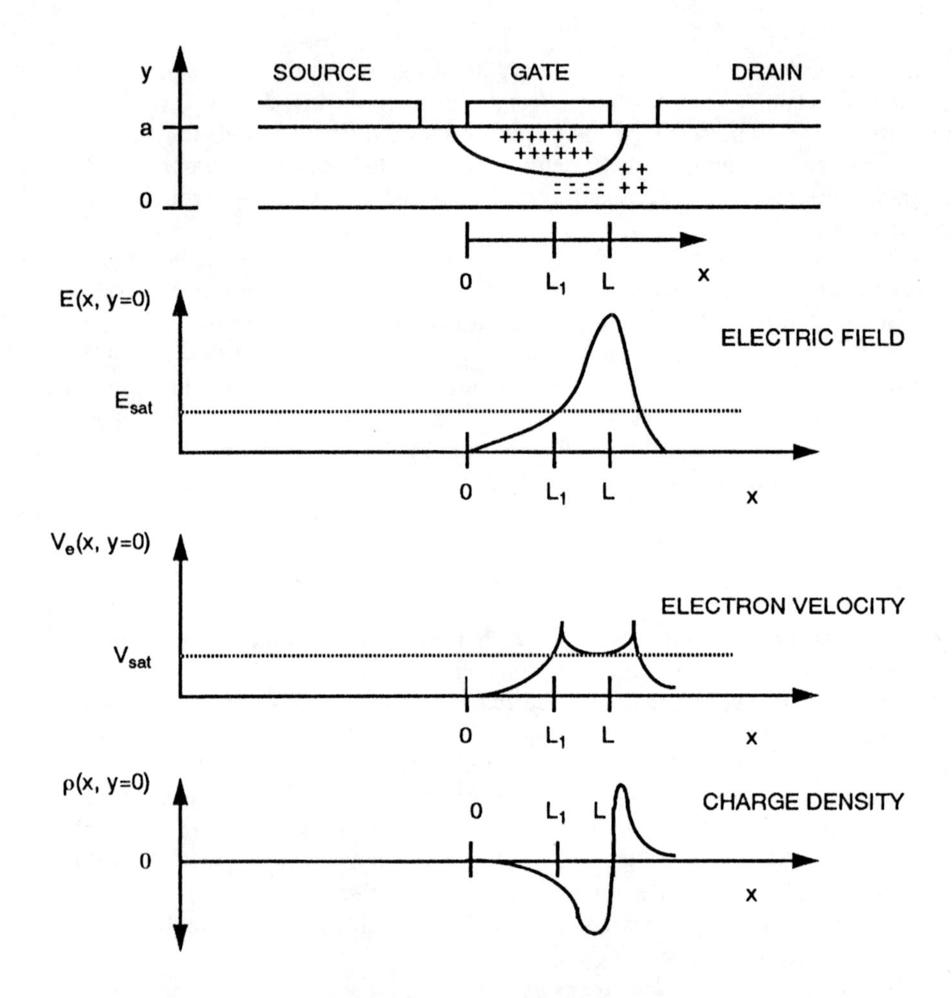

Figure 3.5 Channel electric field, electron velocity, and charge density when the FET is in saturation. L is the gate length; L_1 is the velocity-saturation point.

be expected. First, the edge of the depletion region is not really abrupt. The transition between full depletion and the fully doped channel may be several percent of the channel thickness. Second, the boundary between the conductive channel and insulating buffer layer is never perfectly abrupt, allowing some current to pass through the buffer layer, even when the epitaxial region is completely depleted.

Another phenomenon is nonequilibrium operation. In the preceding discussion, we implicitly assumed that the FET was in equilibrium (i.e., no transient phenomena occurred in the channel). At large electric field strengths, however, which exist in short-gate FETs, electrons can actually exceed saturated drift velocity over part of the channel's length. The reason for this "velocity overshoot" is that velocity saturation occurs only when the electric field is strong enough to excite the electron into a "satellite valley," above the conduction band, where the mobility is relatively low. Transfer of the electron into this valley (called *intervalley scattering*) is a random process, requiring on the average about 1 ps. If the electron transit time is about the same as this scattering time, an appreciable fraction of the channel electrons may not be scattered into this valley, but instead pass under the gate at a very high velocity. There has been some speculation that enhancing this phenomenon may improve a MESFET's performance.

The transconductance of a MESFET varies with gate-to-source voltage because the depletion width is a nonlinear function of that voltage; transconductance is lower at higher reverse biases. The high saturation velocity exhibited by GaAs and the high low-field mobility give the GaAs MESFET higher transconductance than that of a comparable silicon FET.

Naturally, circuits cannot be designed around a qualitative description of MESFET operation; a quantitative model is required. The model must include the FET's nonlinearities, because these are responsible for mixing phenomena. Although other approaches to FET modeling have been taken, the most successful models have been ones consisting of a lumped equivalent circuit that includes a mix of linear and nonlinear elements. Many such models have been proposed; we shall examine a few of them in Section 3.2.

3.1.3 HEMT Structure and Operation

A relatively new FET variant is the HEMT. HEMTs are MESFETs fabricated on heterojunction material; such materials have very high electron mobility and saturation velocity, significantly higher than those of GaAs. Increasing a semiconductor's mobility and saturation velocity invariably increases the gain and reduces the noise figure of devices fabricated with it; thus, HEMTs have higher gain and lower noise than conventional MESFETs.

The structure of heterojunction materials allows for many degrees of freedom in their design; consequently, many different HEMT structures are possible. Figure 3.6 shows one of the most common structures. The "heart" of the device is the longitudinal interface between layers of undoped GaAs and silicon-doped AlGaAs. These layers are very thin; the AlGaAs layer is only 400Å to 500 Å thick, and the

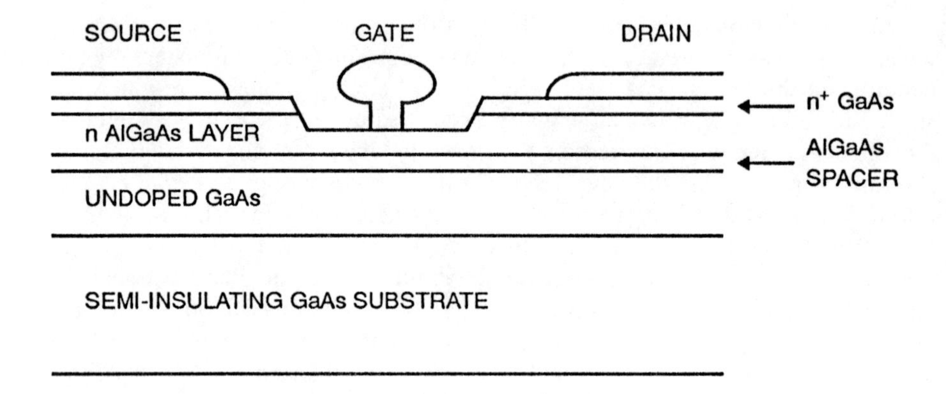

Figure 3.6 Cross section of a conventional AlGaAs HEMT. The AlGaAs layer and spacer are very thin, approximately 400Å and 50Å, respectively.

spacer layer between the GaAs and AlGaAs is only about 50Å thick. These very thin layers (which are, in fact, only a few molecules thick) must be fabricated by molecular-beam epitaxy.

HEMT operates in a manner more like a silicon *metal oxide semiconductor FET* (MOSFET) than a GaAs MESFET. The channel is created at the interface when the AlGaAs material gives up electrons to the GaAs layer; these electrons are trapped in a thin layer near the interface. The density of electrons in this thin layer (sometimes called a *two-dimensional electron gas*) is controlled by the voltage applied to the gate; making the gate more positive increases the electron density and therefore the channel current. Depending on the thickness of the AlGaAs layer (which is controlled by the depth of the recess shown in Figure 3.6) the electron density can be made zero or nonzero at $V_g = 0$, resulting respectively in either an enhancement-mode or depletion-mode device. This is significantly different from MESFETs, which are always operated in a depletion mode.

Because there are no dopant ions in the GaAs to scatter the electrons, they move at high velocity through the material. As a result, the electron saturation velocity and mobility are much greater in HEMTs than in MESFETs: the room-temperature low-field mobility of a HEMT is typically 8,000, compared to 5,000 for a MESFET. Furthermore, the electron mobility increases dramatically as temperature is reduced; for this reason HEMTs are sometimes operated at cryogenic temperatures, where they have achieved amplifier noise temperatures as low as those of masers.

The topology of the HEMT's equivalent circuit is identical to the GaAs MESFET's. However, the *I*/*V* characteristic is much different. HEMTs usually exhibit a "softer" drain I/V characteristic, implying that the drain-to-source resistance is generally lower than in MESFETs. More striking is the difference between GaAs MESFET and HEMT gate *I*/*V* characteristics. The transconductance of a HEMT usually rises sharply with gate voltage and has a distinct peak; this implies that the device is more strongly nonlinear than a GaAs MESFET, and, in general, HEMTs exhibit greater nonlinear distortion. It also implies that relatively low LO power is needed to achieve good conversion loss.

A disadvantage of the HEMT in comparison to the MESFET is that its electron layer is very thin, much thinner than a MESFET's channel, so the number of electrons per unit of channel area is lower, and its maximum drain current is also lower. This implies that the HEMT is inherently a small-signal device. However, it is possible to overcome this limitation through the use of multiple heterojunctions. Good high-frequency power HEMTs have been produced.

3.2 CIRCUIT MODELS FOR GaAs MESFETs

3.2.1 Lumped Large-Signal Models

Analysis of large-signal FET circuits requires an accurate nonlinear FET model. Because FET mixer operation is dependent primarily on the variation of transconductance with gate-to-source (and sometimes gate-to-drain) voltage, it is especially important that the model accurately predict the transconductance (i.e., the first derivative of the gate *I*/*V* characteristic) at any combination of terminal voltages. Mixer intermodulation depends on the higher derivatives of the *I*/*V* characteristic; in many cases these must be expressed accurately as well.

In many types of mixers, it is also important to model the drain *I*/*V* characteristics. In most well-designed active mixers the FET operates in saturation over the entire LO cycle, and it is not necessary to model the "knee" of the drain *I*/*V* characteristic. However, in other types of mixers, the FET is driven into its linear region over at least part of the LO cycle. For example, so-called *drain mixers*, in which the LO is applied to the drain, provide frequency mixing in part because the drain-to-source resistance varies as the FET is driven into its linear region over part of the LO cycle. In these mixers, it is clearly necessary to model the entire drain *I*/*V* characteristic accurately.

Although the gate-source and gate-to-drain capacitance nonlinearities rarely contribute significantly to mixing, it is still important to model them accurately. In most active mixers, the LO is applied to the gate, and the resulting large ac gate voltage causes the gate-to-source capacitance to vary. The input impedance of the device depends strongly on the average value of this capacitance, which can be determined only if the capacitive nonlinearity is well modeled. Paradoxically, it is often the linear properties of the FET that depend on the nonlinear ones.

Because of the complexity of a FET's operation, the strongly nonlinear velocity-field characteristic of GaAs, and the difficulty of quantifying such phenomena as subthreshold conduction, analytical or "physical" MESFET models have been only partially successful. Similarly, two-dimensional analysis of the FET's channel [1] is the most accurate way to model a FET, and this type of analysis produces much valuable information about the operation of this device; however, two-dimensional analysis is far too slow for use in computer-aided design. Empirical models developed specifically for use in computer-aided design have been more successful. We shall concentrate on the latter.

The earliest models of GaAs MESFETs [2]–[7] were developed to improve the general understanding of the operation of these devices. Eventually, the goal of developing models for use in circuit simulation became an important concern, and FET models were developed first for use in the well-known circuit simulator, SPICE [8]–[10]. The limitations of these early models became evident, and much work was devoted to their improvement. At present, the most widely used models are those of Curtice and Ettenberg [11], Materka and Kacprzak [12], and Statz et al. [13].

Figure 3.7 shows the nonlinear equivalent circuit of a MESFET. The equivalent circuit includes five nonlinear elements, C_{gs}, C_{gd}, I_d, and two diodes; these are controlled by two voltages, V_g and V_d, the *internal* gate and drain voltages (i.e., as opposed to the *external* voltages, those measured at the device terminals, which include the voltage drops across resistive parasitics R_s and R_d). All other elements in the equivalent circuit are linear. The FET's transconductance, dI_g/dV_g, dominates the conversion performance in well designed active mixers. R_s and R_d are primarily ohmic-contact resistances, but they also include the resistances of adjacent parts of the epilayer. R_g is the metal resistance of the FET's gate. R_i is the resistance of the ohmic part of the channel between the gate depletion region and the source. The diodes represent the Schottky gate-to-channel junction; they have little effect as long as the voltage across either of the diodes does not exceed its "knee" voltage (approximately 0.5V); if it does, they begin to conduct and clamp the junction voltage at a low positive value.

C_{gs} and C_{gd} are the gate-to-source and gate-to-drain parasitic capacitances. Like those of a Schottky diode, they are defined as incremental capacitances, the change in depletion charge with voltage; unlike the diode, they are functions of both control voltages:

$$C_{gs}(V_g, V_d) = \left.\frac{\partial Q_d}{\partial V_g}\right|_{V_d - V_g = \text{constant}} \tag{3.1}$$

$$C_{gd}(V_g, V_d) = \left.\frac{\partial Q_d}{\partial V_d}\right|_{V_g = \text{constant}} \tag{3.2}$$

where Q_d is the total positive charge in the depletion region.

The dependence of C_{gs} and C_{gd} on drain voltage is illustrated in Figure 3.8. At very low drain-source voltages, the depletion capacitance is nearly equally split between C_{gs} and C_{gd}. As drain voltage is increased and the MESFET approaches current saturation, the dipole layer absorbs increases in drain-to-source voltage, and progressively less change in the depletion width occurs as drain voltage is increased

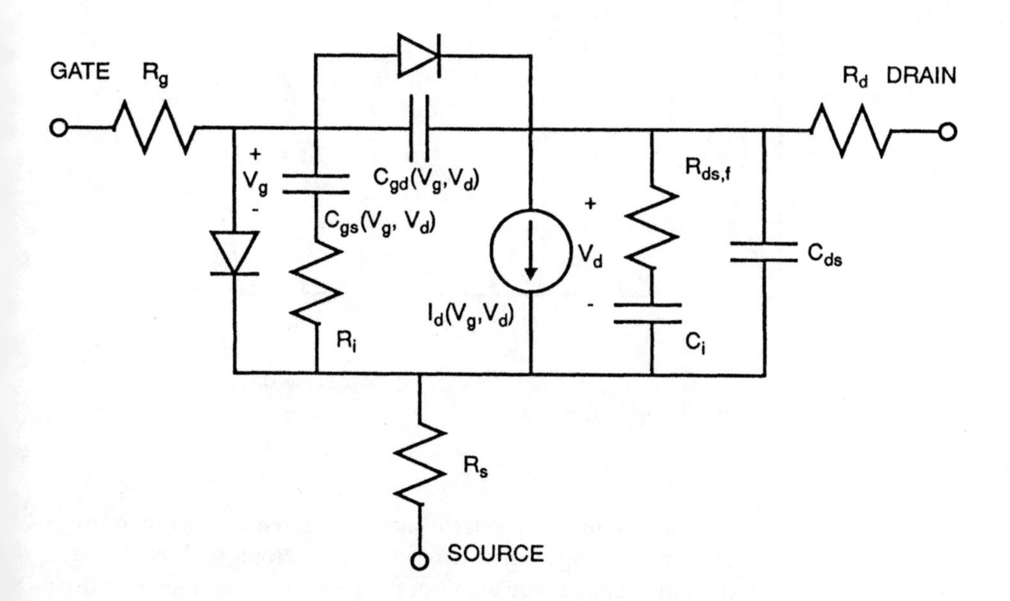

Figure 3.7 Nonlinear equivalent circuit of the MESFET. The diodes are used to model gate-to-channel conduction when the gate junction is forward biased. $R_{ds,f}$ and C_i model the high-frequency drain-to-source resistance; C_i is on the order of 1 μF.

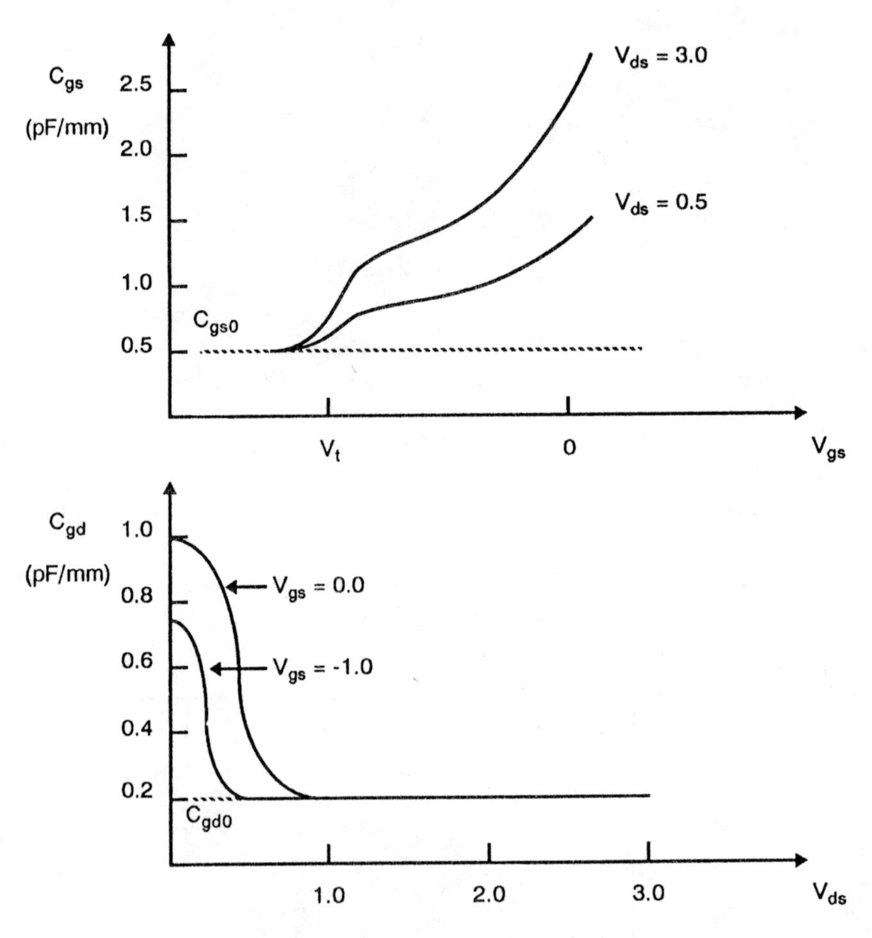

Figure 3.8 Nonlinear capacitances C_{gs} and C_{gd}. The dotted lines show the components of these capacitances from the metallizations.

further. Equation (3.2) indicates that the gate-to-drain capacitance C_{gd} then drops rapidly with voltage. At drain voltages of a few volts, C_{gd} drops so low that it is dominated by the electrostatic capacitance between the gate and drain metallizations C_{gd0}, not the depletion capacitance; consequently, C_{gd} is nearly independent of drain voltage. C_{gs} rises abruptly as the device enters saturation, since C_{gs} then represents the entire gate-to-channel depletion capacitance.

This description illustrates a paradox involving C_{gd}: C_{gd} is, under certain conditions, both the most strongly and most weakly nonlinear of the FET's capacitances. When the drain voltage stays high and relatively constant (as it does in well-designed active FET mixers), the nonlinearity of C_{gd} is negligible. However, when the drain voltage is reduced to the point where the device enters its linear region, C_{gd} is significantly nonlinear and must be modeled as such. Examples of the latter are strongly-driven power amplifiers or, occasionally, FET mixers having the LO applied to the drain.

The dependence of C_{gs} on gate bias is also important. When both V_d and V_g are nearly zero, and the channel is not completely depleted, both C_{gs} and C_{gd} can be described accurately as classical Schottky-junction depletion capacitances. At high drain voltages, C_{gd} is almost completely insensitive to both V_d and V_g, and the dependence of C_{gs} on V_g diverges from that of a classical Schottky diode. As V_g is increased further (in the reverse-bias direction), it eventually reaches a value (the MESFET's *threshold*, or *turn-on*, voltage, V_t) that completely depletes the channel. At this point, the depletion region can expand no further, so its charge varies much more slowly with V_g. Consequently, C_{gs} drops suddenly. This phenomenon is very different from that of the Schottky diode, where the completely depleted epitaxial layer causes the junction capacitance to lose much of its nonlinearity, but not to drop precipitously. The behaviors of the two devices differ because the diode's epitaxial layer is grown on a heavily doped substrate, while the FET's is grown on high-resistivity (undoped) material.

The capacitance C_{ds} is the drain-to-source capacitance. C_{ds} consists of two components: the drain metallization and the capacitance of the dipole layer. Both are relatively small, approximately 0.3 pF/mm of gate width. Although the drain metallization should be described by a separate capacitance connected between the drain and source terminals, including it as part of the dipole capacitance is usually acceptable.

A common variation in FET equivalent circuits is to split C_{gd} into two capacitances and connect the second one between the top of the current source and the bottom of C_{gs}. This variation may give somewhat better agreement with small-signal S parameters, but is rarely worth the complexity it introduces into the process of determining the circuit-element values.

The most important nonlinear element in the MESFET equivalent circuit is the drain current, $I_d(V_g, V_d)$. The transconductance g_m and drain-to-source conductance G_{ds} are very important in the analysis of active mixers. They are linearized quantities. The transconductance describes the dependence of I_d upon V_g; it is

$$g_m(V_g, V_d) = \left.\frac{\partial I_d}{\partial V_g}\right|_{V_d = \text{constant}} \tag{3.3}$$

The drain-to-source conductance requires more effort to model. It consists of two components, a dc conductance $G_{ds,dc}$, which is usually relatively small, and an RF conductance $G_{ds,f}$, which is much greater. The dc component of the conductance is modeled implicitly by the current source:

$$G_{ds,dc} = \left.\frac{\partial I_d}{\partial V_d}\right|_{V_g = \text{constant}} \tag{3.4}$$

The value of $G_{ds,dc}$ obtained from Equation (3.4) is valid only at dc. At high frequencies, the conductance is considerably greater, by a factor of two to five, than indicated by dc measurements. The reason for this phenomenon is not entirely clear, but it is probably related to traps in the channel. These have relatively long time constants; the transition between the low- and high-frequency values of G_{ds} is usually less than a few megahertz. To model this phenomenon, we include a series network consisting of $R_{ds,f}$ (= $1/G_{ds,f}$) and C_i in parallel with the controlled source. This allows the FET to have the measured dc I/V characteristic, yet also to display the correct RF drain-to-source impedance.

Another problem with R_{ds}, which has been related theoretically and experimentally to the vagaries of the dipole layer, is the appearance of a negative resistance region near the knee of the drain I/V curves, especially at high currents. This phenomenon is observed when the ratio of channel thickness to gate length is large, and in extreme cases leads to low-frequency bias-circuit oscillations or high-frequency Gunn oscillations. This phenomenon is well known, and is usually successfully avoided in modern FETs. The primary means of preventing such oscillations is to keep the channel's thickness-to-length ratio relatively small; this ratio should ideally be below 0.25, although in short-gate FETs, this small ratio may be impossible to achieve.

Because of the voltage drops across R_s and R_d, the I/V characteristics and transconductance measured at the accessible gate, drain, and source terminals (the *external* characteristics) are not the same as those defined in terms of V_g and V_d (the *internal* characteristics). At dc, these quantities are related as follows:

$$g_{m_{\text{ext}}} = \frac{g_{m_{\text{int}}}}{1 + R_s g_{m_{\text{int}}}} \tag{3.5}$$

$$V_d = V_{ds} - I_d (R_s + R_d) \tag{3.6}$$

$$V_g = V_{gs} - I_d R_s \tag{3.7}$$

I_d is not a quasistatic function of the external voltages V_{gs} and V_{ds}; thus, I_d is always expressed as a function of the internal voltages V_g and V_d.

The two diodes shown in Figure 3.7 are used to model gate rectification, which occurs whenever the peak value of the gate-to-channel voltage exceeds the knee of the junction's *I*/*V* characteristic. When the excitation is large, these diodes "clamp" the gate-to-source and gate-to-drain voltages to a peak value of approximately 0.5 volts. Thus, they represent an important factor in limiting the excitation level, and therefore the calculation of a mixer's conversion loss as a function of LO level.

3.2.2 Curtice Quadratic Model

This *I*/*V* model, proposed by Curtice [8], is the simplest of those considered here and is probably the least accurate. It is most suitable for digital circuits, where perfect agreement with measured *I*/*V* characteristics is not necessary, and the agreement with measured g_m, in particular, is not important. This model has been included in some versions of the popular circuit-simulation program SPICE2.

The relation for the *I*/*V* characteristic is simply that of a conventional silicon JFET in its current saturation region, with a multiplicative hyperbolic tangent function to account for nonsaturated behavior. $R_{ds,f}$ is not modeled separately, as in Figure 3.7; instead, it is assumed to be equal to $R_{ds,dc}$ and included in I_d. The expression for drain current is

$$I_d = \beta (V_g - V_t)^2 (1 + \lambda V_d) \tanh(\alpha V_d) \tag{3.8}$$

where V_t is the turn-on voltage of the device, and α, β, and λ are constants chosen for best agreement between measured and calculated *I*/*V* characteristics. (When $V_d >> 1/\alpha$, $\lambda \sim 1/R_{ds}$.) C_{gs} is modeled as a Schottky-junction capacitance, and C_{gd} is a constant when the device is saturated.

Although simple quadratic models such as this often give reasonable predictions of I/V characteristics, their transconductance is nearly useless for the analysis of FET mixers. The derivative of Equation (3.8) gives the transconductance:

$$g_m = 2\beta(V_g - V_t)(1 + \lambda V_d)\tanh(\alpha V_d) \tag{3.9}$$

which, as a function of V_g, is a straight line. The transconductance of real devices does not increase monotonically with V_g, but usually becomes approximately constant near $V_g = 0$. Use of this expression in FET mixer analysis would imply that conversion gain can be increased arbitrarily as LO level is increased.

3.2.3 Curtice Cubic Model

The Curtice cubic model [11] is probably the most widely used model of those described here. Although it was developed primarily for use in power FET amplifiers, it has largely overcome the limitations of the quadratic model for use in mixer analysis, and it generally gives good results. The I/V expression is

$$I_d = (A_0 + A_1V_1 + A_2V_1^2 + A_3V_1^3)\tanh(\alpha V_d) \tag{3.10}$$

where A_n and α are constant coefficients and V_1 is an intermediate variable:

$$V_1 = V_g(t - \tau)(1 + \beta(V_{d0} - V_d)) \tag{3.11}$$

The quantity τ (which is proportional to V_d) is the time delay between the drain current and gate voltage (usually a few picoseconds), β is a constant, and V_{d0} is the voltage at which the A_n coefficients were measured. The form of this equation, as well as the use of the tanh function to represent the functional dependence of I_d on V_d, is not physically significant; both are justified entirely on empirical grounds.

The purpose of the intermediate variable V_1 is to account for the shift in pinch-off voltage at high drain voltages. This phenomenon is important in power devices, for which the model was primarily intended, but not in small-signal devices, operated at low drain voltages, that are used for mixers. Thus, it is almost always acceptable in mixer analysis to set $\beta = 0$ and $V_1 = V_g(t - \tau)$. Similarly, in small-signal devices used at the lower microwave frequencies, τ is often negligible.

In [11] there is no attempt to account for the nonlinearity of C_{gs} or C_{gd}. As we mentioned earlier, the nonlinearity of these elements is probably not significant in well-designed mixers, where the gate-to-drain voltage stays above a volt or two throughout the LO cycle. However, implementations of this model in general-purpose circuit-analysis programs often include the nonlinearity of these capacitances.

3.2.4 Materka and Kacprzak Model

This model [12] uses a modified version of the common quadratic expression for the gate *I*/*V* function. Specifically,

$$I_d = I_{dss}\left(1 - \frac{V_g}{V_1}\right)^2 \tanh\left(\frac{\alpha V_d}{V_g - V_1}\right) \tag{3.12}$$

where V_1 is the modified threshold voltage,

$$V_1 = V_t + \gamma V_d \tag{3.13}$$

and I_{dss}, V_t, γ, and α are the parameters of the model. The model includes junction diodes and treats C_{gs} as an ideal Schottky-barrier capacitance; C_{gd} is treated as a linear quantity. It also includes a time delay; that is, $V_g \equiv V_g(t - \tau)$.

3.2.5 Statz et al. Model

This model [13] is based on the observation that the FET's gate *I*/*V* characteristic appears to be nearly quadratic near $V_g = V_t$ and nearly linear near $V_g = 0$. An empirical expression having this characteristic is

$$I_d = \frac{\beta(V_g - V_t)^2}{1 + b(V_g - V_t)} f(V_d) \tag{3.14}$$

This model gives a nearly "square" curve for the transconductance as a function of gate voltage. This is quite different from the observed transconductance of real FETs, and thus the model is probably not as accurate for mixer use as some of the others. Two consequences of this square transconductance function are low predictions of LO power requirements and unrealistically optimistic predictions of harmonic conversion efficiency.

The tanh function used in the other models to express the drain *I*/*V* characteristic is a transcendental function. Thus, quite a bit of computer time is needed to evaluate it, probably more time than it is worth. To circumvent this problem, the following expression is used instead of the tanh function:

$$f(V_d) \sim 1 - \left(1 - \frac{\alpha V_d}{n}\right)^n \tag{3.15}$$

where n is in the range of 2 to 3. When $V_d > n/\alpha$, $f(V_d) = 1$.

This model originally included a much more inventive approach to the modeling of the gate-to-channel capacitances than the previous models. However, it turned out that the model suffered from charge nonconservation. This is a serious problem in time-domain circuit simulators, since they often compute both charge and capacitance waveforms, and these must be completely consistent.

3.2.6 Maas-Neilson Model

The author reserves the right to include a description of his own model, regardless of its ultimate acceptance or practical value. This expression [14] was chosen to fit not only the FET's gate I/V characteristic and transconductance function dI_d/dV_g, but its second and third derivatives as well. The latter two derivatives are not important in modeling conversion performance, but they are dominant in modeling mixer intermodulation and spurious responses up to the third order. The I/V expression is

$$I_d = (A_0 x + A_1 \sin(x) + A_2 \sin(2x) + A_3 \sin(3x))\, f(V_d) \tag{3.16}$$

The intermediate variable x is

$$x = \pi \left(\frac{V_g - V_t}{V_f - V_t} \right) \tag{3.17}$$

where V_f is a parameter of the model. It is typically the maximum forward gate voltage that the device will experience; $V_f = 0$ is a common choice. Users of the model are free to substitute their favorite expression for $f(V_d)$.

The expression for I_d treats the gate I/V characteristic as a linear function plus a variation from linearity expressed by a Fourier series. Because it consists of a linear combination of basis functions, the I_d expression and its derivatives can be fit to a set of data by means of any least-squares fitting process; we prefer the use of singular-value decomposition. If the higher derivatives are not of interest, this expression can be fit to the measured transconductance very easily and almost exactly.

3.2.7 Modeling FET Resistive Mixers

The possibility of using the resistive channel of a FET for mixing has generated considerable interest recently [15]. Such *FET resistive mixers* exhibit very low intermodulation and high 1-dB compression points. They are operated with the LO applied to the gate along with dc gate bias, but without drain bias. The RF and IF ports are connected to the drain through appropriate filtering and matching networks.

Because the FET mixer operates at low drain voltages, the electron velocity-field characteristic is nearly linear, and the channel can be modeled by Shockley's theory [16]. The I/V characteristic is

$$I_d = I_1 (3 (u^2(V_g, V_d) - u^2(V_g, 0)) - 2 (u^3(V_g, V_d) - u^3(V_g, 0))) \tag{3.18}$$

where $u(V_g, V_d)$ is the normalized depletion depth,

$$u(V_g, V_d) = \sqrt{\frac{-V_g + V_d + \varphi}{-V_t}} \tag{3.19}$$

and ϕ is defined as a positive quantity. The channel conductance g_d can be found by differentiating Equation (3.18):

$$g_d(V_g) = 3\frac{I_1(1 - u(V_g, 0))}{-V_t} \tag{3.20}$$

Equation (3.18) is adequate for determining conversion performance, but it does not appear to be adequate for modeling intermodulation properties.

Because the drain is unbiased, the gate-to-channel capacitance is divided approximately equally between the gate-to-source and gate-to-drain capacitances. these are well modeled as depletion capacitances:

$$C_{gs} = C_{gd} = \frac{C_{g0}}{\sqrt{1 - \frac{V_g}{\varphi}}} \tag{3.21}$$

The equivalent circuit of the FET having an unbiased drain is almost the same as that of the biased FET; because the channel is inactive, the elements C_i, C_{ds}, R_i, and R_{dsf} need not be included. The *I/V* characteristics and element values, given by Equations (3.18) through (3.21), are, of course, very different from those of the active device.

3.2.8 Modeling HEMT Devices

Most MESFET models are not easily applied to HEMTs. The main problems are that the tanh function used to model the drain *I/V* characteristic is too flat in the saturation region, and the function that models the gate *I/V* does not reproduce the transconductance peak. It is certainly possible to include this peak by modifying the parameters of certain models. By adding fourth- or fifth-degree terms, the *I/V* expression of the Curtice model (Equation (3.10)) can be made to reproduce the transconductance peak. The Maas-Neilson model is capable of reproducing this peak accurately; however, it is necessary to treat the parameter V_f in Equation (3.17) as a variable that must be fit to measured *I/V* data. The drain *I/V* function can be modified as in Equation (3.8) by the inclusion of a $(1 + \lambda V_d)$ term, or through the use of an entirely different function.

Golio [17] describes a method for modifying the gate *I/V* expression in any of the above models for use with HEMTs. This method is particularly appropriate for use in mixer analysis, because it is based primarily on achieving the correct transconductance characteristic. The characteristic is modified by the addition of a second, subtractive term:

$$g_{m_H} = g_{m_M} - \xi (V_g - V_{g_p})^{\psi} \tag{3.22}$$

where ξ and ψ are empirical constants, g_{m_M} is the transconductance function of the MESFET model, g_{m_H} is the transconductance of the HEMT model, and V_{g_p} is the voltage at which the peak transconductance occurs. The *I/V* characteristic is then found by integrating over V_g:

$$I_{dg}(V_g) = \int_{V_p}^{V_g} g_{m_H}(V)\,dV \tag{3.23}$$

where V is a variable of integration representing the gate voltage. Finally, the complete I/V expression is obtained by multiplying I_{dg} by the drain-current term, $f(V_d)$.

3.2.9 Comparison of Models

There is, in one sense, not much difference between FET models: all use the same or almost the same circuit topology, and all attempt to reproduce the same set of I/V and C/V parameters. Indeed, when we speak of a FET "model," we usually refer not to the equivalent-circuit topology, but to the mathematical expression that models the FET's I/V characteristic. This is the focus of most FET-modeling efforts.

The merits of different FET models is probably the most contentious subject in the area of nonlinear-circuit analysis. Yet, in spite of very strong opinions about the advantages of various models, almost all seem to work at least acceptably well for predicting a mixer's basic parameters, conversion loss, and port impedances. The reason for this paradoxical situation is that in well-designed active FET mixers (and we do not care much about poorly-designed mixers), the fundamental-frequency component of the pumped FET's transconductance waveform is by far the dominant factor affecting conversion gain or loss. Virtually all other elements in the FET's equivalent circuit can be acceptably approximated as single-valued, linear ones.

What affects the fundamental-frequency component of the transconductance waveform? First, we recognize that a FET mixer is usually biased near $V_g = V_t$, so the FET conducts in pulses, and the transconductance waveform is a train of approximately half-cosine-shaped pulses. A few minutes examining Fourier-transform expressions should convince the reader that only two parameters of this pulse train affect the fundamental-frequency component strongly: the duty cycle and the peak value. The shape of the individual pulses affects the higher harmonics relatively strongly, but not the lower ones. Therefore, any model that expresses the peak transconductance and current accurately (and, of course, includes accurate values for the linear-circuit elements) is likely to produce good results in mixer analysis.

Modeling intermodulation in mixers, saturation levels in strongly driven mixers and amplifiers, and the modeling of certain other nonlinear phenomena are more complicated problems, which depend far more strongly than simple conversion analysis on the properties of the device model, especially the higher derivatives of the gate *I/V* characteristic. Many of these phenomena are difficult to model accurately for other reasons, which are related to subtle limitations of the various methods of analysis that might be employed (we shall examine these further in Chapter 4). However, for many purposes, especially mixer analysis, the problem of modeling the device is not very severe.

3.2.10 Determining Equivalent-Circuit Parameters

Physical models of the MESFET are rarely accurate enough for circuit simulation and never have adequate numerical efficiency. Therefore, it is inevitably necessary to use an empirical model and to measure some or all of the elements that make up the FET's equivalent circuit; the process of measuring those parameters is called *parameter extraction*. The most popular methods are to fit the element values and dc *I/V* parameters to measured S-parameter and *I/V* data, or to measure the element values directly by means of impedance- or capacitance-measuring equipment. The latter may involve measurements of either a biased or a "cold" device.

A practical way to determine the elements of the equivalent circuit of Figure 3.7 is to derive them from S-parameter measurements. The process is straightforward in concept and involves varying the element values empirically until the measured S parameters and those calculated from the equivalent circuit are the same. General-purpose computer programs for analyzing and optimizing microwave circuits can be used for this purpose. The voltage dependencies of the nonlinear elements are determined by repeating the process at a number of bias-voltage values.

A problem arises from the large number of variables—the values of the circuit elements—that must be optimized. It invariably happens that good agreement with measured S parameters can be obtained with both sensible element values and values that clearly bear no relation to the correct ones. This situation exists because S parameters are not affected uniquely by individual element values; for example, reducing R_s or R_g will increase S_{21} as effectively as increasing g_m. Also, increasing the source inductance has much the same effect on S_{11} as increasing R_s, R_g, or R_i.

One of the best ways to circumvent this problem is to measure as many elements as possible at dc or low frequencies. One method is to measure S parameters at frequencies and to convert them to Y parameters. At frequencies of a few megahertz, most of the capacitances are negligible (but the output conductance has its RF value), and the resistive elements can be calculated easily from the Y parameters of the linearized circuit. At somewhat higher frequencies (a few hundred megahertz), the capacitive elements are no longer negligible, but their effect on Y parameters is fairly simple, and their values can be determined easily.

At frequencies of a few hundred megahertz, the low-frequency Y parameters of the FET equivalent circuit are approximately

$$y_{11} = \frac{j\omega C_{gs}}{j\omega C_{gs}(R_g + R_i + R_s) + 1 + g_m R_s} \tag{3.24}$$

$$y_{22} = \frac{1 + j\omega C_{ds} R_{ds}}{(1 + g_m R_s) R_{ds}} + j\omega C_{gd} \tag{3.25}$$

$$y_{21} = \frac{g_m}{1 + g_m R_s + R_s / R_{ds}} \tag{3.26}$$

$$y_{12} = -j\omega C_{gd} \tag{3.27}$$

From these, the element values can be estimated as follows:

$$C_{gs} = \frac{y_{11}(1 + g_m R_s)}{1 - y_{11}(R_g + R_i + R_s)} \approx y_{11}(1 + g_m R_s) \tag{3.28}$$

$$g_m = \frac{y_{21}}{1 - y_{21} R_s} \tag{3.29}$$

$$g_{ds} = \frac{1}{R_{ds}} = \frac{(1 + g_m R_s) y_{22}}{1 + y_{22} R_s} \tag{3.30}$$

$$C_{gd} = \frac{-y_{12}}{j\omega} \tag{3.31}$$

$$C_{ds} = (1 + g_m R_s) \left(\frac{\text{Im}\{y_{22}\}}{\omega} - C_{gd} \right) \qquad (3.32)$$

Equations (3.28) through (3.32) can be used to estimate several important element values from Y parameters measured at a single low frequency. However, better accuracy can be obtained by fitting over a range of low frequencies. Note that Equations (3.24), (3.25), and (3.27) indicate that the magnitudes of y_{11}, y_{22}, and y_{12} vary approximately linearly with frequency; thus, these quantities can be plotted as functions of frequency, and C_{gs}, C_{gd}, and C_{ds} can be derived from the plots. This process provides much better accuracy than the use of data measured at a single frequency.

C_{gs}, C_{gd}, and C_{ds} can also be measured directly at dc or low frequencies with a sensitive capacitance bridge. Measuring capacitances in this manner is more practical for FETs than for Schottky diodes, because the capacitances are much greater.

One of the more difficult matters is to separate g_m and R_s; it is possible to extract only the product $g_m R_s$, but not the individual values, from conventional gate and drain *I/V* measurements. A number of methods have been developed to separate these quantities. A simple and reasonably accurate method for measuring R_s is to apply a current through the gate-to-source junction and to measure the voltage between the source and the open-circuit drain; this voltage is approximately the voltage across R_s.

R_s, R_d, and R_g are difficult to determine either from S parameters or dc measurements. They can be found at least approximately from forward-bias measurements of the gate-to-channel junction in a manner similar to that described in Chapter 2 for diodes. It is important to recognize, however, that forward-bias measurements may give values of these parameters that are somewhat different from the actual reverse-bias quantities because the shape of the depletion region is different under forward bias from that under reverse bias. Especially in devices having gate lengths below 0.5 μm, the junction current distribution under forward bias also may not be uniform across the gate because of the voltage drop across R_g. Finally, under forward bias, the gate current is injected along the edge of the depletion region closest to the source and bypasses the region comprising R_i; R_i is therefore not measured. Nevertheless, dc measurements give good starting values for the S-parameter fitting and may act as a "sanity check" on the results.

The values of R_s, R_d, and R_g can be determined by three separate measurements. The theory behind the procedure has been described by Fukui [18] and, somewhat modified, by Weinreb [19]. The procedure is as follows:

1. Measure the *I/V* characteristic of the gate-to-source junction with the drain open-circuited. Determine the series resistance as described in Section 2.2; this series resistance is $R_s + R_g$. Record also the voltage V_{gs} corresponding to the current I_j at which the series resistance was determined.
2. Bias the gate-drain junction, with the source open-circuited, to I_j, the same current used to determine the series resistance in Step 1. Record the voltage V_{gd}; it is not necessary to measure the entire *I/V* characteristic.
3. Bias the gate-channel junction with the drain and source connected together to the same current as in Step 2 and record the voltage, V_{gsd}.
4. Calculate the resistances

$$R_1 = (V_{gs} - V_{gsd}) / I_j \tag{3.33}$$

$$R_2 = (V_{gd} - V_{gs}) / I_j \tag{3.34}$$

5. Find R_s and R_d:

$$R_s = R_1 + (R_1^2 + R_1 R_2)^{1/2} \tag{3.35}$$

$$R_d = R_2 + R_s \tag{3.36}$$

6. R_g is found by subtracting R_s from the series resistance found in Step 1.

Clearly, before performing the S-parameter fitting process, it is necessary to measure the FET's S parameters accurately (by accurately de-embedding the test fixture and compensating for test-equipment errors). Although the calibration processes are sometimes more difficult, it is best to make on-wafer measurements or to measure unpackaged chips, rather than packaged devices, so that uncertainties in modeling the package parasitics can be avoided. Losses in the test fixture, if not removed by calibration, increase the measured values of R_s and R_g and are therefore particularly serious. Considerable variability is observed between devices of the same type, even among devices produced on the same wafer, so manufacture's published data is rarely an adequate source of device S parameters. S parameters must be measured over a bandwidth wide enough that the S-parameter phases vary substantially.

3.3 DUAL-GATE FETs

3.3.1 Description

Dual-gate FETs are not new to high-frequency electronics. Dual-gate silicon MOSFETs have been used as amplifiers and mixers in low-noise VHF and UHF receiver front ends since the 1960s. Dual-gate MOSFETs provide lower intermodulation distortion and better automatic gain control than bipolar transistors, although with slightly higher noise figures.

A dual-gate MESFET is similar in structure to a single-gate device, except that it includes a second gate between the first gate and the drain. This gate has several effects on the device's operation. Its primary use is to control the small-signal transconductance of the first gate and thereby the RF gain of the device; this makes the second gate useful for automatic or manual gain control. This capability is particularly valuable for the input amplifier of a communication receiver; this amplifier may have to accommodate a very large range of RF input levels. Second, because the second gate is usually grounded at the RF frequency, it acts as a shield between the drain and the first gate, reducing the feedback capacitance C_{gd} to a very low value. The low value of this feedback capacitance ensures good stability and high maximum available gain. Unfortunately, however, the noise figure of a dual-gate FET is usually about 1 dB worse than that of a comparable single-gate device.

The fact that the dual-gate FET's transconductance can be varied by the voltage on the second gate makes the dual-gate FET useful as a mixer. In mixer applications, the local oscillator voltage is usually applied to the second gate (the one closest to the drain), and the RF input is applied to the first gate. Approximately 20 dB of LO-to-RF isolation is thereby achieved, often allowing the use of a single-ended mixer instead of a balanced mixer, or at least simplifying the RF input filter. This simplification of the receiver circuitry can be quite valuable for miniature receivers and GaAs integrated circuits.

3.3.2 Modeling Dual-Gate Devices

Dual-gate MESFETs are usually modeled as two single-gate FETs in series, as shown in Figure 3.9. This configuration is similar to a cascode amplifier using discrete single-gate devices, and the dual-gate FET is often described as such. However, in order for the second FET to control the transconductance of the first, the lower FET must be operated in its linear region. This situation is very different from that of the cascode amplifier, where both FETs are operated in current saturation.

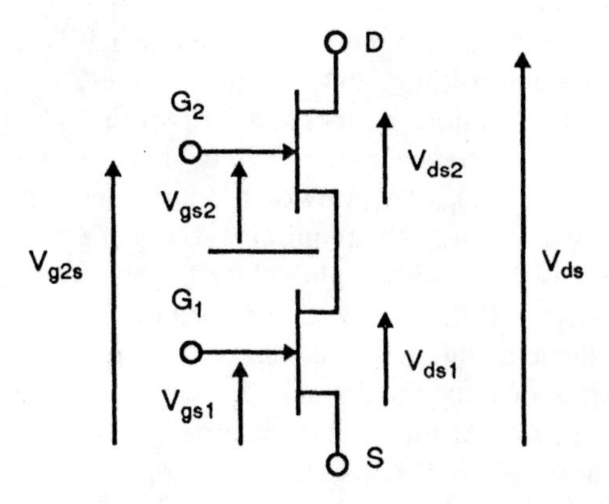

Figure 3.9 The dual-gate MESFET modeled as two single-gate FETs in series.

Virtually all of the models in Sections 3.2.1 to 3.2.6 may be used to describe the individual FETs of Figure 3.9. The *I/V* characteristics of the dual-gate FET can be found from those of the single-gate devices by observing three rather obvious constraints: the channel current in both devices must be equal, $V_{ds1} + V_{ds2} = V_{dd}$, and $V_{gs2} = V_{g2s} - V_{ds1}$. The applied voltages are V_{dd}, V_{g2s}, and V_{gs1}. To determine the drain current for any set of applied voltages, V_{ds1} is treated as an independent variable and is varied between zero and V_{dd} until a value is found that gives the same drain current for both single-gate FETs.

For mixer analysis, the dual-gate FET, like the single-gate device, must be described by a lumped-element equivalent circuit. A single equivalent circuit for the dual-gate FET can be created by substituting the single-gate equivalent circuits of the individual FETs into Figure 3.9. This is the process that has been followed in published models [20]-[21]. However, today the availability of general-purpose nonlinear circuit simulators using harmonic-balance methods makes such "single-device" models unnecessary, and, in any case, modeling the dual-gate device as a series connection of single-gate devices is more practical and more accurate than attempting to model the dual-gate FET as a single unit.

3.3.3 I/V Characteristics

The *I/V* behavior of the dual-gate FET is considerably more complicated than that of the single-gate device. The current in the upper FET is controlled by its gate-to-source voltage, as is the current in the lower FET. The gate-to-source voltage in the lower FET, V_{gs1}, is the applied gate voltage. In the upper FET, however, the applied voltage, V_{g2s}, is not the controlling gate-to-source voltage, V_{gs2}, which depends on the voltage at the inaccessible node connecting the two FET channels and, therefore, on the way V_{dd} is divided between the two devices. The voltage at the node connecting the two devices, V_{ds1}, is allowed to "float," so that it takes on a value between zero and V_{dd} that satisfies the requirement of equal channel currents in both devices. A stable value of V_{ds1} is invariably close to zero or to V_{dd}.

The curves of Figure 3.10 show how the *I*/*V* characteristics of the dual-gate FET can be derived from those of the individual single-gate FETs. Figure 3.10(a) shows the drain characteristics of two devices superimposed when $V_{dd} = 5$V. The curves are arranged so that the sum of the drain-to-source voltages in both devices equals V_{dd}. The drain current of the dual-gate device is the point where the curves for the two devices intersect. If, for example, the upper FET is operated at $V_{gs2} = -1$V and the lower FET has $V_{gs1} = -2$V, the drain current is 14 mA and the upper FET has approximately zero volts across its channel. Figure 3.10(b) shows an equivalent set of curves for the same device, also at $V_{dd} = 5$V, except the current for the second device is graphed in terms of V_{g2s}, the voltage between gate 2 and the source of the lower FET. Figure 3.10(b) was obtained directly from Figure 3.10(a) by determining points on the two sets of curves that represent contours of constant V_{g2s}. The latter graph illustrates the effect of gate voltages on the operating points more clearly than the former, because the graphs are made in terms of observable quantities (the voltage applied to gate 2) rather than a voltage between gate 2 and an inaccessible point (the source node of the upper device). It is evident that, in the dual-gate FET, a substantial change in the transconductance from V_{g1s} to I_d, due to changes in V_{g2s}, occurs only when the lower FET is biased in its linear region and the upper FET is in saturation. The shaded area in Figure 3.10(b) shows the region in which the operating points of the two devices must be located for successful mixer or gain-controlled amplifier operation.

The fact that the mixing FET is in saturation over a large part of the LO cycle has great implications for the performance of dual-gate FET mixers: it is the primary reason why dual-gate FETs invariably exhibit lower conversion loss and higher noise figure than single-gate FET mixers. During the time that the FET is in its linear region, its transconductance and output conductance are reduced; these parameters are critical to the performance of a FET mixer. We shall examine this point further, as well as other differences between single- and dual-gate mixers, in Chapter 9.

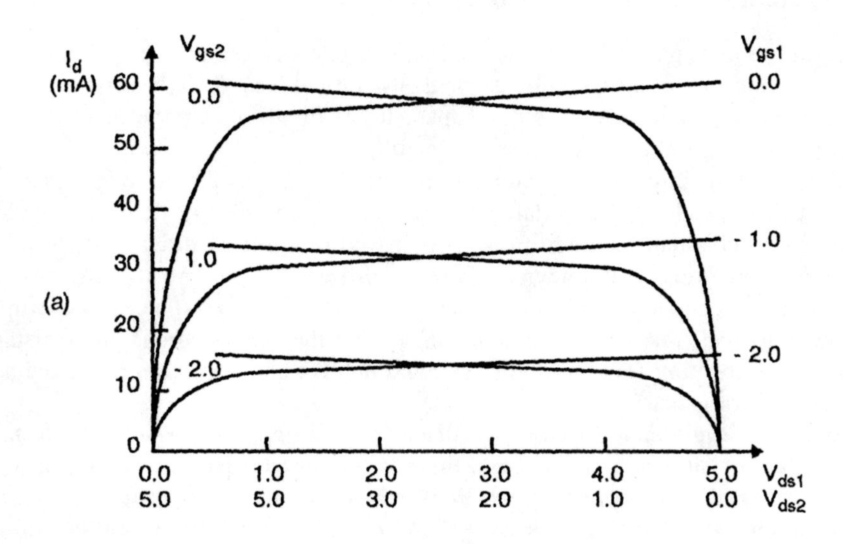

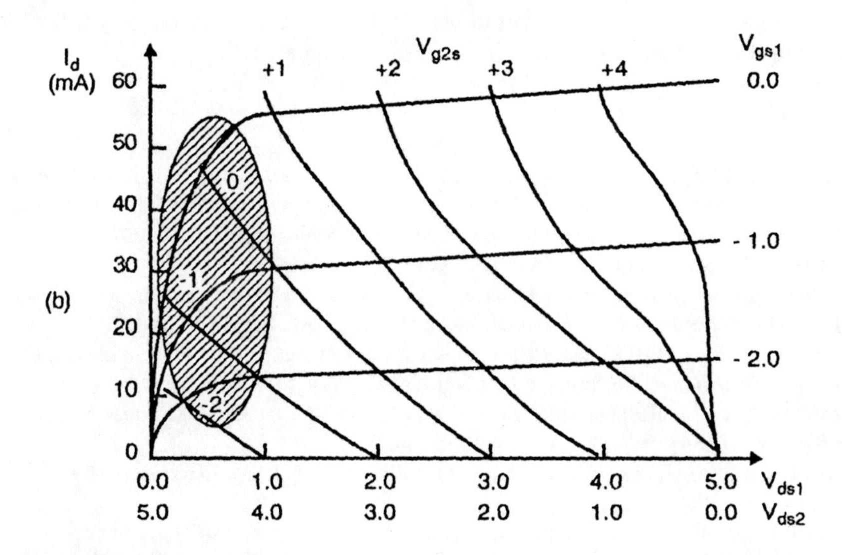

Figure 3.10 I/V characteristic of a dual-gate FET, in terms of the I/V characteristics of the individual devices: (a) in terms of V_{gs2}; (b) in terms of V_{g2s}. The shaded area represents the region in which the dual-gate FET operates when used as a mixer.

3.3.4 Parameter Extraction for Dual-Gate Devices

Modeling the dual-gate FET as a pair of single-gate devices allows one the use of the single-gate modeling techniques for dual-gate devices. This has the advantage that new techniques need not be developed. It has the disadvantage that individual devices may not be available for measurement.

In monolithic integrated circuits, it is not difficult to create single-gate test devices that have channel structures nearly identical to those of dual gate devices. These can be characterized and used, perhaps with minor modifications, to model the "equivalent" dual-gate device. (The modifications are mostly obvious: for example, it is necessary to eliminate the source and drain resistances of the upper and lower devices, respectively, replace them with the estimated channel resistance of the interconnecting section of epilayer, and include an interelectrode capacitance between the two gates.)

Obtaining equivalent circuits of discrete dual-gate devices is much more difficult. It is usually impractical to fit them to measured S parameters, because the number of variables is very great. Methods have been proposed for creating equivalent circuits of dual-gate devices. Scott and Minasian [20] present one model; however, it is valid only to approximately 6 GHz. Tsironis et al. present another which is considerably more difficult to use. Unfortunately, there are as yet few good solutions to the problem of modeling discrete dual-gate FETs.

3.4 REFERENCES

[1] Kennedy, D. P., and R. R. O'Brien, "Computer-Aided Two-Dimensional Analysis of the Junction Field-Effect Transistor," *IBM J. Research and Development*, Vol. 14, 1970, p. 95.

[2] Lehovec, K., and R. Zuleeg, "Voltage-Current Characteristics of GaAs JFETs in the Hot-Electron Range," *Solid-State Electronics*, Vol. 13, 1970, p. 1415.

[3] Yamaguchi, K., S. Asai, and H. Kodera, "Two-Dimensional Numerical Analysis of Stability Criteria of GaAs FETs," *IEEE Trans. Electron Devices*, Vol. ED-23, 1976, p. 1283.

[4] Yamaguchi, K., and H. Kodera, "Drain Conductance of Junction Gate FETs in the Hot Electron Range," *IEEE Trans. Electron Devices*, Vol. ED-23, 1976, p. 545.

[5] Madjar, A., and F. Rosenbaum, "A Large-Signal Model for the GaAs MESFET," *IEEE Trans. Microwave Theory Tech.*, vol. MTT-29, 1981, p. 781.

[6] Shur, M. S., "Analytical Model of GaAs MESFETs," *IEEE Trans. Electron Devices*, Vol. ED-25, 1978, p. 612.

[7] Pucel, R., H. A. Haus, and H. Statz, "Signal and Noise Properties of GaAs Microwave Field-Effect Transistors," in *Advances in Electronics and Electron Physics*, L. Martin, ed., Vol. 38, New York: Academic Press, 1975, p. 195.

[8] Curtice, W. R., "A MESFET Model for Use in the Design of GaAs ICs" *IEEE Trans. Microwave Theory Tech.*, Vol. MTT-28, 1980, p. 448.

[9] Shur, M. S., and L. F. Eastman, "*I/V* Characteristics, Small-Signal Parameters and Switching Times of GaAs FETs," *IEEE Trans. Electron Devices*, Vol. ED-25, 1978, p. 606.

[10] Sussman-Fort, S. E., S. Narasimhan, and K. Mayaram, "A Complete GaAs MESFET Model for SPICE," *IEEE Trans. Microwave Theory Tech.*, Vol. MTT-32, 1984, p. 471.

[11] Curtice, W. R., and M. Ettenberg, "A Nonlinear GaAs FET Model for Use in the Design of Output Circuits for Power Amplifiers," *IEEE Trans. Microwave Theory Tech.*, Vol. MTT-33, 1985, p. 1383.

[12] Materka, A., and T. Kacprzak, "Computer Calculation of Large-Signal GaAs FET Amplifier Characteristics," *IEEE Trans. Microwave Theory Tech.*, Vol. MTT-33, 1985, p. 129.

[13] Statz, H., P. Newman, I. W. Smith, R. A. Pucel, and H. A. Haus, "GaAs FET Device and Circuit Simulation in SPICE," *IEEE Trans. Electron Devices*, Vol. ED-34, 1987, p. 160.

[14] Maas, S. A., and D. Neilson, "Modeling MESFETs for Intermodulation Analysis of Mixers and Amplifiers," *IEEE Trans. Microwave Theory Tech.*, Vol. MTT-38, 1990, p. 1964.

[15] Maas, S. A., "A GaAs MESFET Mixer with Very Low Intermodulation," *IEEE Trans. Microwave Theory Tech.*, Vol. MTT-35, 1987, p. 425.

[16] Shockley, W., "A Unipolar Field-Effect Transistor," *Proc. IRE*, Vol. 40, 1952, p. 1365.

[17] Golio, J. M., ed., *Microwave MESFETs and HEMTs*, Norwood, MA: Artech House, 1991.

[18] Fukui, H., "Determination of the Basic Device Parameters of the GaAs MESFET," *Bell Syst. Tech. J.*, Vol. 58, 1979, p. 771.

[19] Weinreb, S., "Low-Noise Cooled GASFET Amplifiers," *IEEE Trans. Microwave Theory Tech.*, Vol. MTT-28, 1980, p. 1041.

[20] Scott, J. R., and R. A. Minasian, "A Simplified Microwave Model for the GaAs Dual-Gate MESFET," *IEEE Trans. Microwave Theory Tech.*, Vol. MTT-32, 1984, p. 243.

[21] Tsironis, C., and R. Meirer, "Microwave Wide-Band Model of GaAs Dual-Gate MESFETs," *IEEE Trans. Microwave Theory Tech.*, Vol. MTT-30, 1982, p. 243.

Chapter 4
Diode Mixer Theory

The application of semiconductor diodes to microwave mixers was pioneered by the MIT Radiation Laboratories as part of the development of radar in the early 1940s. The earliest theory of semiconductor diode mixers appeared in 1939 [1], although the major early work, covering mixer circuits and diode design as well as theory, appeared considerably later, in 1948 [2]. Improvements in circuits, diodes, and theoretical understanding in the ensuing years [3]–[5] resulted in major improvements in mixer performance. At present, mixer theory is considered mature, and design approaches at frequencies well into the terahertz region (>1000 GHz) have been reported.

This chapter is concerned with the calculation of conversion performance and noise figure (or temperature) of diode mixers. Although some of the early theoretical work was completely analytical, accurate simulation of practical diode mixers, without simplifying assumptions, requires a fully numerical approach. We assume in this analysis that the LO voltage serves only to vary the diode's small-signal junction conductance and capacitance, and that frequency conversion occurs via these linear, time-varying, small-signal elements. This assumption is valid as long as the applied RF signal is small compared to the LO voltage, a condition usually satisfied in practice. It is not satisfied when the RF input level is large, so this approach is not directly useful for determining intermodulation or saturation levels.

A complete analysis requires knowledge of the impedances presented to the diode (the set of *embedding impedances*) at all significant LO harmonics and mixing frequencies and all diode parameters. This is a fairly large set of data, and for high-frequency mixers it often is not readily available. For this reason, the full-scale mixer analysis described in this chapter may not, in many cases, be directly useful as a design tool, but it does give insights that are invaluable for optimizing practical mixers. A practical approach to the design of diode mixers, which can be used alone or in conjunction with this analysis, will be the subject of Chapter 6.

An assumption inherent in the large-signal analysis is that only a single solution exists; that is, the pumped diode is stable under large-signal and small-signal conditions. It is possible, however, for diodes to undergo parametric oscillation, or to exhibit other types of instabilities. This analysis can, in principle, be used to predict certain types of instability; however, in unstable mixers the large-signal algorithm may not converge to a solution. Such behavior is fortunately unusual and is always characteristic of a poor design. Hence, it is usually not of great practical interest.

4.1 CURRENTS AND VOLTAGES IN THE PUMPED DIODE

A diode mixer is sensitive to many frequencies as well as those at which it is designed to operate. The most famous of these is the *image* frequency, which is found at the LO sideband opposite the input, or *RF* frequency. The mixer is also sensitive to similar sidebands on either side of each LO harmonic. These responses are usually undesired; the exception is the *harmonic mixer*, which is designed to operate at one or more of these sidebands.

When a small-signal voltage is applied to the pumped diode at any one of these frequencies, currents and voltages are generated in the junction at all other sideband frequencies. These frequencies are called the *small-signal mixing frequencies,* ω_n, and are given by the relation:

$$\omega_n = \omega_0 + n\omega_p \tag{4.1}$$

where ω_p is the LO frequency and

$$n = \ldots -3, -2, -1, 0, 1, 2, 3, \ldots \tag{4.2}$$

These frequencies are shown in Figure 4.1. The frequencies are separated from each LO harmonic by ω_0, the difference between the LO frequency and the RF frequency (in conventional, downconverting mixers, ω_0 is the IF). It does not matter which mixing frequency is the excitation, or even if the mixer is excited at more than one of these frequencies; a voltage, a current, or both exist in the diode at each mixing frequency ω_n. It is impossible to eliminate any frequency completely. Short-circuiting the diode at one frequency (e.g., by connecting it in parallel with a series resonant circuit) removes the voltage at that frequency, but currents may still exist. An open circuit removes the current component, but the voltage remains.

An apparent oddity of Equation (4.1) is that the lower sidebands are represented as negative frequencies. It will become clear later that this representation allows

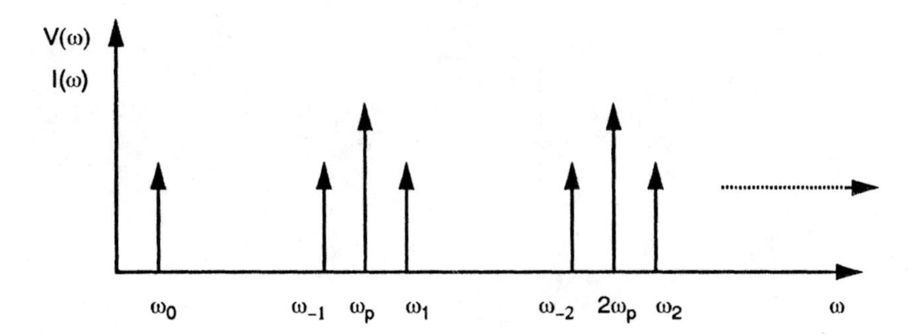

Figure 4.1 Small-signal mixing frequencies ω_n and LO harmonics, $n\omega_p$. Voltage and current components exist in the diode at these frequencies.

considerable simplification of the small-signal analysis. However, since the voltages and currents at the mixing frequencies are represented as phasors, and phasor quantities are traditionally positive frequencies, some manipulation will be necessary to allow for the use of negative frequencies.

In theory, an infinite number of mixing frequencies exist. In practice, n must be limited to some maximum value, N, implying that a total of $2N + 1$ small-signal mixing frequencies and, as we shall see, $N + 1$ positive-frequency LO harmonics (including dc) are significant. The question naturally arises as to the number of LO harmonics that must be retained. N should be chosen to allow enough harmonics for a meaningful analysis, yet not so many that excessive computer time and memory are required. We shall examine this dilemma further in Section 4.2.

4.2 LARGE-SIGNAL ANALYSIS

4.2.1 The Diode Model

We examined the Schottky barrier diode in Chapter 2. The equivalent circuit of the diode, including the nonlinear junction capacitance and fixed, linear series resistance, is shown in Figure 4.2. The junction I/V characteristic is

$$I(V_j) = I_0\left(\exp\left(\frac{qV_j}{\eta KT}\right) - 1\right) \tag{4.3}$$

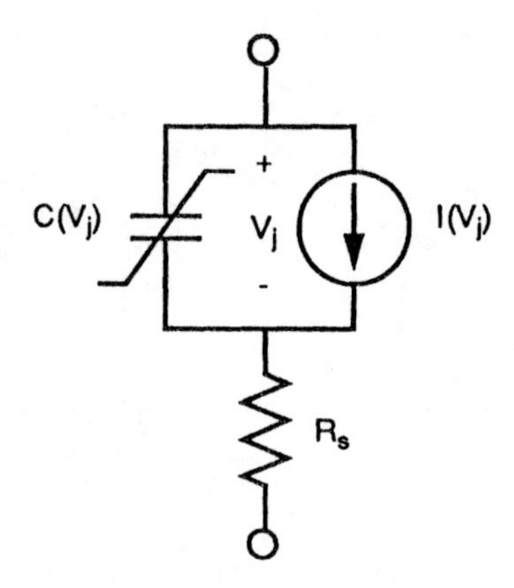

Figure 4.2 Equivalent circuit of the Schottky-barrier diode.

and the small-signal junction conductance $g(V_j)$ is

$$g(V_j) = \frac{d}{dV_j}I(V_j) = \frac{q}{\eta KT}\exp(\frac{qV_j}{\eta KT}) \approx \frac{q}{\eta KT}I(V_j) \tag{4.4}$$

where I_0 is the reverse saturation current, q is the electron charge (1.6×10^{-19} coul), V_j is the junction voltage (not including the voltage drop across the series resistance R_s), η is the ideality factor (usually less than 1.25), K is Boltzmann's constant (1.37×10^{-23} J/K), and T is absolute temperature. The junction capacitance $C(V_j)$ is

$$C(V_j) = \frac{C_{j0}}{\left(1 - \frac{V_j}{\phi_{bi}}\right)^{\gamma}} \tag{4.5}$$

where C_{j0} is the zero-voltage junction capacitance and ϕ_{bi} is the junction's built-in voltage. The exponent $\gamma = 0.5$ when the diode's expitaxial layer is uniformly doped; if the doping is not uniform, the exponent may deviate somewhat from $\gamma = 0.5$. The capacitance $C(V_j)$, is defined as the incremental change in depletion-region charge Q_d that results from a change in junction voltage:

$$C(V_j) = \frac{dQ_d}{dV_j} \tag{4.6}$$

R_s, the series resistance, is normally treated as a linear quantity, and may be found from the diode's dc *I*/*V* curve. In (4.5), ϕ_{bi} is a positive quantity and V_j is positive when the junction is forward-biased.

There is often confusion regarding the voltage and current relations for the nonlinear capacitor under large-signal and small-signal excitation. The large-signal expression is defined as an *incremental* capacitance. The large-signal current in the capacitor, $I_c(t)$, is

$$\begin{aligned} I_c(t) &= \frac{dQ_d}{dt} = \left.\frac{dQ_d}{dV}\right|_{V = V_j(t)} \frac{d}{dt} V_j(t) \\ &= C(V_j(t)) \frac{d}{dt} V_j(t) \end{aligned} \tag{4.7}$$

where $V_j(t)$ is the large-signal junction voltage. In small-signal analysis, however, the capacitor is not treated as a nonlinear element, but as a *linear*, time-varying capacitance. Then

$$C(t) = C(V_j(t)) \tag{4.8}$$

$$\begin{aligned} i_c(t) &= \frac{d}{dt} C(t) v(t) \\ &= C(t) \frac{d}{dt} v(t) + v(t) \frac{d}{dt} C(t) \end{aligned} \tag{4.9}$$

where $v(t)$ is the small-signal junction voltage. Equation (4.7) is used in the large-signal analysis, and (4.9) is used in the small-signal analysis.

4.2.2 The Harmonic-Balance Equations

The goal of the large-signal analysis is to determine the voltage $V_j(t)$ (or, as an alternative, its Fourier components $V_j(n\omega_p)$), the control voltage of the nonlinear junction conductance and capacitance. When $V_j(t)$ is known, it is substituted into

Equations (4.4) and (4.9) to determine the time-varying conductance and capacitance waveforms; these are subsequently used to determine conversion loss and input or output impedances. Thus, our task is first to find $V_j(t)$. In doing so, we note that $v(t) << V_j(t)$, so we can neglect $v(t)$.

A diode mixer consists of one or more diodes and a set of filtering and matching circuits. These circuits can be described in the frequency domain by Y parameters evaluated at the LO frequency ω_p and its harmonics. There must also be voltage or current sources at the fundamental LO frequency and dc, representing the excitation and dc bias. Because the diode is nonlinear, however, it must be described by a lumped-element model in the time domain; it is rarely possible to describe nonlinear elements adequately in the frequency domain. We are therefore faced with a fundamental problem: the diode and its external circuit require inherently incompatible analyses.

A practical solution to this problem is to use a type of harmonic balance analysis. This analysis can be implemented in a number of ways, but all have in common the following: a time-domain junction voltage is estimated, and the junction current is calculated from the diode equations (4.3) and (4.5); similarly, a frequency-domain solution is found that satisfies the external circuit equations. If the voltage is indeed estimated correctly, the resulting diode and linear-circuit currents will be equal. If the voltage is not correct, however, they will be unequal; in this case the voltage can be modified and tested again in both sets of equations. Eventually that modification process, if implemented appropriately, will cause the voltage to converge to a solution after a number of iterations.

The key to implementing this process successfully is to find two things: an initial estimate of the voltage that is close enough to the solution to allow convergence, and a criterion for modifying it–embodied in a computer algorithm–that produces fast, reliable convergence. In this section, we shall develop an expression for the quantity, called the *current-error vector*, that must be solved. In later sections, we shall examine iterative methods for solving it.

Figure 4.3(a) shows the circuit of a single-diode mixer. In Figure 4.3(b), the fixed elements of Figure 4.3(a) have been absorbed into the linear network to create a two-port. The diode's series resistance R_s, the source impedance $Z_s(\omega_p)$, and the terminated RF and LO ports have been included in the linear subcircuit; only the diode's junction conductance and capacitance are included in the nonlinear subcircuit. At each LO harmonic, this two-port has the admittance equations

$$\begin{bmatrix} I_1(n\omega_p) \\ I_2(n\omega_p) \end{bmatrix} = \begin{bmatrix} Y_{11}(n\omega_p) & Y_{12}(n\omega_p) \\ Y_{21}(n\omega_p) & Y_{22}(n\omega_p) \end{bmatrix} \begin{bmatrix} V_j(n\omega_p) \\ V_s(n\omega_p) \end{bmatrix} \tag{4.10}$$

The variables are defined in the figure. Usually, $V_s(n\omega_p) = 0$ when $n > 1$.

The circuit of Figure 4.3(b) can be further reduced to that of the Norton equivalent in Figure 4.3(c), where

$$I_s(n\omega_p) = Y_{12}(n\omega_p)V_s(n\omega_p) \tag{4.11}$$

and the linear circuits in the mixer have been reduced to a simple two-terminal network consisting of an excitation and a set of admittances at each LO harmonic. These admittances are called the *embedding admittances*, and they represent the admittance at each harmonic "seen" by the diode at its terminals. Finally, the current in Port 1 of the linear subcircuit, $I_l(n\omega_p)$, is

$$I_l(n\omega_p) = I_s(n\omega_p) + Y_e(n\omega_p)V_j(n\omega_p) \tag{4.12}$$

where Y_e represents the embedding admittances,

$$Y_e(n\omega_p) \equiv Y_{11}(n\omega_p) \tag{4.13}$$

If the linear network is lossless, it is possible to obtain a practical form of this embedding network without evaluating the full admittance network. It is necessary only to calculate $Y_e(n\omega_p)$ and create a source I_s that has the same available power as the original source:

$$|I_s(\omega_p)| = |V_s(\omega_p)|\sqrt{\frac{Re\{Y_e(\omega_p)\}}{Re\{Z_s(\omega_p)\}}} \tag{4.14}$$

Unfortunately, the lossless assumption is not valid if R_s is absorbed into the linear network. However, in many cases (e.g., if the reflection algorithm, described below, is used), it is practical to treat R_s as part of the nonlinear subcircuit, and Equation (4.14) may be valid. The phase of I_s is usually of no concern.

The harmonic voltage components $V_j(n\omega_p)$ are the variables that define the system; if these are known, all the voltages and currents in the circuit can be determined*. These voltages must satisfy the condition

---

* Borrowing from thermodynamics, we sometimes use the term *state variables* to represent a set of variables that is sufficient to define a system.

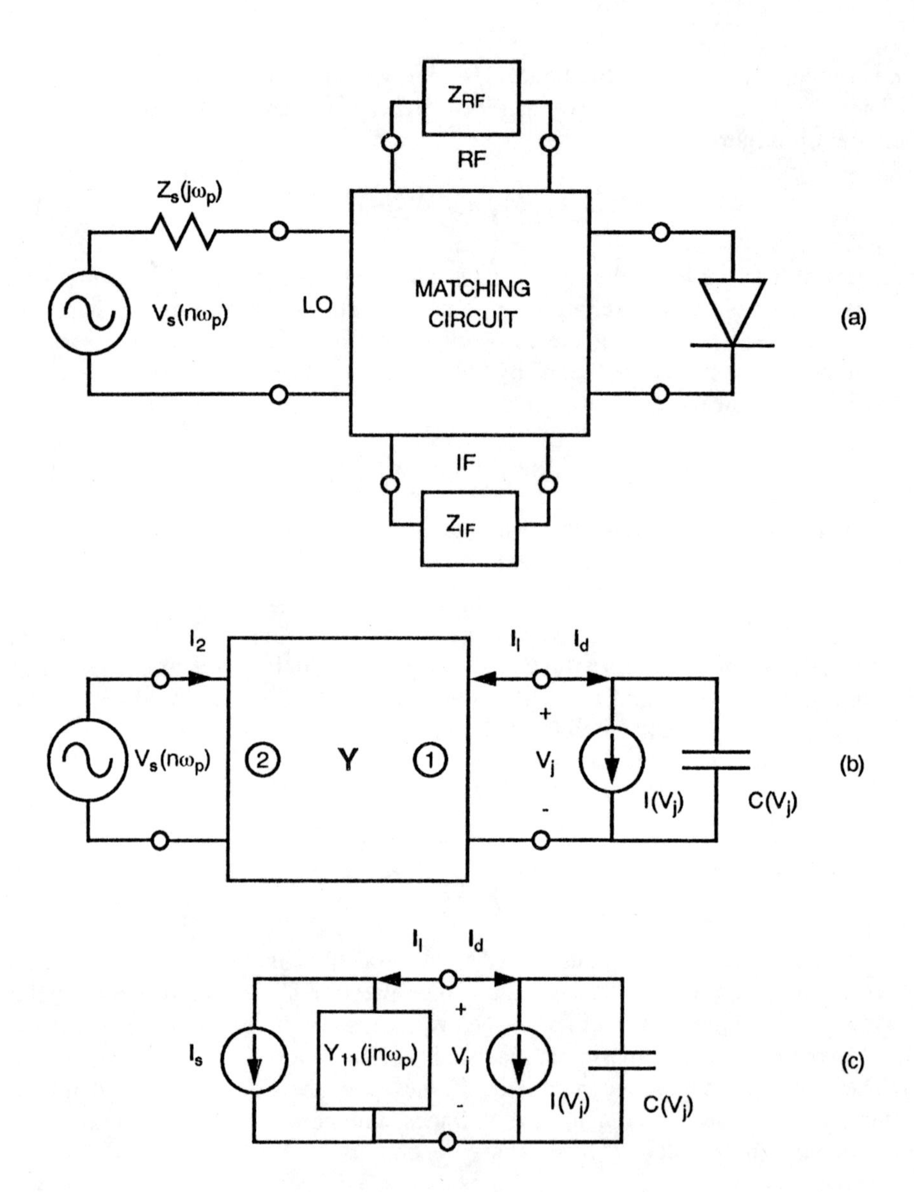

Figure 4.3 Large-signal (LO) equivalent circuit of a single-diode mixer: (a) the mixer; (b) R_s and load impedances are absorbed into the matching network; (c) the Y matrix is reduced to a set of excitation sources and an embedding admittance, Y_{11}.

$$I_l(n\omega_p) = -I_d(n\omega_p) \qquad n = 0, 1, 2, \ldots, N \tag{4.15}$$

or, in vector form,

$$\begin{bmatrix} I_l(0) \\ I_l(\omega_p) \\ I_l(2\omega_p) \\ \cdots \\ \cdots \\ I_l(N\omega_p) \end{bmatrix} = -\begin{bmatrix} I_d(0) \\ I_d(\omega_p) \\ I_d(2\omega_p) \\ \cdots \\ \cdots \\ I_d(N\omega_p) \end{bmatrix} \tag{4.16}$$

where $I_d(n\omega_p)$, the current in the diode junction, is the sum of the resistive current $I_g(n\omega_p)$ and reactive current $I_c(n\omega_p)$:

$$I_d(n\omega_p) = I_g(n\omega_p) + I_c(n\omega_p) \tag{4.17}$$

The fundamental problem is to find a set of harmonic voltage components $V_j(n\omega_p)$ that satisfy Equation (4.16). Using boldface characters to represent vectors, we can express (4.16) simply as

$$\mathbf{I}_l + \mathbf{I}_d = 0 \tag{4.18}$$

Before the iterative process has converged to the correct $V_j(n\omega_p)$,

$$\mathbf{F}(\mathbf{V}_j) = \mathbf{I}_l + \mathbf{I}_d \tag{4.19}$$

$\mathbf{F}(\mathbf{V}_j)$ is the current-error vector. Its magnitude is a measure of how far the solution process is from convergence.

We now examine the nonlinear part of the circuit. I_d(t) consists of two components, the current in the junction capacitance and the current in the resistive junction. The latter, I_g(t), is found by substituting $V_j(t)$ into (4.3), and the frequency-domain components are found by Fourier transforming. We obtain a current vector $\mathbf{I}_g$ having the same form as the vectors in Equation (4.16).

The capacitive current vector $\mathbf{I}_c$ requires an extra step. We first express the charge waveform as a function of junction voltage:

$$Q_d(t) = f(V_j(t)) = C_{j0}\phi(1 - V_j(t)/\phi)^{\gamma} \tag{4.20}$$

The frequency-domain charge components $Q_d(jn\omega_p)$ are found by Fourier-transforming $Q_d(t)$. Note that

$$I_c(t) = \frac{d}{dt}Q_d(t) \tag{4.21}$$

Because differentiation in the time domain corresponds to multiplication by $j\omega$ in the frequency domain,

$$I_c(jn\omega_p) = jn\omega_p Q_d(jn\omega_p) \qquad n = 0, 1, 2\ldots, N \tag{4.22}$$

(One could also use Equation (4.7), which involves time derivatives, to find $I_c(t)$ directly, and then Fourier transform to obtain $I_c(jn\omega_p)$. Time derivatives, however, can be relatively difficult to estimate accurately in numerical calculations, so frequency-domain methods are usually preferred.) Finally, Equation (4.22) can be written

$$\mathbf{I}_c = j\Omega\mathbf{Q}_d \tag{4.23}$$

where Ω is a diagonal matrix having $n\omega_p$ at each position along the diagonal, with $n = 0$ at the top and $n = N$ at the bottom. The charge vector $\mathbf{Q}_d$ has the same form as the vectors in Equation (4.16).

Substituting Equations (4.12), (4.23), $\mathbf{I}_s$, and $\mathbf{I}_g$ into (4.19), we obtain for the current-error vector

$$\mathbf{F}(\mathbf{V}_j) = \mathbf{I}_s + \mathbf{Y}_e\mathbf{V}_j + \mathbf{I}_g + j\Omega\mathbf{Q}_d \tag{4.24}$$

where $\mathbf{Y}_e$ is the diagonal matrix of embedding admittances (Equation (4.12)):

$$\mathbf{Y}_e = \begin{bmatrix} Y_e(0) & 0 & 0 & \ldots & 0 \\ 0 & Y_e(\omega_p) & 0 & \ldots & 0 \\ 0 & 0 & Y_e(2\omega_p) & \ldots & 0 \\ \ldots & \ldots & \ldots & \ldots & \ldots \\ 0 & 0 & 0 & \ldots & Y_e(N\omega_p) \end{bmatrix} \tag{4.25}$$

This is the form of $\mathbf{F}(\mathbf{V}_j)$ most frequently cited. Our final task is to obtain $\mathbf{V}_j$ such that

$$\mathbf{F}(\mathbf{V}_j) = 0 \tag{4.26}$$

Methods for finding $\mathbf{V}_j$ are the subject of the next section.

4.2.3 Solution Algorithms

The fundamental problem we must address is to find a vector $\mathbf{V}_j$, consisting of the voltage components $V_j(jn\omega_p)$, that solves the equation $\mathbf{F}(\mathbf{V}_j) = 0$. A number of ways to do this have been proposed; we examine the most popular of these below.

Optimization

The problem of finding a set of voltage components $V_j(jn\omega_p)$ can be treated as an optimization problem. The process begins with the selection of a trial function for $V_j(t)$ (or alternatively $V_j(jn\omega_p)$), and $\mathbf{F}(\mathbf{V}_j)$ is evaluated. Clearly, $|\mathbf{F}(\mathbf{V}_j)|$ may be selected as an error function, $V_j(jn\omega_p)$ as variables of optimization, and conventional optimization techniques [6], [7] can be used to perform the minimization.

Although optimization may at first seem to be the most logical way to solve Equation (4.24), it is clearly not the best. The reason is that much of the information inherent in $\mathbf{F}(\mathbf{V}_j)$ is lost when $|\mathbf{F}(\mathbf{V}_j)|$ is formed; specifically, we lose the information about the way the *individual* components of $\mathbf{F}$ depend on the *individual* voltage components of $\mathbf{V}_j$. The practical result is that the optimization process converges very slowly and often unreliably. If we were to retain the information that is discarded by optimization, we could make the solution process far more efficient. This is, in fact, the main idea behind Newton's method, which we shall examine in due course.

The advantage of optimization is that many optimization routines are readily available in computer subroutine libraries, so this critical part of the computer programming is often available without additional effort. Thus, optimization may satisfy an occasional need for a "fast and dirty" mixer simulation. Like most of the methods we shall examine, optimization is not self-starting; we must begin with an initial estimate of the solution.

Optimization Algorithm

1. Select an initial estimate of $V_j(t)$: $V_j(t) = A_0 + A_1 \cos(\omega_p t)$ is often a good estimate.
2. Use Equations (4.3) and (4.5) to obtain $Q_d(t)$ and $I_g(t)$.
3. Fourier-transform to obtain $\mathbf{Q}_d$ and $\mathbf{I}_g$.
4. Use Equations (4.11) and (4.12) to obtain $\mathbf{I}_l$.
5. Form $|\mathbf{F}(\mathbf{V}_j)|$ (Equation (4.24)).
6. If $|\mathbf{F}(\mathbf{V}_j)|$ is small enough, end the process.
7. Modify $V_j(jn\omega_p)$, $n = 0, 1, 2, \ldots, N$ according to the chosen method of optimization.
8. Form the time waveform $V_j(t)$ and continue from Step 2.

Relaxation Method ("Splitting Method")

In 1980, Hicks and Khan [8] proposed a relaxation method for solving Equation (4.26) that has been very widely accepted. Although it is probably not suitable for general-purpose harmonic-balance analysis, it works well for simple problems such as the one at hand, finding the LO waveforms in a diode mixer.

As with most of the other algorithms, we begin with an initial estimate of $\mathbf{V}_j$. This is Fourier-transformed to obtain $V_j(t)$, and $I_d(t)$ is found from the nonlinear elements' equations, (4.3) and (4.7). It is then assumed that $I_l(t) = -I_d(t)$, and after Fourier-transforming $I_l(t)$ to obtain $\mathbf{I}_l$, an estimate of the junction voltage $\mathbf{V}_j'$ is found from the linear-circuit equations:

$$\mathbf{V}'_j = \mathbf{Y}_e^{-1}(\mathbf{I}_l - \mathbf{I}_s) \tag{4.27}$$

The new estimate of $\mathbf{V}_j$, $\mathbf{V}_j^{(2)}$, is a quantity geometrically between the previous two:

$$\mathbf{V}_j^{(2)} = s\mathbf{V}'_j + (1 - s)\,\mathbf{V}_j \tag{4.28}$$

where $0 < s < 1$. The coefficient s must be found by trial and error; usually $s \sim 0.1$ to 0.3.

The value of the coefficient s has a strong effect on the convergence properties of the process. Small values of s result in relatively slow but reliable convergence; large values result in much greater modification of $\mathbf{V}_j$ per iteration, but increase the chance that the process will diverge. The process is also very sensitive to the value of the embedding impedances. Even under the best of circumstances, it is possible for the process to diverge even when the nonlinearities are weak and the initial estimate is arbitrarily close to a solution.

Even with its precarious convergence, its simplicity has made this method quite popular. Furthermore, the process does not require knowledge of derivatives of the device's I/V or Q/V characteristics, and therefore can be used with devices requiring numerical models that are fairly complex, and hence difficult to differentiate.

Splitting Algorithm

1. Estimate $\mathbf{V}_j$.
2. Inverse Fourier-transform $\mathbf{V}_j$ to obtain $V_j(t)$.
3. Calculate $I_d(t)$ from the diode-junction equations.
4. Set $I_l(t) = -I_d(t)$, and Fourier-transform $I_l(t)$ to obtain $\mathbf{I}_l$.
5. Calculate $\mathbf{V}_j'$ from the linear-circuit equations.
6. Form a new estimate of $\mathbf{V}_j$ from Equation (4.28), and repeat from Step 2.

Reflection Algorithm

Another popular technique is that of Kerr [9], which we shall call the *reflection algorithm.* As with the previous algorithms, the circuit is divided into linear and nonlinear parts, and an iterative process is used to solve the relevant equations. The most significant difference between the reflection algorithm and other methods is that it uses propagating voltage waves (or scattering variables) instead of port voltages and currents. Another significant difference is that this method is self-starting: it does not require an initial estimate of the solution.

The reflection algorithm is a type of relaxation method. In fact, it can be viewed as a version of the splitting algorithm implemented with scattering variables instead of voltages and currents, and, as originally described, with $s = 1$. A method using other values of s developed by Hwang and Itoh [10] had, as might be expected, better performance.

Figure 4.4 illustrates the algorithm. The LO source and linear subcircuit are separated from the diode by a section of ideal transmission line having characteristic impedance Z_C. The transmission line is assumed to be an integral number of wavelengths long at the fundamental LO frequency, and therefore the steady-state voltages and currents at the diode terminals are the same as those at the linear subcircuit. Because of the presence of the line, however, the circuit can be divided into the two equivalents shown in Figure 4.4(b); the fictitious transmission line thus facilitates the separation of the linear and nonlinear calculations*.

The algorithm mimics the transient response of the circuit of Figure 4.4(a) from the time it is first turned on until it converges to steady-state conditions. At $t = 0$, LO and dc-bias voltages are applied to the left end of the transmission line, exciting an incident wave, $V_i^1(t)$. The equivalent circuit of the input end of the line is shown in Figure 4.4(b); the incident voltage wave is

$$V_i^1(t) = \frac{V_{LO} Z_c}{\sqrt{Z_c^2 + |Z_e(j\omega_p)|^2}} \cos(\omega_p t + \theta) + \frac{V_b Z_c}{Z_c + Z_e(0)} \tag{4.29}$$

where

$$\theta = \operatorname{atan}\left(\frac{\operatorname{Im}\{Z_e(j\omega_p)\}}{\operatorname{Re}\{Z_e(j\omega_p)\} + Z_c} \right) \tag{4.30}$$

and the superscripts indicate the iteration number. V_{LO}, V_b, and Z_e are as shown in Figure 4.4(a).

This incident wave propagates toward the diode. The equivalent circuit of the diode-to-transmission-line interface is shown in Figure 4.4(c), where the diode is excited by the incident wave through the transmission line impedance, Z_c. This circuit consists entirely of lumped linear and nonlinear resistive and capacitive elements; therefore, we must develop a set of time-domain nonlinear differential equations and integrate them numerically to obtain the diode's terminal voltage and current waveforms, $V_d^1(t)$ and $I_d^1(t)$. The equation to be integrated is

$$(R_s + Z_c)\left[C(V_j)\frac{d}{dt}V_j(t) + I_g(V_j(t))\right] + V_j(t) - 2V_i^1(t) = 0 \tag{4.31}$$

In order to reach the steady-state condition, it may be necessary to integrate the diode equation through several LO cycles until the transient response has died out.

---

* Of course, scattering variables can be defined in the absence of a transmission line, so the transmission line is not strictly necessary. (See, for example, Balabian and Bickart, *Electrical Network Theory*, Wiley, New York, 1969.)

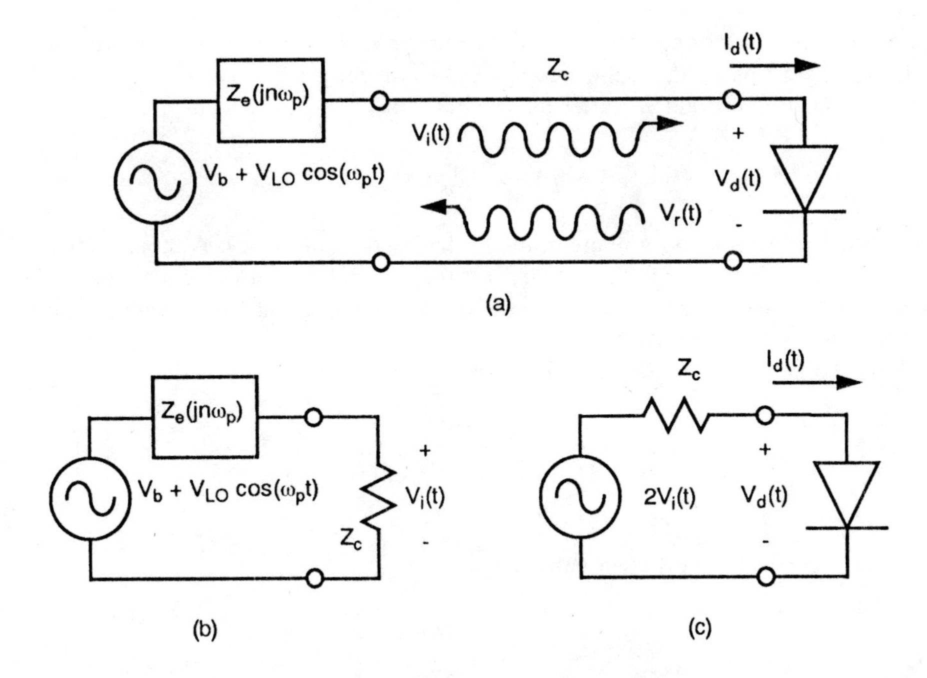

Figure 4.4 Large-signal equivalent circuit for use with the reflection algorithm: (a) a fictitious transmission line, $n\lambda$ long, is placed between the diode and the embedding network; (b) equivalent circuit of the input end of the line; (c) equivalent circuit of the diode end. The diode includes the series resistance, R_s.

The reflected wave is then given by

$$V_r^1(t) = \frac{V_d^1(t) - I_d^1(t)Z_c}{2} \tag{4.32}$$

and it propagates back toward the LO source. Because the diode is a nonlinear device, its terminal voltage and current, and consequently the reflected wave, contain harmonic components. At the LO source, the reflected wave is Fourier-transformed by means of the *fast Fourier transform* (FFT) algorithm, yielding its frequency-domain voltage V_{rn} at each harmonic $n\omega_p$ of the LO frequency. The new incident wave is found by multiplying V_{rn} by the source reflection coefficients at each harmonic, inverse Fourier-transforming to obtain the time-domain representation, and adding them to the original incident wave. In Hwang's modification of this method, the frequency-domain reflected wave is only partially updated at each iteration (i.e., after the pth iteration):

$$V_{rn}^{p+1} = (1-s)\,V_{rn}^{p} + s\bar{V}_{rn} \tag{4.33}$$

where $\bar{V}_{rn}$ is the reflected voltage at $n\omega_p$ calculated from (4.32). Although Hwang states that a value of $s \cong 0.5$ usually results in fastest convergence, his work involved FET mixers, not diode mixers, and it is likely that the diode's stronger nonlinearity may require lower values of s.

After the pth iteration, the new incident wave is given by

$$V_i^{p+1}(t) = V_i^1(t) + \sum_{n=-N}^{N} \Gamma_{sn} V_{rn}^{p} \exp(jn\omega_p t) \tag{4.34}$$

where the LO source coefficient at $n\omega_p$, Γ_{sn}, is

$$\Gamma_{sn} = \Gamma_s(jn\omega_p) = \frac{Z_e(jn\omega_p) - Z_c}{Z_e(jn\omega_p) + Z_c} \tag{4.35}$$

This process is repeated until convergence is achieved. Convergence is indicated by minimal variation in diode terminal voltage or current waveforms between iteration cycles. The simplest criterion is to determine the rms difference in the waveforms between two cycles; reducing this quantity to a few millivolts is usually adequate for practical purposes. Kerr [9] presents an alternative frequency-domain criterion.

The value of Z_c has a strong effect upon the speed of convergence. Convergence time can be reduced if Z_c is chosen to minimize the reflection coefficients Γ_{sn}, especially at the lowest harmonics. In particular, $Z_e(0)$, the dc embedding impedance, should be set equal to Z_c instead of its usual value of zero. This is necessary to prevent large changes in the dc component of $V_i(t)$ at each iteration, when the reflected wave encounters the source reflection coefficient $\Gamma_{s0} = -1$. This artificially high value of $Z_e(0)$ offsets the bias voltage at the diode, but it can be compensated easily by raising the bias-source voltage.

Z_c also affects the efficiency of the solution of Equation (4.31). Increasing the value of Z_c increases the time required for the integration of the differential equation to reach steady-state conditions. If Z_c is made too small, numerical instability may result, especially if the diode's C_{j0} and R_s are small. For most applications, a value of $Z_c = 25\Omega$ to 100Ω works well.

Reflection Algorithm

1. Determine the initial incident wave, $V_i^1(t)$ from Equation (4.29).
2. Solve Equation (4.31) numerically to obtain $V_j^1(t)$.
3. Find V_d:

$$V_d^1(t) = V_j^1(t) + R_s I_d^1(t) \tag{4.36}$$

4. Find the reflected wave from Equation (4.32).
5. Fourier-transform $V_r^1(t)$ to obtain V_{rn}, $n = 0, 1, 2,...,N$.
6. Use Equation (4.34) to find the new incident wave.
7. Repeat from Step 2.

Newton's Method

Newton's method (more properly, *Newton-Raphson iteration*) is an iterative technique for finding the zeros of a function. The zero is estimated at each iteration as the straight-line extrapolation to the abscissa; by performing this estimate several times, the process converges quadratically to the zero. Although the one-dimensional form of Newton's method, illustrated in Figure 4.5, is best known, the method can be extended to the multidimensional case very easily. Clearly, solving Equation (4.26) is such a problem.

Newton's method is most useful when the derivatives of the function in question are easily evaluated, and (because its global convergence is often poor) a reasonably good initial estimate of the solution can be made. The former condition is well satisfied in diode mixers, although the latter may not be. Even so, Newton's method has become the most commonly used method for solving harmonic-balance equations in all types of nonlinear circuits, and is the solution method most frequently used in general-purpose harmonic-balance simulators.

The iterative process is described in terms of the frequency-domain voltage vector, $\mathbf{V}_j$; after the pth iteration,

$$\mathbf{V}_j^{p+1} = \mathbf{V}_j^p - \mathbf{J}^{-1}\mathbf{F}(\mathbf{V}_j^p) \tag{4.37}$$

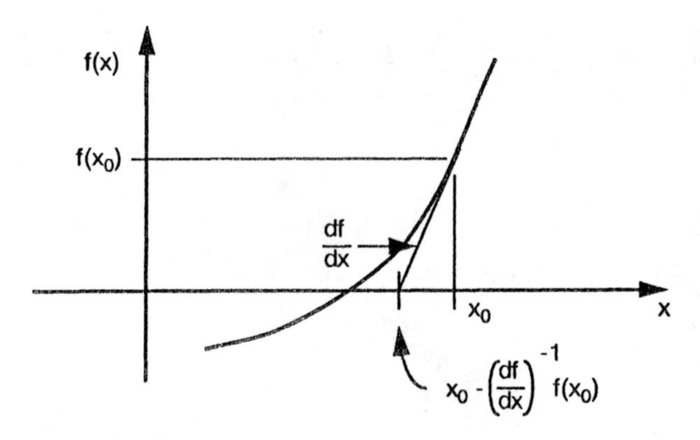

Figure 4.5 Newton's method in one dimension. The zero of $f(x)$ is estimated iteratively from the function's value and derivative. One iteration is illustrated.

where the superscript p indicates the iteration number, $\mathbf{F}(\mathbf{V}_j)$ is described by Equation (4.24), and **J** is the Jacobian matrix,

$$\mathbf{J} = \frac{\partial}{\partial \mathbf{V}_j}\mathbf{F}(\mathbf{V}_j) = \mathbf{Y}_e + \frac{\partial \mathbf{I}_g}{\partial \mathbf{V}_j} + j\Omega\frac{\partial \mathbf{Q}_d}{\partial \mathbf{V}_j} \tag{4.38}$$

The partial derivatives in (4.38) are matrices; their elements (see [11] for a derivation) are

$$\frac{\partial Q_d(k\omega_p)}{\partial V_j(l\omega_p)} = \frac{1}{T}\int_{-T/2}^{T/2} \frac{\partial Q_d(t)}{\partial V_j(t)} \exp(-j\,(k-l)\,\omega_p t)dt \tag{4.39}$$

and

$$\frac{\partial I_g(k\omega_p)}{\partial V_j(l\omega_p)} = \frac{1}{T}\int_{-T/2}^{T/2} \frac{\partial I_g(t)}{\partial V_j(t)} \exp(-j\,(k-l)\,\omega_p t)dt \tag{4.40}$$

and $\mathbf{Y}_e$ is given by Equation (4.13). The components of **J**, other than $\mathbf{Y}_e$, are Fourier-series components of the derivative waveforms. For further information on the mathematical underpinnings of Newton's method, and a few subroutines for implementing it, see [7].

4.2.4 Convergence and Accuracy

Three critical properties of any iterative analysis are its speed, the reliability of its convergence, and the accuracy of the solution. Each algorithm, of course, has its own properties; however, parameters that are common to all the algorithms–the number of harmonics retained in the analysis and the sampling process–affect efficiency as well. The relative merits of the different algorithms is probably the most contentious issue in this field of endeavor; we shall wade into these piranha-infested shallows momentarily. First, we consider harmonics and the sampling process.

Harmonics and Time Samples

The number of harmonics N that must be retained in a nonlinear analysis inevitably involves a trade-off between speed and accuracy. A large N improves the accuracy at the expense of slower analysis, while a small N improves speed at the expense of accuracy. In general, the optimum N depends on the level of excitation and the strength of the nonlinearity; in diode mixers, the nonlinearity of the junction is relatively strong, and good performance usually requires strong pumping of the diode. Furthermore, an FFT requires that the number of harmonics be a power of two; thus, $N = 8$ is usually selected, and is adequate for most purposes.

It is important to note that the time required to perform an FFT with $N = 8$ is rarely a large fraction of the time required to perform an analysis. This is especially true in Newton's method, where the dimension of the Jacobian is $N + 1$, and the time required to invert the Jacobian, not the FFT, is dominant in establishing the time required for the analysis.

An N-harmonic FFT requires $2N$ time samples. As we shall see, it is often worthwhile to "oversample" the time waveform (i.e., to use more samples than necessary, and discard the extra frequency components that emerge from the FFT). This improves the accuracy of all the Fourier components, but beyond some point may not improve the lower harmonics significantly. The number of harmonics that must be retained, and the need for oversampling, must be established empirically.

Properties of Algorithms

The trade-off between solution algorithms is fundamentally between speed per iteration and degree of convergence per iteration. Algorithms that are relatively fast per iteration invariably require many iterations, while those that are slow improve the error function much more on each iteration.

Newton's method is quadratically convergent: near the solution, the number of significant digits in the components of the solution vector approximately doubles at each step. However, each step requires the inversion of a large matrix, a process that is notoriously slow. Although ways exist for speeding the process, it is inevitably slower than other methods. The compensation for the use of Newton's method is that, with a good initial estimate of the solution, its convergence is fairly reliable and its convergence is complete and accurate.

The splitting method is relatively fast per iteration, but convergence is at best linear; thus, it requires many more iterations to reach a solution. Although both may diverge under certain conditions, there is no question that Newton's method is more robust. The reflection algorithm, especially as modified in [10], represents a good trade-off between speed and reliability of convergence.

Optimization methods are distinctly inferior to the other methods, and should be considered obsolete. They may be useful on occasion when the ease of implementing them outweighs the inefficiency.

Accuracy of the Large-Signal Analysis

The accuracy of a harmonic-balance analysis is dominated by three factors: the number of frequency components retained, the properties of the Fourier transforms, and the accuracy of the nonlinear-device model. We can dispense with the latter very easily: the Schottky barrier diode is probably the only microwave device in common use that can be described well enough for virtually all purposes by the simple set of closed-form equations (4.3) and (4.5). Furthermore, the parameters of these equations can be measured at dc or low frequencies, and are valid at frequencies well into the millimeter-wave region.

In each step of the harmonic-balance analysis, the time waveform $V_j(t)$ is substituted into Equations (4.3) and (4.5), and the diode current $I_d(t)$ is found. The current includes all the frequency components of $V_j(t)$, plus more harmonics, generated by the nonlinearity, that are outside the range of the FFT. When the current is Fourier-transformed, these higher harmonics are neglected; neglecting them represents an error in $I_d(t)$, which, in turn, generates an error in all the components of $\mathbf{I}$. The critical question is this: how bad are the errors in $\mathbf{I}$ that result from errors in $I_d(t)$?

These errors can be estimated by considering the properties of the transform. The FFT is simply an algorithm for minimizing repeated multiplications in a *discrete Fourier transform* (DFT). The DFT can be described by a matrix,

$$\mathbf{i} = \mathbf{A}\mathbf{I} \tag{4.41}$$

where **i** is the vector of sampled time-domain currents at the sampling intervals t_1, $t_2,\ldots,t_K$, and **I** is the vector of frequency-domain currents. The elements of **A** are

$$a_{k,n} = \exp(jn\omega_p t_k) \tag{4.42}$$

In a discrete Fourier transform, the rows of **A** are orthogonal, and the matrix is said to be *well conditioned.* The conditioning of the matrix is described by the *condition number*, κ(**A**), where

$$\kappa(\mathbf{A}) = \|\mathbf{A}\|\|\mathbf{A}^{-1}\| \tag{4.43}$$

and $\|\mathbf{A}\|$ is the infinite norm,

$$\|\mathbf{A}\| = \max_i \sum_j |a_{i,j}| \tag{4.44}$$

The significance of all this is that the errors in $I_d(t)$–which consist primarily of the ignored frequency components $n > N$–are effectively multiplied by the condition number, and distributed among all the components of **I**. Thus,

$$\frac{\|\delta\mathbf{I}\|}{\|\mathbf{I}\|} \le \kappa(\mathbf{A})\frac{\|\delta\mathbf{i}\|}{\|\mathbf{i}\|} \tag{4.45}$$

and the worst-case fractional error in the components of **I** is bounded by the condition number times the worst-case fractional error in $I_d(t)$. Fortunately, the condition number is lowest when the rows of **A** are orthogonal, as they are in the FFT, and the errors incurred by truncating the transform are minimal. However, conditioning may be much worse when the frequency components are not harmonics; this situation arises in intermodulation analysis or when harmonic-balance analysis alone (rather than the large-signal/small-signal analysis we describe below) is used for conversion-loss calculations.

Therefore, it is usually better to use a large number of frequency components and discard unwanted components than simply use fewer components in the FFT. The former requires oversampling of the time waveform, or, in practice, evaluating the diode equations more frequently than the latter. However, the resulting increase in computation time caused by oversampling is often justified by the improvement in accuracy.

4.2.5 Improving Speed and Convergence

A number of methods can be used to improve the speed of a nonlinear circuit analysis. We have already examined one: use of fewer frequency components in the Newton iterations than in the FFT. Because the efficiency of Newton's method (in terms of both speed and memory use) is related most strongly to the size of **J**, considerable improvement can be achieved in this manner. The time required to solve (4.26) is proportional to the cube of the dimension of **J**; memory occupancy is proportional to its square*.

An effective way to improve the performance of Newton's method is to recycle the Jacobian. After the first few iterations, **J** rarely changes much on each iteration; thus, it need not be reformulated until the improvement in the solution at each iteration begins to degrade. Delaying the reformulation of **J** reduces the improvement in **F**(**V**) in each iteration, but speeds the iterations enormously. The overall effect is a significant improvement in speed. Newton iterations performed in this manner are sometimes called *Samanskii iterations*.

Improving convergence is somewhat more problematical. All of the above methods except splitting will converge if the initial estimate of the solution is sufficiently close to the actual solution. Unfortunately, in many cases, it is difficult or even impossible to make such an estimate accurately. In those cases, the process may fail to converge, and other methods may be necessary.

Source stepping is a technique that can be used to achieve convergence in difficult cases, especially when it is impossible to make a good initial estimate of the solution. It works only with methods which converge more reliably as the circuit becomes more linear; thus, it may not work well in algorithms such as the splitting method, whose reliability does not improve dramatically in nearly linear circuits.

Source stepping is a very simple idea: begin by setting the LO to a relatively low level, much lower than the level that would be used in practice. At a low enough LO level the circuit is nearly linear; only the fundamental frequency is significant, and the solution is easily found. That solution is used as the initial estimate for the second trial, where the LO level is slightly greater. By increasing the LO level stepwise in this manner until the desired level is reached, the initial trial function for each step is close enough to the solution for convergence to be achieved easily. The obvious disadvantage of this process is that the circuit must be analyzed several times.

---

* We assume that a method such as LU decomposition is used to solve Equation (4.26); although sparse-matrix techniques are theoretically capable of reducing this strong dependence of speed on matrix dimension, the Jacobian of a single- (or even multiple-) diode mixer is relatively dense, and thus sparse-matrix techniques are often not applicable in practice.

Convergence failure in the reflection algorithm often occurs in the solution of the nonlinear differential equation, (4.31), describing the circuit in Figure 4.4. Increasing the value of Z_c or reducing the sampling interval often provides successful convergence; however, both options slow the solution process.

4.2.6 LO Power and Matching

In the design of any mixer, it is important to minimize the LO power requirements. It is particularly important above 100 GHz, where adequate LO power is often very difficult and expensive to obtain. In a linear circuit, LO power requirements are minimum when the source and load are conjugate matched; however, since the diode is a nonlinear element, its impedance cannot be precisely defined. Fortunately, we can define a "quasi-impedance" of the diode as the ratio of the fundamental-frequency voltage to the fundamental-frequency current:

$$Z_d(j\omega_p) = \frac{V_d(\omega_p)}{I_d(\omega_p)} \tag{4.46}$$

and for optimum power transfer from the LO source to the diode,

$$Z_e(j\omega_p) = Z_d^*(j\omega_p) \tag{4.47}$$

This quantity is a function of the LO level and diode bias. It also depends on the embedding impedances at the fundamental and higher harmonics of ω_p.

The LO power dissipated by the diode is given in the time domain by

$$P_d = \frac{1}{T}\int_0^T V_d(t) I_d(t)\,dt - V_{dc} I_{dc} \tag{4.48}$$

where $T = 2\pi/\omega_p$, and the available power from the LO source is

$$P_{\mathrm{LO}} = \frac{V_{\mathrm{LO}}^2}{8\mathrm{Re}\{Z_e(j\omega_p)\}} \tag{4.49}$$

The ratio of these, P_d/P_{LO}, is the LO efficiency. If the diode is matched in the sense of Equation (4.47), the LO efficiency is unity. In many practical mixers, especially if the LO is variable in frequency, little effort is made to match the LO port. The primary design effort is applied to the optimization of the mixing frequency terminations in order to minimize conversion loss and noise. Local oscillator inefficiency is overcome simply by using more power. For millimeter-wave mixers, this luxury is usually not allowable, and mixer conversion performance is often limited primarily by available LO power, rather than by diode quality or other aspects of the circuit's design.

4.2.7 Multiple-Diode Mixers

Multiple-diode mixers (usually balanced mixers, treated in Chapters 7 and 8) are more commonly used in microwave systems than the single-diode mixer we have considered up to now. There are two ways to approach the subject of multiple-diode mixers. The first is to generate a single-diode equivalent circuit of the multiple-diode mixer. The single-diode equivalents of different types of mixers will be examined, as the subject arises, throughout this book.

The second way is to extend Equations (4.10) through (4.26) to accommodate multiple diodes. This approach is relatively straightforward, and is illustrated in Figure 4.6, where port $K + 1$ of the linear subcircuit is the excitation, and diodes are connected to ports 1 through K. As before, the diodes' series resistances are absorbed into the admittance matrix of the linear subcircuit. The voltage and current vectors in (4.19) become vectors of vectors:

$$\mathbf{I}_l = \begin{bmatrix} \mathbf{I}_{l,1} \\ \mathbf{I}_{l,2} \\ \cdots \\ \cdots \\ \mathbf{I}_{l,K} \\ \mathbf{I}_{l,K+1} \end{bmatrix} \tag{4.50}$$

where each subvector $\mathbf{I}_{l,k}$ has the same form as $\mathbf{I}_l$ in Equation (4.16). Similarly,

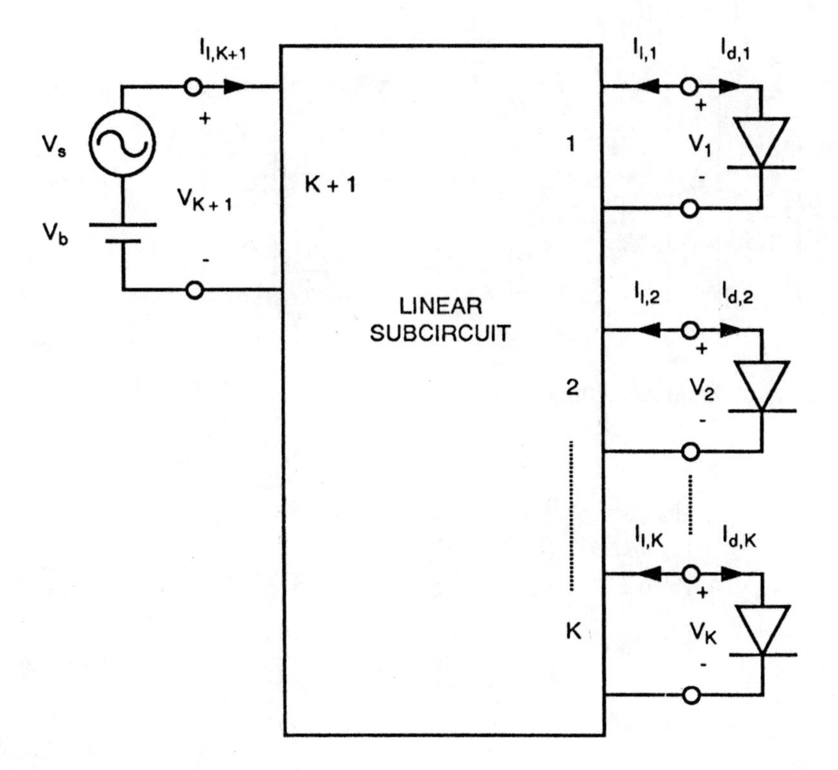

Figure 4.6 Large-signal equivalent circuit of a multiple-diode mixer.

$$\mathbf{I}_d = \begin{bmatrix} \mathbf{I}_{d,1} \\ \mathbf{I}_{d,2} \\ \cdots \\ \cdots \\ \mathbf{I}_{d,K} \end{bmatrix} \quad \text{and} \quad \mathbf{V} = \begin{bmatrix} \mathbf{V}_1 \\ \mathbf{V}_2 \\ \cdots \\ \cdots \\ \mathbf{V}_K \\ \mathbf{V}_{K+1} \end{bmatrix} \tag{4.51}$$

$\mathbf{V}_{K+1}$ is the excitation, and includes only V_s and V_b. The admittance equations of the linear subcircuit, corresponding to Equation (4.10), are

$$\begin{bmatrix} \mathbf{I}_{l,1} \\ \mathbf{I}_{l,2} \\ \cdots \\ \cdots \\ \mathbf{I}_{l,K} \\ \mathbf{I}_{l,K+1} \end{bmatrix} = \begin{bmatrix} \mathbf{Y}_{11} & \mathbf{Y}_{12} & \cdots\cdots & \mathbf{Y}_{1,K+1} \\ \mathbf{Y}_{21} & \mathbf{Y}_{22} & \cdots\cdots & \mathbf{Y}_{2,K+1} \\ \cdots & \cdots & \cdots\cdots & \cdots \\ \cdots & \cdots & \cdots\cdots & \cdots \\ \mathbf{Y}_{K,1} & \mathbf{Y}_{K,2} & \cdots\cdots & \mathbf{Y}_{K,K+1} \\ \mathbf{Y}_{K+1,1} & \mathbf{Y}_{K+1,2} & \cdots\cdots & \mathbf{Y}_{K+1,K+1} \end{bmatrix} \begin{bmatrix} \mathbf{V}_1 \\ \mathbf{V}_2 \\ \cdots \\ \cdots \\ \mathbf{V}_K \\ \mathbf{V}_{K+1} \end{bmatrix} \tag{4.52}$$

where the $\mathbf{Y}_{p,q}$ are diagonal submatrices,

$$\mathbf{Y}_{p,q} = \begin{bmatrix} Y_{p,q}(0) & 0 & 0 & \cdots & 0 \\ 0 & Y_{p,q}(\omega_p) & 0 & \cdots & 0 \\ 0 & 0 & Y_{p,q}(2\omega_p) & \cdots & 0 \\ \cdots & \cdots & \cdots & \cdots & \cdots \\ 0 & 0 & 0 & \cdots & Y_{p,q}(N\omega_p) \end{bmatrix} \tag{4.53}$$

The matrix corresponding to $\mathbf{Y}_e$ in (4.12) is found by eliminating the last row and column of the admittance matrix in (4.52); $\mathbf{I}_s$ is $\mathbf{V}_{K+1}$ times the first K elements of last column of the admittance matrix. The generalization of the splitting method and reflection algorithm are straightforward: iterations must be performed separately at each port. If Newton's method is used, Equations (4.38) through (4.40) still apply at each port. Note that the use of multiple diodes increases the size of the Jacobian substantially, from dimension $N + 1$ to $K(N + 1)$, and makes it far more sparse. This represents a considerable increase in both memory occupancy and computation time.

4.3 SMALL-SIGNAL ANALYSIS

4.3.1 Diode Conductance and Capacitance Waveforms

Frequency mixing occurs in the diode because its small-signal junction conductance and capacitance vary in time with the LO voltage waveform. The conductance arises from the exponential *I*/*V* characteristic of the junction, and the capacitance from the junction depletion layer. The diode's exponential *I*/*V* characteristic is clearly a stronger nonlinearity than the inverse square-root characteristic of its capacitance, and it is clearly the dominant element in determining the mixer's conversion loss and noise temperature. Indeed, the junction resistance usually varies over the LO cycle from a few ohms to an open circuit, while the capacitance varies by a factor of at most three to four.

The mixer's performance depends more strongly on the average value of the capacitance than on its variation with time. However, in order to determine the average junction capacitance, it is necessary to model its nonlinearity accurately. The variation of the LO capacitance is nevertheless responsible for a number of curious phenomena, mostly undesirable: increase in the diode's effective noise temperature as conversion loss decreases, parametric instability, and adding surprisingly large reactive components (sometimes inductive!) to the RF input and IF output impedances.

4.3.2 Conversion Matrices

The time waveforms for the junction capacitance and conductance are found by substituting $V_j(t)$ from the large-signal analysis into (4.4) and (4.5). To determine their frequency-conversion properties, we must first examine frequency conversion in the individual circuit elements, and finally show how the conversion properties of the diode and its embedding circuit can be found.

Figure 4.1 and Equation (4.1) define the mixing frequencies of the diode's small-signal voltage and current. At each of these frequencies, phasors represent the voltage across the element and the current in it. The sum of these voltage phasors is

$$\hat{v}(t) = \sum_{n=-\infty}^{\infty} V_n \exp(j(\omega_0 + n\omega_p)t) \tag{4.54}$$

where V_n is the voltage component at the mixing frequency ω_n. Note that, despite its superficial similarity to a Fourier series, (4.54) is not a Fourier series, but a sum of phasor components. Expressed in this way, $\hat{v}(t)$ is complex, not a real function of time, and the circumflex is used to emphasize this point. For the currents, the same holds true:

$$\hat{i}(t) = \sum_{n=-\infty}^{\infty} I_n \exp(j(\omega_0 + n\omega_p)t) \tag{4.55}$$

Although the summations in (4.54) and (4.55) are over an infinite number of terms, the summations must eventually be limited to $n = -N,\ldots,N$, where the highest significant mixing frequencies are those near the Nth LO harmonic. This truncation is obviously necessary, since computers are notoriously poor at handling infinite summations, and acceptable, since only a limited number of LO harmonics can be significant.

The Fourier series representation for a time-varying resistor having fundamental frequency ω_p is

$$r(t) = \sum_{n=-\infty}^{\infty} R_n \exp(jn\omega_p t) \tag{4.56}$$

The voltage and current in the resistor are related by Ohm's law,

$$\hat{v}(t) = r(t)\hat{i}(t) \tag{4.57}$$

Substituting (4.54) through (4.56) into (4.57) gives the relation:

$$\sum_{l=-\infty}^{\infty} V_l \exp(j(\omega_0 + l\omega_p)t) = \sum_{m=-\infty}^{\infty} \sum_{n=-\infty}^{\infty} R_m I_n \exp(j(\omega_0 + (n+m)\omega_p)t) \tag{4.58}$$

Clearly, the same frequencies exist on both sides of the equation. Separating the terms at the same frequency on each side of (4.58), equating them, and expressing them in matrix form gives

$$
\begin{bmatrix} V_{-N} \\ V_{-N+1} \\ \cdots \\ V_{-1} \\ V_0 \\ V_1 \\ \cdots \\ V_{N-1} \\ V_N \end{bmatrix} = \begin{bmatrix} R_0 & R_{-1} & R_{-2} & \cdots\cdots & R_{-N} & \cdots & R_{-2N} \\ R_1 & R_0 & R_{-1} & \cdots\cdots & R_{-N+1} & \cdots & R_{-2N+1} \\ \cdots & \cdots & \cdots & \cdots\cdots & \cdots & \cdots & \cdots \\ R_{N-1} & R_{N-2} & R_{N-3} & \cdots\cdots & R_{-1} & \cdots & R_{-N-1} \\ R_N & R_{N-1} & R_{N-2} & \cdots\cdots & R_0 & \cdots & R_{-N} \\ R_{N+1} & R_N & R_{N-1} & \cdots\cdots & R_1 & \cdots & R_{-N+1} \\ \cdots & \cdots & \cdots & \cdots\cdots & \cdots & \cdots & \cdots \\ R_{2N-1} & R_{2N-2} & R_{2N-3} & \cdots\cdots & R_{N-1} & \cdots & R_{-1} \\ R_{2N} & R_{2N-1} & R_{2N-2} & \cdots\cdots & R_N & \cdots & R_0 \end{bmatrix} \begin{bmatrix} I_{-N} \\ I_{-N+1} \\ \cdots \\ I_{-1} \\ I_0 \\ I_1 \\ \cdots \\ I_{N-1} \\ I_N \end{bmatrix} \tag{4.59}
$$

The terms in the matrix are the Fourier-series components of the resistance waveform between dc and the $2N$th harmonic. Since the time waveform is obviously real, the positive and negative frequency terms are complex conjugates. Although this matrix represents a resistance, some of its terms may be complex. This does not imply that the resistor is reactive; it is a consequence of the choice of time axis and asymmetry in the shape of the large-signal junction-current waveform. Voltage and current components at the same frequency are related by R_0, the dc component of the Fourier series, which is obviously real. The matrix in Equation (4.59) is called the *impedance-form conversion matrix* of the time-varying resistor.

Phasors are conventionally defined as positive frequencies; however, (4.1) shows that the currents and voltages in (4.59), I_n and V_n, have negative frequencies when $n < 0$. This situation will cause trouble when we wish to include the external circuit, so we would do well to correct it at this point. From now on, we shall define the lower-sideband quantities (the voltage and current terms having $n < 0$) as ordinary, positive-frequency phasors, so those terms in the voltage and current vectors must be made conjugate. The resulting conversion matrix in its final form is

$$
\begin{bmatrix} V^*_{-N} \\ V^*_{-N+1} \\ \cdots \\ V^*_{-1} \\ V_0 \\ V_1 \\ \cdots \\ V_{N-1} \\ V_N \end{bmatrix} = \begin{bmatrix} R_0 & R_{-1} & R_{-2} & \cdots\cdots & R_{-N} & \cdots & R_{-2N} \\ R_1 & R_0 & R_{-1} & \cdots\cdots & R_{-N+1} & \cdots & R_{-2N+1} \\ \cdots & \cdots & \cdots & \cdots\cdots & \cdots & \cdots & \cdots \\ R_{N-1} & R_{N-2} & R_{N-3} & \cdots\cdots & R_{-1} & \cdots & R_{-N-1} \\ R_N & R_{N-1} & R_{N-2} & \cdots\cdots & R_0 & \cdots & R_{-N} \\ R_{N+1} & R_N & R_{N-1} & \cdots\cdots & R_1 & \cdots & R_{-N+1} \\ \cdots & \cdots & \cdots & \cdots\cdots & \cdots & \cdots & \cdots \\ R_{2N-1} & R_{2N-2} & R_{2N-3} & \cdots\cdots & R_{N-1} & \cdots & R_{-1} \\ R_{2N} & R_{2N-1} & R_{2N-2} & \cdots\cdots & R_N & \cdots & R_0 \end{bmatrix} \begin{bmatrix} I^*_{-N} \\ I^*_{-N+1} \\ \cdots \\ I^*_{-1} \\ I_0 \\ I_1 \\ \cdots \\ I_{N-1} \\ I_N \end{bmatrix} \tag{4.60}
$$

An analogous development can be used for the time-varying conductance. The resulting conversion matrix is

$$
\begin{bmatrix} I^*_{-N} \\ I^*_{-N+1} \\ \cdots \\ I^*_{-1} \\ I_0 \\ I_1 \\ \cdots \\ I_{N-1} \\ I_N \end{bmatrix} = \begin{bmatrix} G_0 & G_{-1} & G_{-2} & \cdots\cdots & G_{-N} & \cdots & G_{-2N} \\ G_1 & G_0 & G_{-1} & \cdots\cdots & G_{-N+1} & \cdots & G_{-2N+1} \\ \cdots & \cdots & \cdots & \cdots\cdots & \cdots & \cdots & \cdots \\ G_{N-1} & G_{N-2} & G_{N-3} & \cdots\cdots & G_{-1} & \cdots & G_{-N-1} \\ G_N & G_{N-1} & G_{N-2} & \cdots\cdots & G_0 & \cdots & G_{-N} \\ G_{N+1} & G_N & G_{N-1} & \cdots\cdots & G_1 & \cdots & G_{-N+1} \\ \cdots & \cdots & \cdots & \cdots\cdots & \cdots & \cdots & \cdots \\ G_{2N-1} & G_{2N-2} & G_{2N-3} & \cdots\cdots & G_{N-1} & \cdots & G_{-1} \\ G_{2N} & G_{2N-1} & G_{2N-2} & \cdots\cdots & G_N & \cdots & G_0 \end{bmatrix} \begin{bmatrix} V^*_{-N} \\ V^*_{-N+1} \\ \cdots \\ V^*_{-1} \\ V_0 \\ V_1 \\ \cdots \\ V_{N-1} \\ V_N \end{bmatrix} \tag{4.61}
$$

where the G_n are the Fourier-series components for the conductance. It should come as no great surprise that, if both waveforms are finite, the conductance matrix is the inverse of the resistance matrix.

Two more points are even more remarkable. First, the matrix I/V representation of the time-varying resistance or conductance and the scalar representation, Ohm's law, are the same, since the matrix equation or (4.60) can be expressed as $V = R\,I$, where V and I are vectors and R is a matrix. Also, because sinusoids at different frequencies are linearly independent, Kirchoff's law must hold for the voltage and current vectors as well as for scalar quantities. This delightful situation gives us the capability of analyzing time-varying circuits with essentially the same tools as are used for steady-state analysis of time-invariant linear networks; the only difference is that equations are formulated in terms of matrices instead of scalars. For example, the conversion matrix of two time-varying components in parallel is the sum of their admittance-form conversion matrices; if they are in series, their impedance-form conversion matrices are summed. The only extra consideration is that the order of matrix multiplication must be preserved, and it is necessary to invert and multiply instead of divide.

The second point is that the conversion matrix is also analogous to a multiport network, except that the "ports" are different frequencies at a single set of terminals instead of physically separate ports. Consequently, multiport network theory is directly applicable to the conversion matrix, and such operations as determining voltage gain, input impedance, and conversion to another type of matrix, such as an S matrix, involve the same relations as are used for conventional multiports. These properties provide all the tools needed to analyze not only diode mixers, but such relatively complicated mixers as single- and dual-gate FETs.

The conversion matrix for the capacitor has an analogous derivation. The small-signal I/V relation for the capacitor, after (4.9), is

$$\hat{i}(t) = C(t)\frac{d}{dt}\hat{v}(t) + \hat{v}(t)\frac{d}{dt}C(t) \tag{4.62}$$

The Fourier-series representation of $C(t)$ is

$$C(t) = \sum_{n=-\infty}^{\infty} C_n \exp(jn\omega_p t) \tag{4.63}$$

Substituting (4.54), (4.55), and (4.63) into (4.62) gives

$$\sum_{l=-\infty}^{\infty} I_l \exp(j(\omega_0 + l\omega_p)t) = \sum_{m=-\infty}^{\infty}\sum_{n=-\infty}^{\infty} j(\omega_0 + (m+n)\omega_p) \times V_m C_n \exp(j(\omega_0 + (m+n)\omega_p)t) \tag{4.64}$$

Equating terms at the same frequency gives the matrix representation

$$\begin{bmatrix} I^*_{-N} \\ I^*_{-N+1} \\ \cdots \\ I^*_{-1} \\ I_0 \\ I_1 \\ \cdots \\ I_{N-1} \\ I_N \end{bmatrix} = \begin{bmatrix} \omega_{-N}C_0 & \omega_{-N}C_{-1} & \cdots\cdots & \omega_{-N}C_{-2N} \\ \omega_{-N+1}C_1 & \omega_{-N+1}C_0 & \cdots\cdots & \omega_{-N+1}C_{-2N+1} \\ \cdots & \cdots & \cdots\cdots & \cdots \\ \omega_{-1}C_{N-1} & \omega_{-1}C_{N-2} & \cdots\cdots & \omega_{-1}C_{-N-1} \\ \omega_0 C_N & \omega_0 C_{N-1} & \cdots\cdots & \omega_0 C_{-N} \\ \omega_1 C_{N+1} & \omega_1 C_N & \cdots\cdots & \omega_1 C_{-N+1} \\ \cdots & \cdots & \cdots\cdots & \cdots \\ \omega_{N-1}C_{2N-1} & \omega_{N-1}C_{2N-2} & \cdots\cdots & \omega_{N-1}C_{-1} \\ \omega_N C_{2N} & \omega_N C_{2N-1} & \cdots\cdots & \omega_N C_0 \end{bmatrix} \begin{bmatrix} V^*_{-N} \\ V^*_{-N+1} \\ \cdots \\ V^*_{-1} \\ V_0 \\ V_1 \\ \cdots \\ V_{N-1} \\ V_N \end{bmatrix} \tag{4.65}$$

which can be expressed as

$$\mathbf{I} = j\Omega\mathbf{CV} \tag{4.66}$$

where Ω is a diagonal matrix whose elements are the mixing frequencies ω_{-N} to ω_N; that is,

$$\Omega = \begin{bmatrix} \omega_{-N} & 0 & 0 & \dots & 0 \\ 0 & \omega_{-N+1} & 0 & \dots & 0 \\ 0 & 0 & \omega_{-N+2} & \dots & 0 \\ \dots & \dots & \dots & \dots & 0 \\ 0 & 0 & 0 & \dots & \omega_N \end{bmatrix} \tag{4.67}$$

and the Fourier coefficient matrix, C, has the same form as (4.61).

To include them in the mixer analysis, the time-invariant elements in the circuit must also have "conversion matrix" forms. The conversion matrix of a time-invariant impedance is a diagonal matrix having the elements $Z(j\omega_n)$:

$$\mathbf{Z} = \begin{bmatrix} Z(j\omega_{-N}) & 0 & 0 & \dots & 0 \\ 0 & Z(j\omega_{-N+1}) & 0 & \dots & 0 \\ 0 & 0 & Z(j\omega_{-N+2}) & \dots & 0 \\ \dots & \dots & \dots & \dots & 0 \\ 0 & 0 & 0 & \dots & Z(j\omega_N) \end{bmatrix} \tag{4.68}$$

The corresponding admittance-form conversion matrix has the same form, except, of course, the admittances $Y(j\omega_n) = 1/Z(j\omega_n)$ appear along the main diagonal.

4.3.3 Small-Signal Analysis

The task at this point is to generate a conversion matrix representing the entire diode, consisting of two time-varying elements, the junction conductance and capacitance, and a time-invariant element R_s. As suggested in the previous section, Kirchoff's laws are applied to the diode equivalent circuit in Figure 4.2 in a manner analogous to that used in time-invariant circuits. Because the junction conductance and capacitance are in parallel, their admittance-form conversion matrices need only to be added to obtain an admittance conversion matrix for the junction. Their sum is then inverted to obtain the equivalent impedance-form matrix, and the conversion matrix representing R_s is added. The result is the impedance-form conversion matrix of the entire diode, $\mathbf{Z}_d$:

$$\mathbf{Z}_d = (\mathbf{G}_j + j\Omega\mathbf{C}_j)^{-1} + \mathbf{R}_s \tag{4.69}$$

where $\mathbf{G}_j$ and $\mathbf{C}_j$ are the conversion matrices for the junction conductance and capacitance, respectively. $\mathbf{Z}_d$ relates the phasor voltage components of $\mathbf{V}_d$ to those of $\mathbf{I}_d$ as follows

$$\mathbf{V}_d = \mathbf{Z}_d \mathbf{I}_d \tag{4.70}$$

or

$$\mathbf{I}_d = \mathbf{Z}_d^{-1} \mathbf{V}_d = \mathbf{Y}_d \mathbf{V}_d \tag{4.71}$$

At this point we can proceed in a number of different ways. If the embedding circuit is lossless (which is usually the case), a single-diode mixer can be modeled as shown in Figure 4.7, where the embedding network has been reduced to a Thevenin equivalent. It is easy to construct this equivalent circuit: $Z_e(j\omega_n)$ is the set of embedding impedances at the mixing frequencies ω_n, the impedances found by "looking back" into the network from the diode's terminals. The magnitude of the voltage is found from the available power and $Z_e(j\omega_s)$, where ω_s is the excitation frequency:

$$V_s = \sqrt{8P_{av} \operatorname{Re}\{Z_e(j\omega_s)\}} \tag{4.72}$$

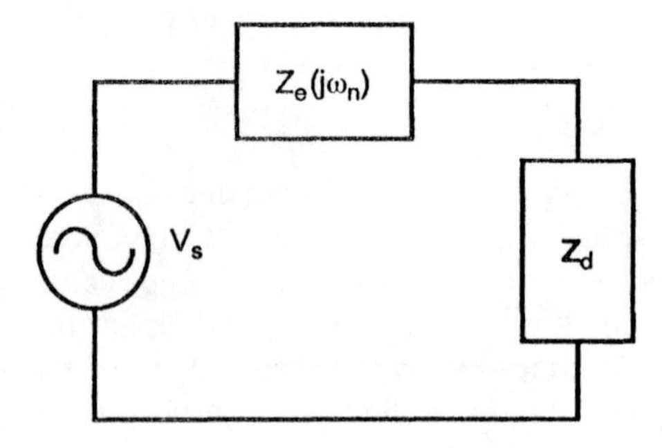

Figure 4.7 Small-signal model of a diode mixer. $Z_e(j\omega_n)$ is the set of embedding impedances, and $\mathbf{Z}_d$ is the diode, represented by an impedance-form conversion matrix.

The current in the loop is

$$\mathbf{I} = (\mathbf{Z}_d + \mathbf{Z}_e)^{-1}\, \mathbf{V}_s \tag{4.73}$$

and the output power at the mixing frequency ω_n is

$$P_n = 0.5\, |I(\omega_n)|^2\, \mathrm{Re}\{Z_e(j\omega_n)\} \tag{4.74}$$

Finally, the conversion loss is

$$L_c = \frac{P_{\mathrm{av}}}{P_n} \tag{4.75}$$

where P_{av} is obtained from (4.72). The input (or output) impedance of the diode at any mixing frequency ω_n can be found by setting the source $V_s(\omega_n)$ to that frequency and calculating

$$Z_{\mathrm{in}}(j\omega_n) = \frac{V_s(\omega_n)}{I(\omega_n)} - Z_e(j\omega_n) \tag{4.76}$$

A second and more general way to perform the small-signal analysis is to treat the diode as a multiport; we alluded to this possibility in Section 4.3.2. The subscripts of the components of the $\mathbf{Z}_d$ or $\mathbf{Y}_d$ matrix are chosen so that Z_{dmn} is the transfer impedance between V_m and I_n. This notation may seem odd, compared with the conventional, where the elements in the top left corner of the matrix is always (1,1); however, it will simplify the notation in subsequent relations considerably.

Figure 4.8 shows the pumped diode's multiport representation. As before, the "ports" represent voltage and current components at different frequencies at the diode terminals, not physically separate ports. One of these ports is the RF port, in this case at ω_1; another, the IF port, at ω_0, is the output. All the other ports are terminated in their embedding impedances $Z_e(j\omega_n)$; although these affect the mixer performance, they are not used as input or output ports. Thus, we can reduce this terminated $2N + 1$-port network to a two-port, with only the RF and IF ports remaining, and with all the embedding impedances "absorbed" into the multiport network.

To convert a terminated multiport network described by an impedance matrix to a two-port, we must first include the embedding impedances in the diode's conversion matrix. To do this, we simply add the impedances to the main diagonal;

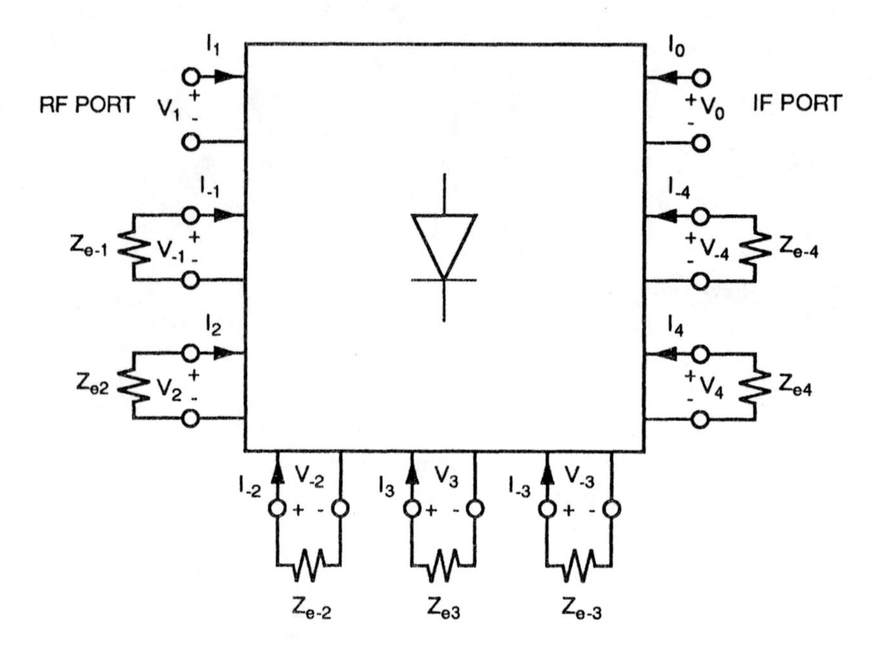

Figure 4.8 Multiport representation of a diode mixer. The ports are currents and voltages at the mixing frequencies ω_n, and all ports except the RF and IF are terminated at the embedding impedance at that frequency. We leave the RF and IF unterminated so that we can reduce this circuit to a two-port. In this example, $N = 4$.

this operation puts the embedding impedances in series with their respective ports. To connect the impedances to each port, we invert this augmented matrix to obtain an admittance form and set the port voltages to zero at all frequencies other than the RF and IF. The augmented Z matrix is

$$\mathbf{Z}_d^a = \mathbf{Z}_d + \hat{\mathbf{Z}}_e \tag{4.77}$$

where the superscript a indicates an augmented matrix, and the circumflex over $\mathbf{Z}_e$ indicates that the embedding impedances at ω_1 and ω_0 are not included; they are artificially set to zero. By eliminating these impedances, which are in fact the source and load impedances, we are free to select them independently. The impedance

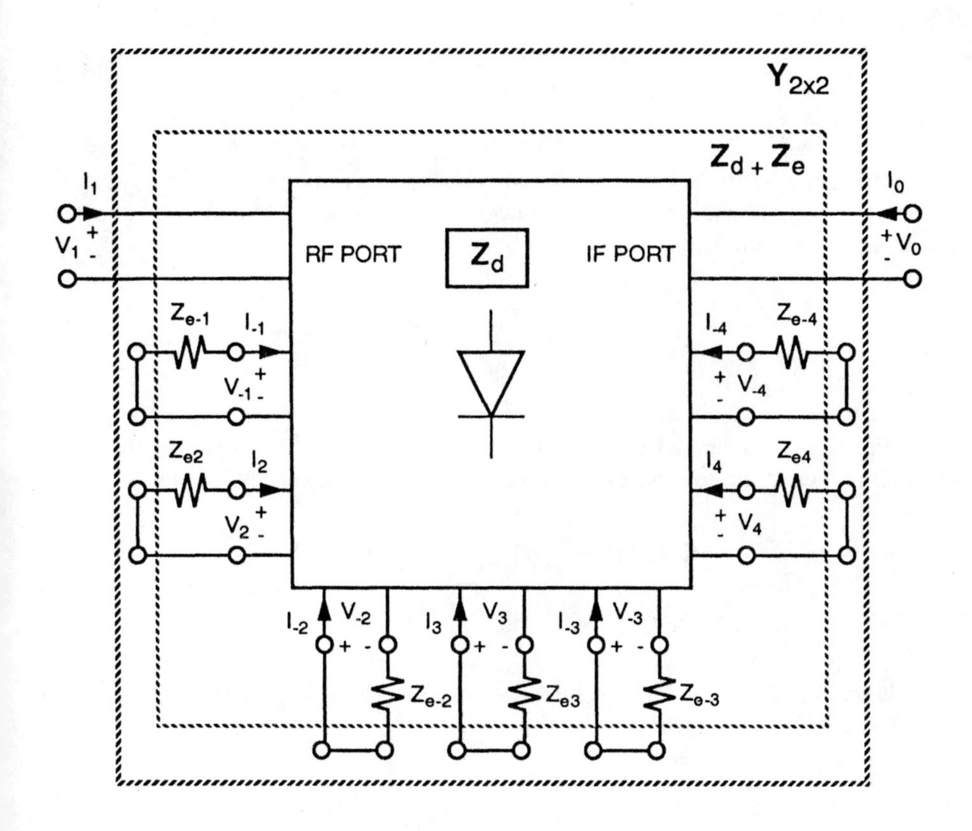

Figure 4.9 Steps in the realization of the 2 x 2 conversion Y matrix. $\mathbf{Z}_d$, the conversion matrix of the diode, is augmented with the embedding impedances yielding $\mathbf{Z}_d + \mathbf{Z}_e$. This matrix is inverted, and the voltages at the ports other than the RF and IF are set to zero. The remaining two ports are those of $\mathbf{Y}_{2x2}$, the Y matrix in Equation (4.78).

matrix $\mathbf{Z}_d^a$ is then inverted to obtain an admittance matrix, and, finally, the voltages corresponding to the undesired mixing frequencies (ω_n, $n \neq 1, 0$) are set to zero. This eliminates all the rows except for those corresponding to the RF and IF. These operations are illustrated in Figure 4.9. The resulting matrix relating the RF and the IF port voltages and currents is

$$\begin{bmatrix} I_0 \\ I_1 \end{bmatrix} = \begin{bmatrix} y_{00} & y_{01} \\ y_{10} & y_{11} \end{bmatrix} \begin{bmatrix} V_0 \\ V_1 \end{bmatrix} \tag{4.78}$$

This matrix may be treated in the same manner as any conventional, time-invariant two-port admittance matrix. The transducer conversion gain (which, of course, is invariably less than unity) is defined as

$$G_t = \frac{P_{\text{IF}}}{P_{\text{RF}}} \tag{4.79}$$

where P_{IF} is the power delivered to the output load and P_{RF} is the input power available power of the source. The conversion gain between ω_1 and ω_0 is

$$G_t = \frac{4\,\text{Re}\{Y_e(j\omega_1)\}\ \text{Re}\{Y_e(j\omega_0)\}\ |y_{01}|^2}{|(y_{11} + Y_e(j\omega_1))\ (y_{00} + Y_e(j\omega_0)) - y_{10}y_{01}|^2} \tag{4.80}$$

which is the standard relationship for the two-port. The reverse gain is found from (4.80) by substituting y_{10} for y_{01}.

The input admittance at ω_1 is

$$Y_{\text{in}}(j\omega_1) = y_{11} - \frac{y_{01}y_{10}}{Y_e(j\omega_0) + y_{00}} \tag{4.81}$$

and the output impedance is

$$Y_{\text{out}}(j\omega_0) = y_{00} - \frac{y_{01}y_{10}}{Y_e(j\omega_1) + y_{11}} \tag{4.82}$$

The source and load admittances that provide a simultaneous conjugate match can also be found. The source admittance is

$$\mathrm{Im}\{Y_e(j\omega_1)\} = -\mathrm{Im}\{y_{11}\} + \frac{\mathrm{Im}\{y_{01}y_{10}\}}{2\,\mathrm{Re}\{y_{00}\}} \tag{4.83}$$

$$\mathrm{Re}\,\{Y_e(j\omega_1)\} = \frac{\sqrt{(2\,\mathrm{Re}\{y_{11}\}\,\mathrm{Re}\{y_{00}\} - \mathrm{Re}\{y_{01}y_{10}\})^2 - |y_{01}y_{10}|^2}}{2\,\mathrm{Re}\{y_{00}\}} \tag{4.84}$$

and the load admittance $Y_e(j\omega_0)$ is found from (4.83) and (4.84) after interchanging y_{01} with y_{10}, y_{11} with y_{00}.

A stability factor can also be found for the mixer. This is more important than it might at first appear; although oscillation is rarely observed in diode mixers, the pumped junction capacitance sometimes causes diode mixers (especially lower sideband upconverters) to be only conditionally stable. Like any active circuit, diode mixers do not have simultaneous conjugate-match impedances when they are conditionally stable, and under these conditions it is impossible to match both the RF and IF ports. The Linvill stability factor c is

$$c = \frac{y_{01}y_{10}}{2\,\mathrm{Re}\{y_{11}\}\,\mathrm{Re}\{y_{00}\} - \mathrm{Re}\{y_{01}y_{10}\} + T_y} \tag{4.85}$$

where T_y is the numerator of Equation (4.84). If $c < 1$ the circuit is unconditionally stable; that is, no combination of passive RF source and IF load admittances can cause oscillation, and a simultaneous conjugate match at the RF and IF ports is possible.

By describing the RF source and IF load as embedding admittances or impedances, we have tacitly assumed that the input circuits are lossless. At relatively low frequencies or in low-loss transmission media such as waveguide, a lossless assumption may be valid, but in the higher millimeter-wave region, it may not be. In these cases it is usually adequate to preserve the lossless assumption, and to add a few tenths of a decibel loss to the calculated conversion loss to compensate for the input and output circuits' inefficiencies.

An admittance representation of the mixer can be used to generate an S matrix. The standard relation is applicable:

$$\mathbf{S} = (1 + \mathbf{Y}_n)^{-1}(1 - \mathbf{Y}_n) \tag{4.86}$$

where $\mathbf{Y}_n$ is the normalized 2 × 2 Y matrix (Equation (4.78)).
S is defined by the relation

$$\begin{bmatrix} b_0 \\ b_1 \end{bmatrix} = \begin{bmatrix} S_{00} & S_{01} \\ S_{10} & S_{11} \end{bmatrix} \begin{bmatrix} a_0 \\ a_1 \end{bmatrix} \tag{4.87}$$

where a_n and b_n are the scattering variables at the mixing frequency ω_n. The interpretations of these quantities are the same as for conventional two-ports: S_{11} and S_{00} are the input and output reflection coefficients when the opposite port is terminated in the normalizing impedance, and $|S_{01}|^2$ and $|S_{10}|^2$ are the forward and reverse transducer gains, respectively, under the same conditions. In a similar manner, it is possible to generate conversion matrices in the form of Z, G, H, or even transfer (ABCD) or transfer-scattering matrices.

The interpretation of the conversion S matrix is somewhat problematical when the input or output is at a lower sideband mixing product. Suppose, for example, that the input is at ω_{-1} (this is the image frequency of the conventional downconverter we have been considering). Then the S matrix is

$$\begin{bmatrix} b^*_{-1} \\ b_0 \end{bmatrix} = \begin{bmatrix} S_{-1,-1} & S_{-1,0} \\ S_{0,-1} & S_{0,0} \end{bmatrix} \begin{bmatrix} a^*_{-1} \\ a_0 \end{bmatrix} \tag{4.88}$$

and, for example,

$$S_{-1,-1} = \left. \frac{b^*_{-1}}{a^*_{-1}} \right|_{a_0 = 0}$$
$$S_{0,-1} = \left. \frac{b_0}{a^*_{-1}} \right|_{a_0 = 0} \tag{4.89}$$

$S_{-1,-1}$ is clearly the conjugate of the input reflection coefficient, and $|S_{0,-1}|^2$ is the transducer gain. However, the phase of $S_{0,-1}$ has no direct, physically meaningful interpretation. Fortunately, it is rarely a quantity of great interest.

4.4 NOISE ANALYSIS

4.4.1 Noise Sources

The history of mixer noise analysis is remarkable in that the properties of noise in mixers were well understood early in their history, but a workable, accurate noise model was not generated until the mid-1970s. The correlation properties of shot noise in mixers were identified early in the development of the theory [12]–[14], which led to the correct conclusion that the effective noise temperature of the ideal mixer diode (one having no junction capacitance or series resistance and is appropriately terminated) is $\eta T/2$, where η is the ideality factor (Chapter 2) and T is the diode's physical temperature in kelvins. Unfortunately, however, measured noise temperatures were generally closer to twice this value, and at very low temperatures the agreement was even worse. This led to the assumption of the existence of an anomalous noise source within the diode [15], and somewhere in the confusion the effect of noise correlation, which in all fairness is rather subtle, was forgotten. In fact, no "anomalous" source of noise existed; it was only necessary to resurrect the earlier understanding of the correlation properties of shot noise and to include the effects of series resistance, nonlinear junction capacitance, and arbitrary embedding impedances. The present theory, as given by Held and Kerr [5], agrees well with experimental results.

The dominant noise sources in Schottky-barrier diodes are thermal noise generated in the series resistance and shot noise arising from carrier emission across the junction. Other effects, such as hot-electron noise and intervalley scattering noise in GaAs, may dominate in poorly designed or poorly fabricated diodes, or at very high frequencies (hundreds of gigahertz) [16]. The following noise analysis is concerned exclusively with shot and thermal noise, because these alone are usually adequate to provide an accurate description of noise in diode mixers.

The key to understanding mixer noise is not simply to understand the shot and thermal noise processes, which are relatively simple, but to appreciate the more subtle correlation properties of the noise, and the effect of the time-varying elements, particularly the junction capacitance, upon them. The correlation properties of a dc-biased diode are simply those of any white noise process: components at different frequencies are not correlated. In a pumped diode, however, the components downconverted to the IF from different mixing frequencies are partially correlated, and their correlation raises the noise level. Furthermore, the time-varying junction capacitance prevents minimum conversion loss and minimum noise temperature from occurring simultaneously: mixers having low conversion loss often do not have commensurately low noise temperature. In fact, very low-loss mixers often are relatively noisy.

4.4.2 Mixer Noise Theory

A Schottky junction creates shot noise because its current consists of a series of pulses that occur as each electron crosses the junction. If the transit time is very short (as it is in a Schottky diode), the current waveform can be treated as a series of random impulses. Although the average number of such pulses in each second is constant, and is proportional to the dc current, the random nature of the impulses causes the instantaneous current to vary with time. The resulting fluctuations in the junction current are a noise process; its mean-squared magnitude is proportional to the dc current. Because it is a major source of noise in vacuum tubes as well as solid-state devices, shot noise has been studied extensively [17]. Through any one of several different approaches, one can show that the mean-square shot-noise current in a forward-biased diode is

$$\overline{i^2} = 2qI_jB \tag{4.90}$$

where q is the electron charge, I_j is the current in the resistive part of the junction, and B is the bandwidth. An assumption inherent in (4.90) is that the electron transit time across the junction is short compared to the inverse of the frequency at which the noise is evaluated.

The other significant noise source is thermal noise in the series resistance. Thermal noise is present in any power-dissipative medium having a temperature above absolute zero. It arises from the random, thermal agitation of electrons, and is closely related to blackbody radiation. Thermal noise is frequency dependent. However, at frequencies below the submillimeter and temperatures above a few kelvins the noise power available from a resistor depends on bandwidth and temperature and not frequency.

A resistor behaves as if it were noiseless and has a mean-square open-circuit noise voltage of

$$\overline{v^2} = 4KTBR \tag{4.91}$$

so the available noise power in a bandwidth B is simply

$$P_n = KTB \tag{4.92}$$

where R is the resistance, K is Boltzmann's constant, and T is absolute temperature. Equations (4.91) and (4.92) are valid for temperatures and frequencies f where

$$hf \ll KT \tag{4.93}$$

and h is Planck's constant, 6.63×10^{-34} J·s. The equivalent circuit of the noisy diode, showing thermal and shot noise, is shown in Figure 4.10(a). In Figure 4.10(b) the voltage noise source has been converted to a current source via Thevenin's theorem. This representation is more practical for noise analysis.

Both thermal and shot noise can be treated as white Gaussian noise processes at frequencies and temperatures where the above assumptions are valid. Therefore, if

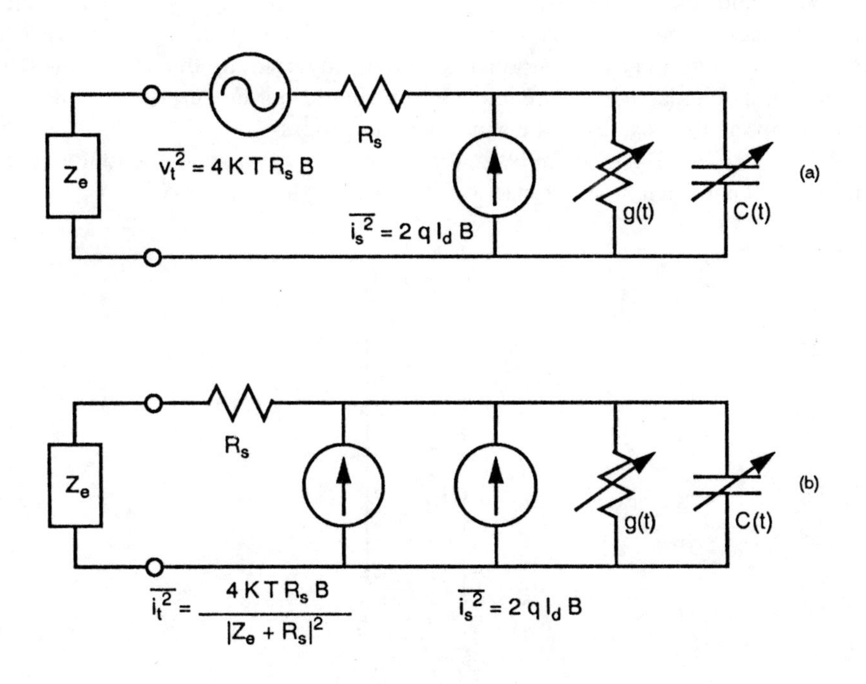

Figure 4.10 Noise equivalent circuit of the diode. Thermal noise in R_s is represented in (a) as a voltage source; in (b), it is converted to a current source.

the diode were to be dc biased and two noise waveforms created by narrowband filtering the diode's noise at two different frequencies, the resulting noise processes would be uncorrelated. Specifically, the narrowband noise processes in the dc-biased diode at the mixing frequencies $\omega_0 + n\omega_p$ are independent and therefore uncorrelated, and if any two were mixed down to the IF frequency ω_0, the components would still be uncorrelated. This situation is illustrated in Figure 4.11(a).

The thermal noise remains constant under LO excitation because the diode's series resistance R_s is time invariant. However, the effect of the LO on the shot noise is very different. When the LO voltage is applied, the diode current is a pulse waveform, and the magnitude of the shot noise is also a pulse waveform at the LO frequency. As shown in Figure 4.11(b), each shot noise component at each mixing frequency is converted to the other mixing frequencies by the fundamental LO frequency and its harmonics. The noise components at any mixing frequency therefore include upconverted and downconverted components from every other mixing frequency. Those components, downconverted to the IF frequency, are correlated. The existence of this correlation should not be surprising, since all the noise components arose from the same random process.

As with the small-signal analysis, the noise currents in the diode junction at each mixing frequency can be expressed as a vector:

$$I_s = \begin{bmatrix} I_{s,-N} \\ I_{s,-N+1} \\ \cdots \\ I_{s,1} \\ I_{s,0} \\ I_{s,1} \\ \cdots \\ I_{s,N-1} \\ I_{s,N} \end{bmatrix} \quad (4.94)$$

where $I_{s,n}$ are the shot-noise currents at the mixing frequency ω_n, $n = -N,...,0,...,N$. The goal is to find the IF output voltage resulting from these noise sources, each of which may be treated as an excitation source.

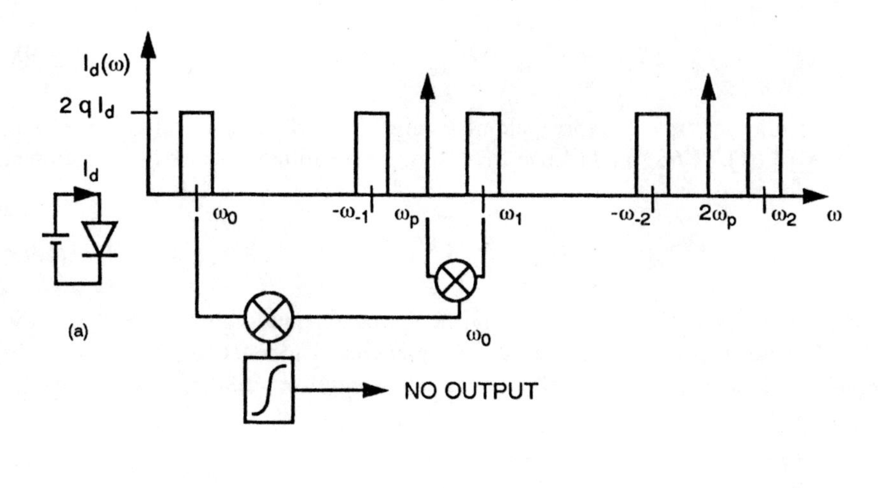

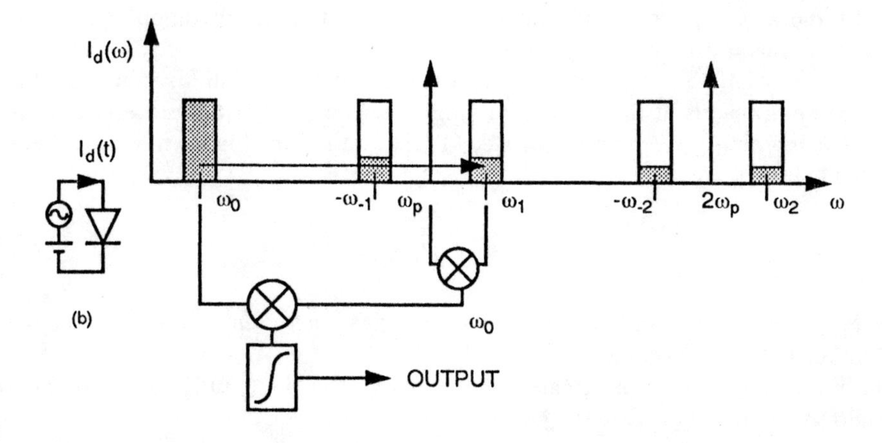

Figure 4.11 Noise correlation in a pumped diode: (a) when the diode is dc biased, noise components at different mixing frequencies are uncorrelated. However, when an LO signal is applied, (b) a portion of the noise power at any mixing frequency is converted to every other mixing frequency; therefore, the downconverted components are correlated.

The impedance-form conversion matrix of the terminated diode is

$$\mathbf{Z}_{cd} = (\mathbf{G}_j + j\Omega\mathbf{C}_j + \mathbf{Y}_t)^{-1} \tag{4.95}$$

where $\mathbf{Y}_t$ is a diagonal matrix whose elements are $1/(R_s + Z_e(j\omega_n))$, and $\mathbf{G}_j$, Ω, and $\mathbf{C}_j$ are given by (4.61), (4.65), and (4.67). The shot-noise voltage across the junction at mixing frequency ω_m is

$$V_{sm} = \mathbf{Z}_m \mathbf{I}_s \tag{4.96}$$

where $\mathbf{Z}_m$ is the row of the conversion matrix $\mathbf{Z}_{cd}$ corresponding to ω_m, and $\mathbf{I}_s$ is the vector of shot-noise currents at all mixing frequencies. The most common case is, of course, $m = 0$. The mean-square value of the junction noise voltage at ω_m is

$$\overline{V_{sm}^2} = \mathbf{Z}_m \langle \mathbf{I}_s \mathbf{I}_s^{*\mathrm{T}} \rangle \mathbf{Z}_m^{*\mathrm{T}} \tag{4.97}$$

where *T indicates the conjugate transpose of the vector, and the angular brackets represent the scalar product of the terms in the $\mathbf{I}_s$ vector.

The term $<\mathbf{I}_s\mathbf{I}_s^{*\mathrm{T}}>$ is a matrix representing the correlations between the shot-noise components at the various mixing frequencies. It is therefore called the *correlation matrix of the pumped diode*, $\mathbf{C}_s$. The terms in this matrix have been evaluated [5]; they are

$$C_{sm} = 2qI_{m-n}B \tag{4.98}$$

where I_{m-n} is the $m - n$th-harmonic Fourier-series coefficient of the large-signal junction-current waveform $I_g(t)$.

The thermal noise can be treated analogously. The magnitude of the thermal noise current source at ω_n (Figure 4.10(b)) is

$$\overline{i_t^2} = \frac{4KTR_sB}{|Z_e(j\omega_n) + R_s|^2} \tag{4.99}$$

Because the thermal noise arises in R_s, which does not vary with LO excitation, the thermal noise components at different frequencies are not correlated, and the correlation matrix is simply a diagonal. Therefore, the correlation matrix of the thermal noise is

$$C_{tmn} = \begin{cases} \dfrac{4KTR_sB}{|Z_e(j\omega_m) + R_s|^2} & m = n \\ \dfrac{4KTR_sB}{|Z_{\text{out}}|^2} & m = n = IF \\ 0 & m \neq n \end{cases} \tag{4.100}$$

where Z_{out} is the IF output impedance of the terminated diode (Equation (4.76)).

The voltage at ω_m is the sum of the shot-noise and thermal-noise components:

$$\overline{V_m^2} = \mathbf{Z}_m (\mathbf{C}_s + \mathbf{C}_t) \mathbf{Z}_m^{*T} \tag{4.101}$$

The noise power dissipated in the output termination, $Z_e(j\omega_m)$, is

$$P_m = \frac{\overline{V_m^2}\,\text{Re}\{Z_e(j\omega_m)\}}{|Z_e(j\omega_m) + R_s|^2} \tag{4.102}$$

and the single-sideband noise temperature (see Chapter 5) is

$$T_{ssb} = \frac{P_m}{KBG_{tmn}} \tag{4.103}$$

where G_{tmn} is the transducer conversion gain between the input at ω_n and the output ω_m.

It is important to note that this model includes only the noise sources within the diode, and not those of the embedding network. In particular, thermal noise from the real parts of the embedding impedances at frequencies other than the RF and IF are not included. Generally, it should be expected that the terminations at the high-order mixing products are reactive, and even if one or two have a substantial real part, the conversion gain between the high-order mixing products and the IF is likely to be small, so the noise of these terminations probably will not be significant. (The one important exception to this generalization is the noise from the image-frequency termination, which may be substantial if the mixer has a significant image response. This noise is invariably treated as an external, not internal, noise source.) If this model were adapted to components such as parametric amplifiers, where the noise from the high-order terminations is usually significant, these noise sources would have to be included. Fortunately, since the noise components from the embedding impedances are uncorrelated, they can be included easily. The additional noise at the output is

$$\Delta P_m = \sum_{\substack{i=-N \\ i \neq m,n}}^{N} K T_i B G_{tmi} \tag{4.104}$$

where T_i is the noise temperature of the real part of the embedding impedance $Z_e(j\omega_i)$. Another source of noise is the mixer's output load itself. If the IF port of the mixer is not well matched, the noise from the terminating impedance may be reflected from the IF port and behave very much as if it had originated within the mixer. This phenomenon, and others affecting the noise performance of systems using mixers, will be treated further in the next chapter.

4.5 MIXER ANALYSIS BY GENERAL-PURPOSE HARMONIC-BALANCE SIMULATORS

A relatively new nonlinear circuit-analysis tool is the general-purpose harmonic-balance simulator. Such simulators are capable of performing a harmonic-balance analysis of a diode or transistor mixer having an arbitrary topology or number of devices. The versatility of these simulators has a high cost, however: they are extremely expensive in both their initial cost and in the cost of the computer needed to run them. They are also far more limited in their capabilities than might be expected.

Virtually all of these simulators use Newton's method (Section 4.2.3) as a solution algorithm and accommodate multiple devices in the manner described in Section 4.2.7. To speed the analysis, most "recycle" the Jacobian in the manner described in Section 4.2.5.

One difference between these simulators and the harmonic-balance analysis described in Section 4.2 is that the variables in many of these simulators are the harmonic-voltage components at all the circuit's *nodes*, instead of those at the *ports* connecting the linear and nonlinear subcircuits. Because a nonlinear circuit always has more nodes than ports, the number of variables is much greater than in the "classical" harmonic-balance analysis described earlier in this chapter. Fortunately, this nodal formulation causes the Jacobian matrix to be relatively sparse; if sparse-matrix methods are used to factor the Jacobian, the efficiency is not as bad as might be expected. Furthermore, this formulation solves a very serious problem in the classical formulation: in many types of mixers the admittance matrix of the linear subcircuit often does not exist.

A second difference is that modern simulators allow the use of multiple, noncommensurate excitations (i.e., excitation frequencies that have no harmonic or subharmonic relationship). This is not as much of a complication as it may seem; note that the terms in Equation (4.16) need not be harmonically related. The main complication in the use of noncommensurate excitations is that a time-to-frequency transformation must be developed that allows this; clearly, an FFT is not useful. Unfortunately, such "almost-periodic" Fourier transforms have much less numerical range than the FFT, and small signals (such as the IF or *intermodulation* (IM) components) are often lost in the resulting numerical "noise." Thus, although they are often advertised as products that can simulate IM in mixers, they usually fail to do so. Furthermore, in simulating multitone mixer IM, the number of variables, $2N(K + 1)$, where N is the number of ports (or nodes) and K is the number of frequency components, is extraordinarily high; thus, the analysis requires a very long computation time and very great memory occupancy. The time and memory required is often impractically large.

Fortunately, many simulators include a generalized version of the conversion-loss analysis described in this chapter. This analysis is far more accurate for calculating conversion loss than the multitone harmonic-balance analysis and is much faster. Unfortunately, although the technology is available to include it, as of this writing none provides noise analysis.

One beneficial property of such simulators is that they usually include a set of linear-element models. These allow the linear subcircuit to be described literally, as a set of elements and their interconnections, not as an admittance matrix. Thus, the problem of determining the diode's embedding network is handled invisibly by the simulator. Furthermore, it is not necessary to depend on simplifying assumptions such as a lossless embedding circuit, or identical diodes in a balanced mixer; losses and asymmetries can be accommodated easily.

4.6 PERFORMANCE OPTIMIZATION

In most microwave electronic components, particularly those using transistors, performance is limited most fundamentally by the capabilities of their solid-state devices. Mixers, however, are limited less by the properties of the diode than by those of the circuit, especially the practical impossibility of achieving optimum terminations at a large number of mixing frequencies. It is not surprising that the diode is optimized by reducing its parasitic series resistance and junction capacitance, and by achieving a low ideality factor. Methods of optimizing the external circuit are not as clear; we examine a few of them here. The methods we shall consider are the use of dc bias, optimizing LO power, image enhancement, and the use of simultaneous-conjugate-match source and load impedances.

Table 4.1 shows the parameters of the diode and mixer that is analyzed in this section. The analysis was performed by means of a mixer-analysis program called DIODEMX; this program uses the reflection algorithm and the theory in Sections 4.3 and 4.4 to compute the LO waveforms. Although we have not discussed the theory of intermodulation analysis in mixers, we show the results of such analysis here. This program includes intermodulation analysis; the theory of intermodulation distortion in diode mixers is covered in detail in [18] and [19].

Table 4.1

Mixer Parameters

Mixer Parameters	*Value*
RF frequency	10.05 GHz
LO frequency	10.0 GHz
IF frequency	50 MHz
Diode type	Si Schottky beam lead
C_{j0}	0.15 pF
R_s	6.0Ω
η	1.19
ϕ_b	0.7V
I_0	5.0×10^{-12} A

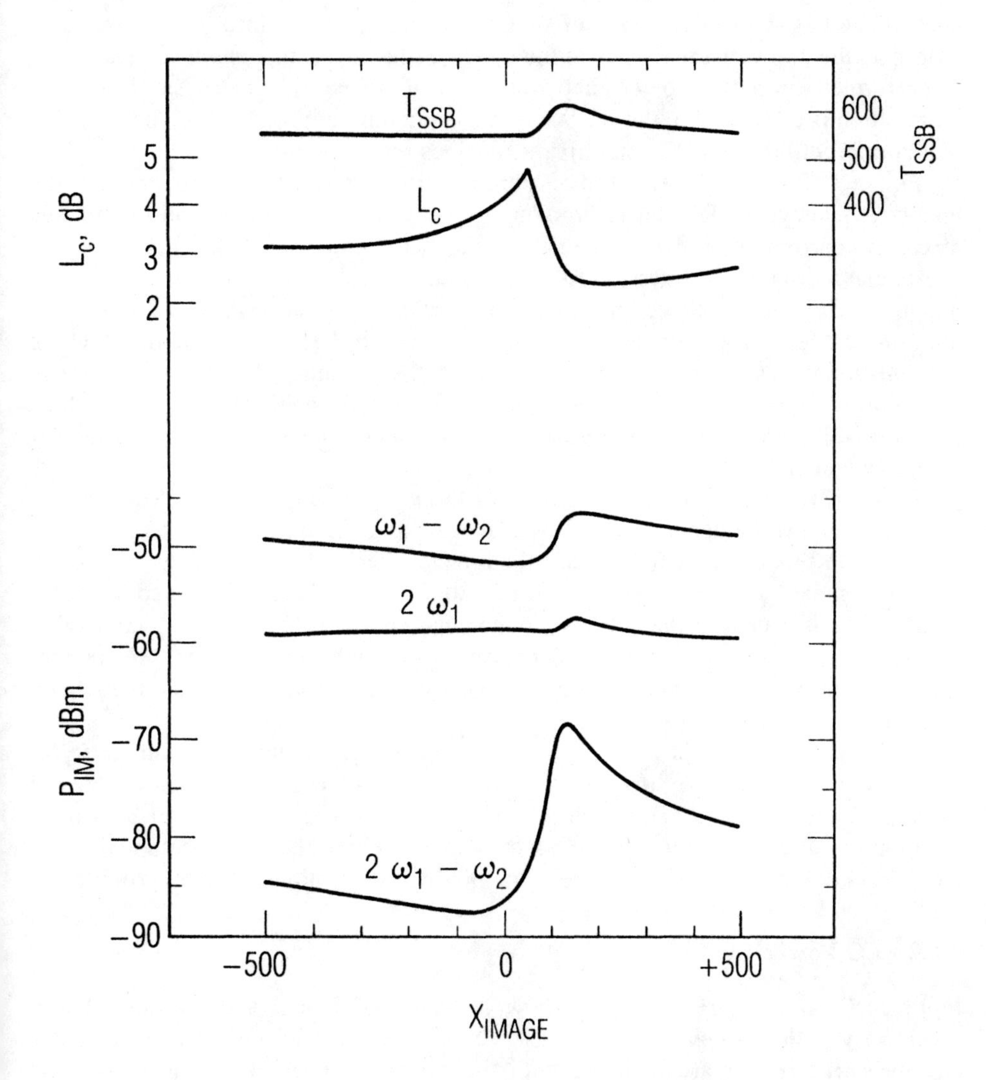

Figure 4.12 Conjugate-match conversion loss, IM levels, and noise temperature as a function of image-termination reactance. All embedding impedances other than the RF, IF, and image are zero. $P_{LO} = 3$ dBm, $V_b = 0$.

4.6.1 Image Terminating Impedance and Image Enhancement

One of the best-known methods of reducing the conversion loss of a mixer is to terminate the diode in a reactance at the image frequency. If this *image enhancement* is performed properly, power that would be otherwise dissipated in the image termination is converted to the IF. The result is not only improved efficiency, but the reduction or elimination of a significant spurious response, the image.

Figure 4.12 shows the effect of varying the diode's terminating impedance at the image frequency; the RF and IF are conjugate matched (the conversion loss of this mixer is approximately 4 dB when the image is conjugate matched). This figure shows clearly that the conversion loss can be minimized through a correct choice of image terminating reactance and, just as importantly, can be maximized through a poor one. Unfortunately, the noise temperature actually *increases* at the point where the conversion loss is minimum! Furthermore, the optimum IF load impedance at this point is approximately 600Ω, a value very difficult to achieve in practice. Thus, the improvement in conversion loss available in theory may be offset in practice by a poor IF match.

Figure 4.12 also shows the IF levels of three important two-tone IM products: second-order IM products at $\omega_1 - \omega_2$ and $2\omega_1$, and the third-order component at $2\omega_1 - \omega_2$. In this case, ω_1 and ω_2 are the downconverted IF frequencies of two RF input tones, each having a power level of –20 dBm. All three of these components exhibit a distinct peak at the point where conversion loss is lowest. The unavoidable conclusion is that best overall performance is achieved not at the image-enhancement point, but over the broad range where the image reactance is low and capacitive.

Image enhancement is provided in practice through the use of an image-reject filter. The filter must have low insertion loss; otherwise, the improvement achieved through image enhancement, which is rarely more than 1 to 2 dB, may be destroyed by filter losses. To allow adjustment of the image-terminating reactance, some empirical adjustment of the distance from the filter to the diode must be provided.

4.6.2 LO Power

Figure 4.13 and Figure 4.14 show how the conversion loss and intermodulation levels vary with LO power. In this case, the source and load impedances are 50Ω, and measured results are included with the calculated ones. The conversion loss decreases monotonically with increasing LO level. Although the conversion loss does not decrease significantly above approximately 6 dBm of LO power, the IM levels continue to decrease. In particular, the third-order IM level continues to drop

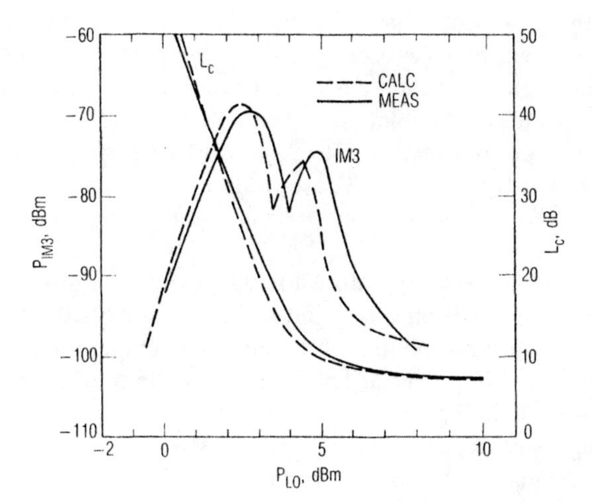

Figure 4.13 Measured and calculated conversion loss and third-order IM levels as a function of LO power. The RF input level is –20 dBm per tone.

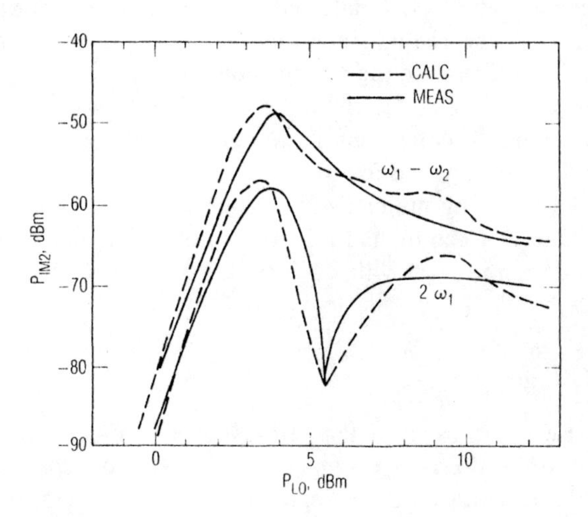

Figure 4.14 Measured and calculated second-order IM levels as a function of LO power. The RF input level is –20 dBm per tone.

dramatically as the LO power is increased. The inescapable conclusion is that relatively high LO power is necessary to achieve good IM performance.

Figure 4.13 and Figure 4.14 show deep nulls in the IM output levels of the third-order and second-harmonic IM components at certain LO levels. Although it may be tempting to exploit these points to achieve good IM performance at low LO power, it is usually not practical to do so. These nulls are very sensitive to virtually all operating parameters of the mixer: temperature, LO frequency, and even RF tone spacing.

The sensitivity of noise temperature to LO power of a mixer using a silicon diode is generally different from one using a GaAs device: silicon diodes exhibit a distinct noise-temperature minimum, while the noise temperatures of GaAs devices do not vary much with LO power at levels that provide minimum conversion loss. The reason may be that the relatively low breakdown voltages of silicon devices cause avalanching—and concomitant high levels of shot noise—to occur at relatively low LO levels.

4.6.3 Minimizing Intermodulation

Intermodulation (both multitone IM and spurious responses) in diode mixers is a strong function of LO level; this is clear from Figure 4.13 and Figure 4.14. A mixer's IM is also sensitive to its embedding impedances. In general, short-circuit embedding impedances at all LO harmonics and unwanted mixing frequencies minimize IM. This is not surprising, since a diode is a strongly nonlinear function of voltage, and short circuits remove the voltage components that generate IM.

Optimizing the diode's characteristics is also important for minimizing IM. Fortunately, diode characteristics that minimize conversion loss and noise also minimize IM: low R_s, low C_{j0}, and low η.

Of all the factors affecting mixer IM, however, the most important is LO level. Unless adequate LO power can be had, optimizing other factors such as embedding impedance and diode parameters will do little to assure low distortion in a diode mixer.

4.6.4 DC Bias Voltage

One of the simplest ways to improve the performance of a mixer is to apply dc bias to the diode. Not only does this reduce the conversion loss and LO power requirements, it also provides an extra degree of freedom for empirical adjustment,

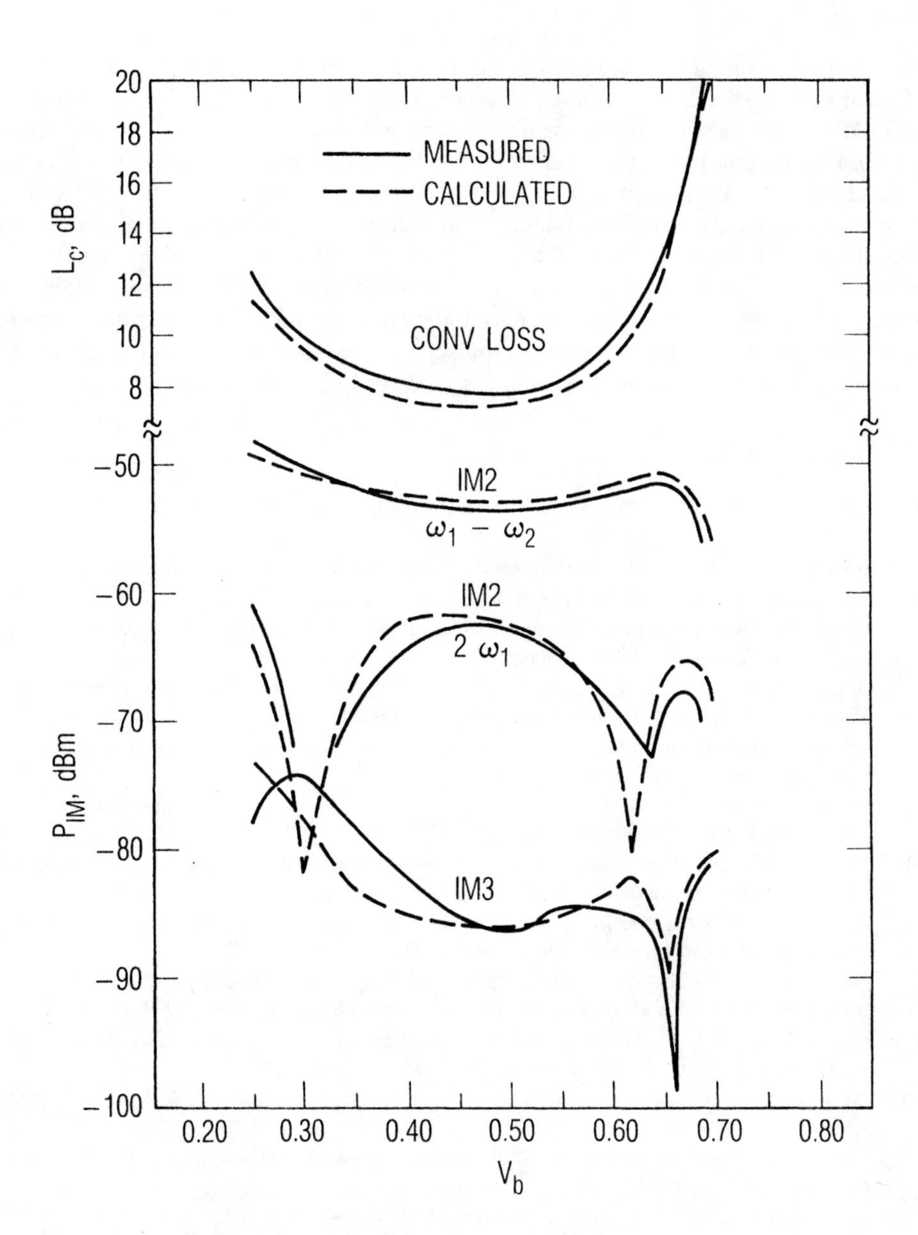

Figure 4.15 Measured and calculated conversion loss and IM level as a function of dc bias voltage, V_b. $P_{LO} = 0$ dBm.

and compensation of manufacturing variations in diodes. Figure 4.15 shows the effect of varying the dc bias voltage when the LO level is relatively low, 0 dBm. It is possible to achieve virtually the same conversion loss at this LO level as at 7 dBm when no dc bias is used. Unfortunately, the IM levels are not as low. There is no substitute for the use of high LO power if good IM performance is to be achieved.

It is usually fairly easy to include dc bias in single-diode mixers. However, in many types of balanced mixers (Chapters 7 and 8), the circuit topology is not well suited to the use of dc bias. Furthermore, the use of bias in balanced mixers always entails the possibility that the bias may unbalance the mixer and thereby degrade some of its most desirable properties: the rejection of LO noise and even-order IM distortion. We shall examine these matters more fully in later chapters.

4.7 REFERENCES

[1] Peterson, E., and L. W. Hussey, "Equivalent Modular Circuits," *Bell Sys. Tech. J.*, Vol. 18, 1939, p. 32.

[2] Torrey, H.C., and C.A. Whitmer, *Crystal Rectifiers*, New York: McGraw-Hill, 1948.

[3] Saleh, A. A. M., *Theory of Resistive Mixer,* Cambridge, MA: MIT Press, 1971.

[4] Egami, S., "Nonlinear, Linear Analysis and Computer-Aided Design of Resistive Mixers," *IEEE Trans. Microwave Theory Tech.*, MTT-22, 1974, p. 270.

[5] Held D. N., and A. R. Kerr, "Conversion Loss and Noise of Microwave and Millimeter-Wave Mixers," *IEEE Trans. Microwave Theory Tech.*, MTT-26, 1978, p. 49.

[6] Dobrowolski, J. A., *Introduction to Computer Methods for Microwave Circuit Analysis and Design*, Norwood, MA: Artech House, 1991.

[7] Press, W., B. P. Flannery, S. A. Teukolsky, and W. T. Vetterling, *Numerical Recipes in C*, Cambridge: Cambridge University Press, 1988.

[8] Hicks, R. G., and P. J. Khan, "Numerical Technique for Determining Pumped Nonlinear Device Waveforms," *Electron. Lett.*, Vol. 16, 1980, p. 375.

[9] Kerr, A. R., "A Technique for Determining the Local Oscillator Waveforms in a Microwave Mixer," *IEEE Trans. Microwave Theory Tech.*, MTT-23, 1975, p. 828.

[10] Hwang, V., and T. Itoh, "An Efficient Approach for Large-Signal Modeling and Analysis of the GaAs MESFET," *IEEE Trans. Microwave Theory Tech.*, MTT-35, 1987, p. 724.

[11] Kundert, K. S., J. K. White, and A. Sangiovanni-Vincentelli, *Steady-State Methods for Simulating Analog and Microwave Circuits*, Kluwer, Boston, 1990.

[12] Uhlir, A., "Shot Noise in PN Junction Frequency Converters," *Bell Sys. Tech. J.*, Vol. 37, 1958, p. 951.

[13] Dragone, C., "Analysis of Thermal Shot Noise in Pumped Resistive Diodes," *Bell Sys. Tech. J.*, Vol. 47, 1968, p. 1883.

[14] Van Der Zeil, A., and R. L. Waters, "Noise in Mixer Tubes," *Proc. IRE,* Vol. 46, 1958, p. 1426.

[15] Weinreb, S., and A. R. Kerr, "Cryogenic Cooling of Mixers for Millimeter and Centimeter Wavelengths," *IEEE J. Solid-State Circ.*, SC-8, 1973, p. 58.

[16] Hegazi, G. M., A. Jelenski, and K. S. Yngvesson, "Limitations of Microwave and Millimeter-Wave Mixers Due to Excess Noise," *IEEE Trans. Microwave Theory Tech.*, MTT-33, 1985, p. 1404.

[17] Van Der Zeil, A., *Noise: Sources, Characterization, and Measurement*, Englewood Cliffs, NJ: Prentice-Hall, 1970.

[18] Maas, S. A., *Nonlinear Microwave Circuits*, Norwood, MA: Artech House, 1988.

[19] Maas, S. A., "Two-Tone Intermodulation in Diode Mixers," *IEEE Trans. Microwave Theory Tech.*, MTT-35, 1987, p. 307.

Chapter 5
System Considerations

In previous chapters we examined Schottky-barrier diodes, their use in mixers, and the calculation of their conversion losses and noise. The next logical step is to examine the use of mixers in systems, and to consider the properties of mixers that affect overall system performance. These include spurious signals and responses, intermodulation, and saturation, as well as other things that we have already addressed to some degree: noise, conversion gain, and port impedances.

In this chapter we are primarily concerned with the use of mixers in microwave and millimeter-wave communication receivers, radar systems, and radiometers; these are the most common and critical applications of microwave mixers. Because the mixer's solid-state device, a diode or transistor, is nonlinear, and is pumped by a large local oscillator waveform, mixers have intermodulation levels that are generally worse than those of nominally linear components. In many cases intermodulation distortion limits a receiver's performance more seriously than noise, and the mixer is the weak link in the receiver system.

In Chapter 4, we treated a mixer as a microwave circuit. In this chapter we shall treat a mixer more as a "black box," a downconverter having at most two responses, the RF and image. The extension to upconverters, harmonic mixers, or other mixer types will usually be straightforward, if not obvious.

5.1 INTERNALLY-GENERATED MIXER NOISE

5.1.1 Noise Temperature and Noise Figure

In preparation for entering one of the more confusing subjects surrounding mixers—the concepts of *single sideband* (SSB) and *double sideband* (DSB) noise characterization—we would be wise to review the concepts of noise figure and temperature. In Chapter 4, the mixer's output noise power was described in terms of the pumped diode's internal noise sources. This theory is fine for determining the noise output power, but it does nothing to show how the mixer's noise affects the operation of a system. To analyze system performance, each component must be treated as a "black box," described only by its external parameters (gain, noise, bandwidth, intermodulation level), instead of a set of circuit equations. In this case, we wish only to know how much noise the mixer generates, not how it arose. Thus, we need another method of noise characterization, one that describes the magnitude of the total noise generated in the mixer. This quantity is the mixer's *noise temperature**; another method of describing a component's noise is its *noise figure*. The persistence of the concept of noise figure to the present is, as we shall see, unfortunate, especially where mixers are concerned; noise temperature is a simpler and more directly useful concept.

In Chapter 4, we saw that the spectral density of the noise power available from a resistor or any other lossy medium, such as a matched waveguide load, is simply KT, where K is Boltzmann's constant and T is its absolute temperature. This "thermal noise" is a frequency-independent, white-noise random process; thus, the power available in a bandwidth B is KTB. Although this relation is approximate, it is extremely accurate at temperatures above a few kelvins and frequencies below a few hundred gigahertz. This remarkable situation—the fact that the available spectral power from a resistor depends only upon its temperature—makes a resistor an ideal noise standard, and the *temperature* of a noise process defines its available spectral power density. Noise temperature is equivalent to available spectral power density; the quantities differ only by Boltzmann's constant.

Noise temperature can be used to specify the noise generated in any two-port, and for mixers, where responses at different mixing frequencies can be treated as separate "ports," the concept can be extended to include multiports. Figure 5.1 shows a linear two-port terminated at its input by a source impedance Z_S, and at its output by a load Z_L. Usually (but not necessarily), these impedances are real and equal, such as the 50Ω coaxial impedance standard used throughout the microwave industry, or a matched waveguide termination. The two-port has a transducer conversion gain G_t, defined as the power delivered to the load Z_L divided by the power available from the source Z_S. G_t may be either greater or less than 1 (in diode mixers, it is invariably the latter). The temperature of the input termination is T_S. The noise power P_L delivered by the input termination and the two-port to the load is

* As we shall see, the concepts of noise temperature and noise figure are applicable to all two-ports, not just mixers.

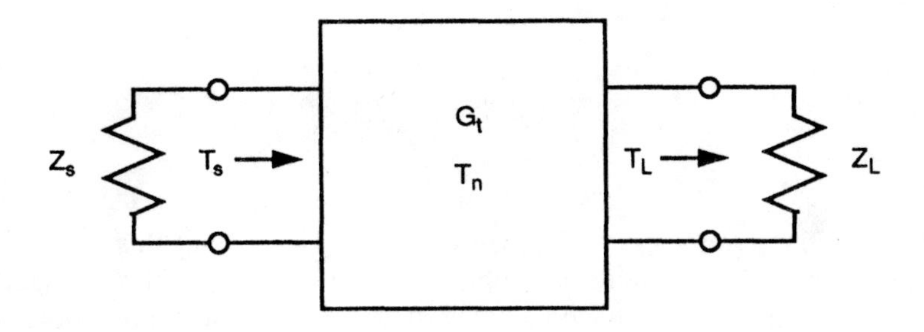

Figure 5.1 Terminated two-port.

$$P_L = KB(G_t T_s + T_{no}) \tag{5.1}$$

or in terms of noise temperature, the noise power delivered to the load T_L is

$$T_L = P_L / (KB) \tag{5.2}$$

The load resistor, of course, generates noise of its own. That noise has been excluded by definition: we are interested in the noise power delivered *to* the load *by* the network; the noise generated by the load is, for this purpose, irrelevant*.

On the right side of (5.1) and (5.2), the term $G_t T_s$ represents the amplified thermal noise from the source. T_{no} is the output noise generated by the noise sources within the two-port. Some of these noise sources may be shot noise in semiconductor junctions, thermal noise from resistors or other losses in the circuit, flicker noise at low frequencies, high-field diffusion noise in FETs, intervalley scattering noise in GaAs, or other mechanisms. The source of the noise or mechanism by which it is generated is not important; it is important only that it exists. The bandwidth has not been specified in (5.2); we assume that the bandwidth is vanishingly small, so the noise spectral density is constant, even if the noise is not white. If the noise is nonwhite, T_{no} and T_L can be made functions of frequency. $T_{no}(f)$ and $T_L(f)$ are then called *spot noise temperatures* at the specified frequency f.

* There are, in fact, ways that the noise of the load can contribute to the noise in the system. We shall examine these later in this chapter; however, the irrelevance of this noise will be our working hypothesis for now.

For a two-port having nonwhite noise, (5.1) becomes

$$P_L = \int_{f_1}^{f_2} K(G_t(f)T_s + T_{no}(f))\,df \tag{5.3}$$

where $G_t(f)$ is the gain; T_L is still given by (5.2). Throughout this chapter, noise temperatures can be interpreted as either noise spectral densities or spot noise temperatures at very small bandwidths. The frequency arguments therefore will be deleted.

Because the two-port is linear*, we can treat it as a noiseless circuit having a single noise source at its input. The noise from this source is added to the noise from the input termination, and the two of these account for all output noise. Thus, we model the noisy two-port as shown in Figure 5.2. The two-port's noise temperature T_n is

$$T_n = \frac{T_{no}}{G_t} \tag{5.4}$$

As long as the source impedance remains equal to Z_S, the representation of Figure 5.2 will be valid. T_n is called the *spot noise temperature of the two-port*, the two-port's noise spectral density, divided by K, referenced to its input.

An older method of characterizing a two-port's noise is the *noise figure* or *noise factor*. Noise figure is not a very useful concept; noise figure cannot be used in calculations without reducing it to a quantity that is proportional to noise temperature. Noise figure was originally conceived as the input signal-to-noise ratio divided by the output signal-to-noise ratio. This definition has several problems, not the least of which is that it is unique only when the input noise level is defined. Noise figure is now defined [1] as the total output noise temperature divided by the output noise temperature due to the input noise source, when the input noise source is a standard load (i.e., of impedance Z_s) at temperature $T_s = T_0 = 290$K. Specifically,

---

* Note that, although it is usually realized by a pumped nonlinear circuit element, the process of shifting a signal from the RF to the IF is linear; thus, a mixer is fundamentally a linear component.

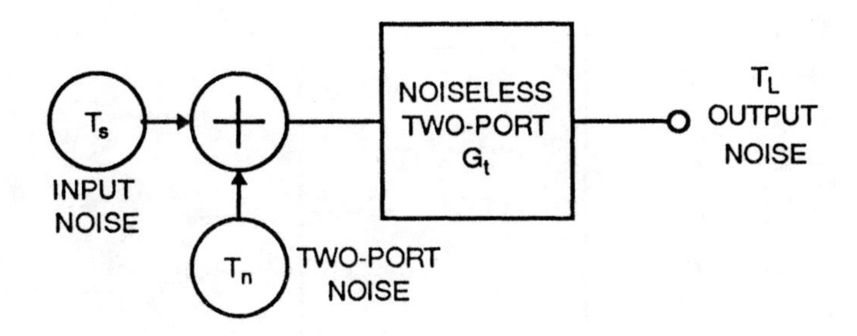

Figure 5.2 The noisy two-port with noise temperature T_n is modeled as a noiseless two-port with an external, additive source of noise temperature T_n at its input.

$$F = \left. \frac{G_t (T_n + T_s)}{G_t T_s} \right|_{T_s = T_0 = 290} = \frac{T_n}{T_0} + 1 \qquad (5.5)$$

This value of T_0 has been adopted as a standard for relating noise figure and temperature. The noise figure F is usually expressed in decibels.

5.1.2 Noise Temperature and Noise Figure of an Attenuator

Because any matched power-dissipating element can be treated for the purpose of noise characterization as an attenuator, an attenuator is a very general and important element in any microwave receiver or other low-noise system. It is even, under most circumstances, a valid model for a diode mixer.

Figure 5.3 shows an attenuator having a resistive source and load of the standard value, all at a temperature T. Because the attenuator and its terminations are in thermal equilibrium, the noise delivered to the load by the combined attenuator and source is equal to the noise that the load delivers to the attenuator and source. This noise level, in terms of noise temperature, is T. From (5.2),

$$T = G_t T + T_{ao} \qquad (5.6)$$

T_{ao} is the noise generated in the attenuator that is delivered to the load, and G_t is the transducer gain (less than one) of the attenuator. This situation is eminently sensible, since there is no fundamental difference between a termination and the output port

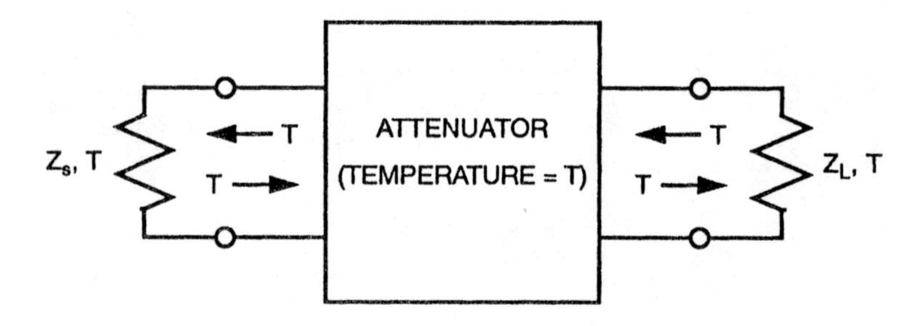

Figure 5.3 A terminated attenuator in thermal equilibrium. The noise delivered to each termination equals the noise that each termination delivers to the network.

of a matched attenuator with its input terminated. By referring T_{ao} to the input, we find the attenuator noise temperature T_{an} to be

$$T_{an} = \frac{T_{ao}}{G_t} = T\left(\frac{1}{G_t} - 1\right) = T(L - 1) \tag{5.7}$$

where $L = 1/G_t$, the loss factor. The noise figure of the attenuator can be found by substituting (5.7) into (5.4). The most significant result is that the noise figure is numerically equal to the loss when $T = T_0 = 290$K.

5.1.3 Noise Temperature of a Cascade of Stages

Figure 5.4 shows a cascade of M two-ports. The noise figure can be determined easily as long as the output impedance of each two-port is equal to the source impedance for which the noise temperature of the next stage is defined (this is rarely a restrictive assumption). The noise temperature and transducer gain of the mth stage in the cascade are T_m and G_m, respectively.

The output noise temperature of the first stage, neglecting the noise of the input termination, is G_1T_1. This is added to the input noise of the next stage, so its output noise is $G_2(G_1T_1 + T_2)$. This process is continued, and the output noise of the Mth stage, the last in the cascade, is

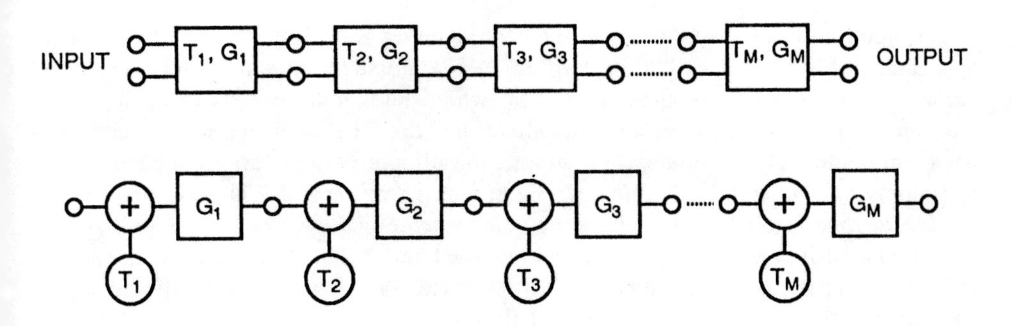

Figure 5.4 Cascade of noisy two-ports and the equivalent representation as noiseless two-ports with input noise sources.

$$T_L = G_M T_M + G_M G_{M-1} T_{M-1} + G_M G_{M-1} G_{M-2} T_{M-2} + \ldots + T_1 \prod_{m=1}^{M} G_m \tag{5.8}$$

Dividing by the gain of the entire cascade (the product of the gains of the individual stages) gives the following result for the noise temperature of the cascade T_n:

$$T_n = T_1 + \frac{T_2}{G_1} + \frac{T_3}{G_1 G_2} + \ldots + T_M \prod_{m=1}^{M-1} \frac{1}{G_m} \tag{5.9}$$

The contribution to the system noise temperature from each stage is the noise of that stage divided by the gain of all the stages ahead of it. An important implication of this result is that, in a cascade of stages, each having gain greater than unity, the noise of the first few stages dominates the overall noise temperature. If the first stage has loss, however, $G_1 < 1$ and the noise temperatures of the subsequent stages become much more significant.

5.1.4 Mixer Noise Temperature

In Chapter 4, the noise temperature of the mixer was determined by calculating the output noise and dividing by the conversion gain between the desired input and the IF output. Thus, the output noise was modeled as if it arose from an additive source at the input, and not at any other mixing frequencies, most notably the image. This approach yields what is called the *single sideband* noise temperature, and is entirely valid for determining signal-to-noise ratio when the signal consists of an input at that one frequency. However, in some situations, the RF and image frequencies are used simultaneously as inputs, for example, in radiometry applications such as radio astronomy and noise measurement. In this case, the mixer noise is described by its *double sideband* noise figure, where the noise is averaged over both sidebands (i. e., is treated as if it arose in both the RF and image bands). The mixer itself is the same in either case; it has an image and an RF response, each having nominally the same conversion loss, and a certain amount of IF output noise power. The only difference is that, in the SSB case, the IF noise is assumed to come entirely from the RF input, and the image "input" is unused and noiseless. In the DSB case, the noise is assumed to come equally (if the image and RF conversion losses are the same) from both. Thus, when the image and RF conversion losses are equal, the DSB noise temperature is precisely half the SSB noise temperature.

These two cases are illustrated in Figure 5.5. Although the RF and image have the same physical ports, in the figure they are shown as two separate ports. This representation is no different from those used in Chapter 4, where the mixer's responses at different frequencies were treated as separate ports.

The output noise temperature of the mixer in Figure 5.5 in the SSB case, is

$$T_L = (T_s + T_{\mathrm{SSB}})\,G_{cr} + T_s G_{ci} \tag{5.10}$$

where G_{cr} and G_{ci} are the transducer conversion gains at the RF and image frequencies, respectively. The SSB noise temperature is the output noise from the mixer, minus that which originated in the input termination, divided by the gain. Therefore, the SSB noise temperature is

$$T_{\mathrm{SSB}} = \frac{T_L - T_s(G_{cr} + G_{ci})}{G_{cr}} \tag{5.11}$$

For the DSB case, T_{DSB} is, by definition, the same at both sidebands. The IF output temperature is

$$T_L = (T_s + T_{\mathrm{DSB}})\,(G_{cr} + G_{ci}) \tag{5.12}$$

which is, of course, numerically equal to the output noise in the SSB case. The DSB noise temperature is

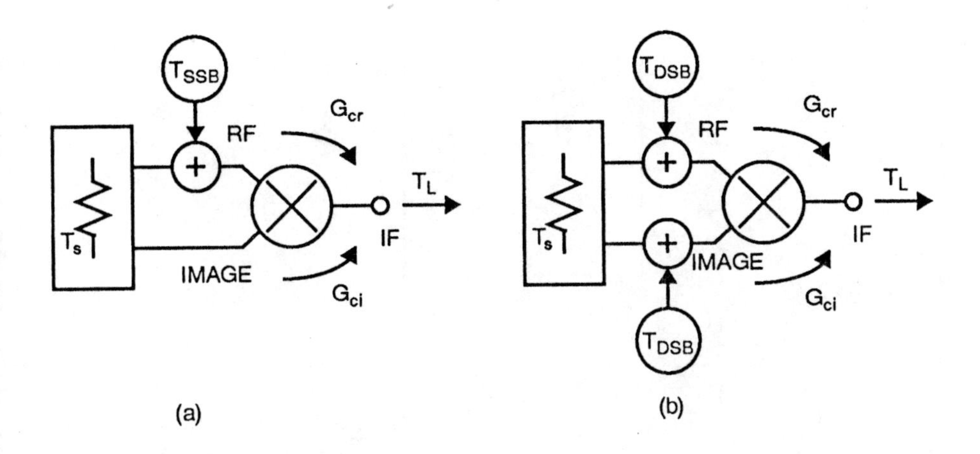

Figure 5.5 Single-sideband (a) and double-sideband (b) noise temperatures. The RF and image bands are shown here as separate ports, although normally they use the same set of terminals.

$$T_{DSB} = \frac{T_L - T_s(G_{cr} + G_{ci})}{G_{cr} + G_{ci}} \tag{5.13}$$

which, if $G_{cr} = G_{ci}$, is half of the SSB noise temperature.

A mixer, for the purpose of noise characterization, is often compared to a three-port attenuator having two inputs and a single output. Such an attenuator is shown in Figure 5.6. Through a derivation nearly identical to that in Section 5.1.2, it is possible to show that the output noise temperature of the attenuator is $T(1 - G_1 - G_2)$. The mixer can be treated in much the same way, but its noise sources are not the same as the simple thermal noise sources of the attenuator; so T may be greater or less than that of the attenuator. To account for these differences, the output noise temperature T_{Lm} of the mixer alone (not including the input termination) is

$$T_{Lm} = T_d(1 - G_{cr} - G_{ci}) \tag{5.14}$$

where T_d is the *effective noise temperature* of the diode. In an ideal diode having no series resistance or junction capacitance, and reactive terminations at all mixing

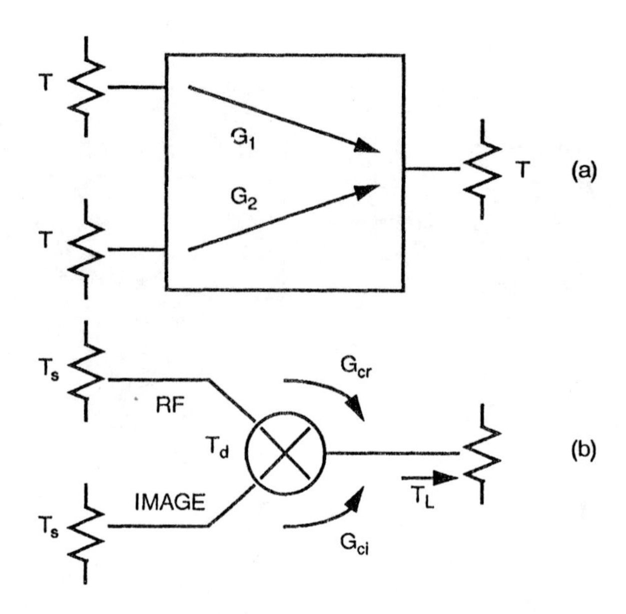

Figure 5.6 The attenuator noise model: (a) three-port attenuator having no coupling between the input ports; (b) a mixer with RF and IF responses treated as separate ports.

products, $T_d = \eta T_s/2$, where η is the diode's ideality factor (Section 2.1.2). This value of T_d represents a lower limit on the obtainable noise temperature. In practice, T_d is usually between 300K and 400K at room temperature, although values as high as 450K to 500K are commonly observed, usually in mixers using image enhancement to achieve very low conversion losses (Chapter 4). In estimating noise temperatures of prosaic mixers, it is often assumed that T_d will be approximately 350K. Most well-designed mixers have T_d very close to that value.

The mixer noise temperature is obtained by substituting (5.14) into (5.11) or (5.13). When G_{cr} and G_{ci} are equal, the noise temperatures are

$$T_{\mathrm{SSB}} = T_d (L - 2) \tag{5.15}$$

$$T_{\mathrm{DSB}} = T_d \frac{(L - 2)}{2} \tag{5.16}$$

where L, the conversion loss, is the inverse of the conversion gain.

Not all mixers have significant image responses. Many use input filters to eliminate the image response. Others achieve image rejection through the use of phase-shift techniques. The SSB and DSB concepts are not applicable to these mixers, and their noise performance can be described by a single unambiguous noise temperature:

$$T_{\mathrm{mxr}} = T_d\,(L - 1) \tag{5.17}$$

which is the same as the noise temperature of an attenuator at physical temperature T_d. The effective diode temperature T_d of an imageless mixer is usually the same as that of a mixer that has an image response (a pathological exception is described in Section 4.6.1). L may be lower, however, because the reactive image-frequency termination usually improves the conversion loss.

5.1.5 Noise Temperature of a Mixer Receiver

Because the mixer may be described by its SSB and DSB noise temperature, Equation (5.9) must be re-examined before it can be applied to a cascade of stages that includes a mixer. A mixer receiver, including an antenna, is shown in Figure 5.7. The antenna in this case is the input termination, and its noise temperature T_a depends on the level of noise it receives. The antenna collects thermal or other noise from the region at which it is pointed, or if it is nondirectional, its environment. Because a directional antenna may be pointed at a warm source (e.g., the earth) or a cold source (e.g., the sky), its noise temperature may be as low as a few degrees or greater than 300K. The noise temperature of a low-frequency nondirectional antenna, which receives primarily atmospheric noise, may be as great as 10^5K. The antenna is a stage in the system, and its noise temperature may be a substantial part of the system's noise temperature. Therefore, the antenna must be included in the receiver's noise temperature.

Figure 5.8 shows the range of temperatures that an antenna can have [2]. At frequencies below 300 MHz atmospheric noise, resulting from thunderstorms, dominates. Between 30 MHz and 10 GHz, galactic noise, which consists primarily of radiation from the galactic center, is dominant, although its level depends on the directivity of the antenna and the direction in which it is pointed. Above 10 GHz, atmospheric absorption noise dominates. At certain frequencies, atmospheric absorption can be very great; for example, oxygen absorption near 60 GHz causes as much as 180 dB loss to a signal traveling straight up through the atmosphere.

The noise temperature of the IF-amplifier chain in Figure 5.7 is T_{IF}; the noise temperatures of individual amplifier stages have been combined according to (5.9),

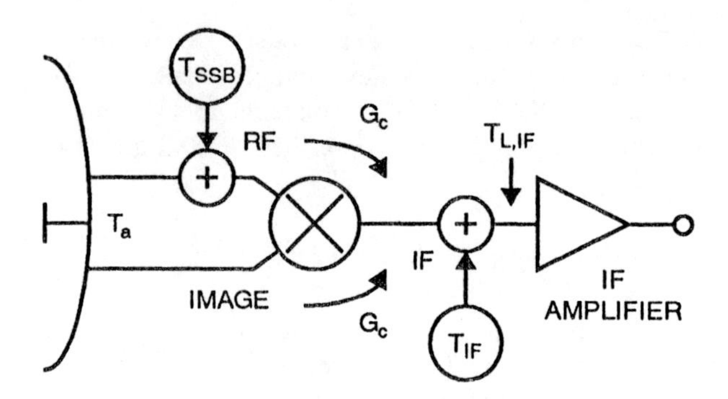

Figure 5.7 SSB noise-temperature model of a cascaded mixer and amplifier. A DSB model would require sources in both the image and RF paths (Figure 5.5).

so the IF amplifier chain can be treated as a single component. From (5.10), the total noise temperature at the input of the IF amplifier, the sum of the amplifier noise temperature and the mixer's output noise, is

$$T_{L,\,\mathrm{IF}} = (2T_a + T_{\mathrm{SSB}})\,G_c + T_{\mathrm{IF}} \tag{5.18}$$

In Equation (5.18), we have assumed that the image and RF conversion gains have the same value, G_c, and that the antenna noise temperature T_a is the same at the RF and image frequencies. Dividing by the gain G_c gives the input SSB noise temperature, including the antenna noise temperature:

$$T_{\mathrm{sys,SSB}} = 2T_a + T_{\mathrm{SSB}} + L\,T_{\mathrm{IF}} \tag{5.19}$$

Where we have again used $L = 1/G_c$. Twice the value of T_a must be used because antenna noise is present at both the image and the RF input frequencies. This is an important detail, and is often forgotten in noise calculations. For the DSB noise temperature, Equation (5.12) gives the IF noise:

$$T_{L,\,\mathrm{IF}} = 2\,(T_a + T_{\mathrm{DSB}})\,G_c + T_{\mathrm{IF}} \tag{5.20}$$

As in deriving (5.13), we divide by the sum of the RF and image gain to obtain, for the DSB noise figure,

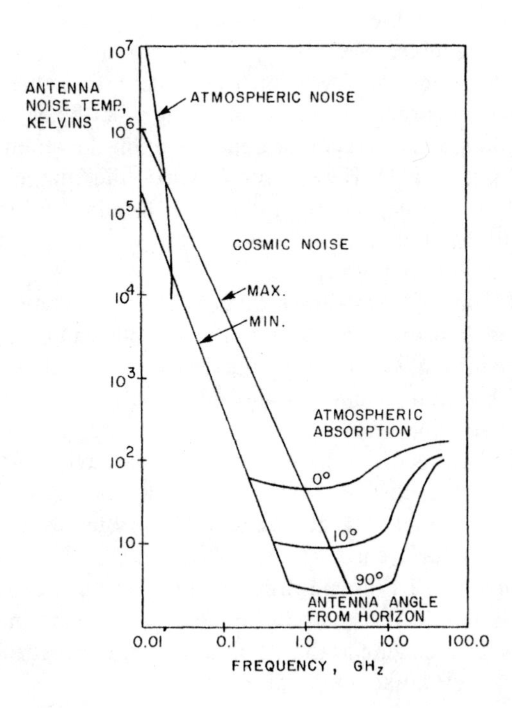

Figure 5.8 Antenna noise temperatures from 10 MHz to 100 GHz. (*Source*: M. Tiuri in J. Kraus, *Radio Astronomy*, Cygnus-Quasar Books, Powell, Ohio, 1982. Reprinted with permission.)

$$T_{\text{sys,DSB}} = T_a + T_{\text{DSB}} + \frac{L\,T_{\text{IF}}}{2} \tag{5.21}$$

which, again, is precisely half the SSB noise temperature.

A mixer without an image response is, in terms of its noise characterization, no different from any other two-port. Therefore, Equation (5.9) can be used directly:

$$T_{\text{sys}} = T_a + T_{\text{mxr}} + L\,T_{\text{IF}} \tag{5.22}$$

It is not immediately obvious from (5.19) to (5.22) and (5.15) to (5.17) whether using a filter to eliminate the image response improves or degrades the noise performance of the receiver. Certainly, for noise measurement or radiometry applications, it is best to use DSB operation and to achieve the DSB noise temperature. However, for receiving sinusoidal signals, the superiority of either an imageless or a double-sideband receiver depends upon the antenna temperature and achievable mixer loss. If a DSB receiver is used, the image response acts as a parasitic source of noise, degrading the signal-to-noise ratio while offering no compensating benefit. In applications where the antenna temperature is very low the added noise is small and may not matter.

Comparing (5.15) and (5.17) shows that, for a given conversion loss, the SSB temperature is lower than the noise temperature of the imageless mixer. However, the imageless mixer may have lower conversion loss and, therefore, a lower noise temperature (notwithstanding some of the qualifications given in Section 4.6.1). Furthermore, if the antenna temperature is equal to or greater than the effective diode temperature (which is often the case), the mixer having the lowest conversion loss clearly achieves the lowest system noise temperature. Finally, it is also necessary to consider the fact that the image is a spurious response, and eliminating it removes a source of interference.

If the mixer follows an RF preamplifier, some care must be exercised. If the RF and image bands are within the passband of the components in front of the mixer, and their gain or loss is the same at the RF and image, the cascade formula (5.9) can be used directly. The SSB noise temperature is

$$T_{\text{sys,SSB}} = 2(T_a + T_1) + \frac{T_{\text{SSB}}}{G_1} + \frac{T_{\text{IF}}}{G_1 G_c} \tag{5.23}$$

where T_1 and G_1 are the noise temperature and gain, respectively, of the stages between the antenna and mixer, and G_c, as before, is the mixer's conversion gain. It is sobering to note that T_1 must be doubled, like T_a, because of the noise in the image band. The DSB noise temperature, as before, is half the SSB noise temperature.

If the gain and noise temperatures of the added stages are not the same at the image and RF frequencies, the SSB noise temperature is

$$T_{\text{sys,SSB}} = T_{aR} + T_{1R} + (T_{aI} + T_{1I})\frac{G_{1I}}{G_{1R}} + \frac{T_{\text{SSB}}}{G_{1R}} + \frac{T_{\text{IF}}}{G_{1R} G_c} \tag{5.24}$$

where T_{1I} and T_{1R} are the image and RF noise temperatures, G_{1I} and G_{1R} are the image and RF gains, and T_{aI} and T_{aR} are the antenna noise temperature at the image and RF frequencies, respectively. Equation (5.24) reduces to the simple cascade formula as G_{1I} approaches zero, and to (5.23) as G_{1I} approaches G_{1R}.

Equation (5.24) shows that, except in situations where the DSB noise temperature is valid, improvement in noise temperature approaching a factor of two can be obtained by eliminating the image response. However, it is necessary to be careful how the filtering is performed. If the filter is installed between the antenna and the first receiver stage, the mixer's image response may still collect noise from the output of the first stage, and the system noise temperature may be nearly the same as it would be with no filter. This may at first seem like a contradiction of (5.24), but it is not, because T_{1I} increases as G_{1I} decreases, so their product remains constant. If the filter is placed between the mixer and preamp, it eliminates the amplifier's image noise, and the desired image-free system is obtained. However, T_{SSB} and L no longer are the same, because the filter terminates the diode in a reactance at the image frequency. In this case the mixer's noise temperature is given by (5.17), not (5.15).

5.1.6 Mixer Noise Figure

Mixers are arguably the best reason to dispense with the antiquated and cumbersome concept of noise figure. However, since the concept is currently alive and well and is recklessly applied to mixers, it is important to examine it, regardless of the confusion it engenders. Part of the blame for the problem can be attributed to the IEEE definition of noise figure. The rest arises from the fact that the calculation of noise figure is in reality little more than an academic exercise, because noise figure must be reduced to a noise temperature (or to a quantity proportional to noise temperature) before it can be used to calculate a system's noise figure. Therefore, it matters little how noise figure is defined, as long as it is a meaningful figure of merit, and the route from noise figure to temperature is the correct one.

The IEEE standard [1] defines the noise figure as the ratio of the total output noise temperature divided by the portion of the output noise engendered by the input termination, when the input termination has the standard temperature of 290K. However, it goes on to say, in the case of mixers, that a noise figure is defined for both the image and RF responses, and the output noise engendered by the input termination *includes only the noise arising from the principal frequency transformation of the system.* In other words, when a SSB noise figure at the RF input is to be determined, the output noise arising from the input termination, at the image frequency, *is not included.* Many engineers are rightly uncomfortable with this definition, as it seems to be an attempt to define away a noise source that can in no way be eliminated. Furthermore, it is impossible to measure directly the noise figure thus defined, because noiseless image terminations are notoriously difficult to obtain. The use of a filter to eliminate the image response does not help, because it changes the image-frequency embedding impedance, and therefore the noise temperature. Finally, it is not at all clear whether the "total output noise" in the definition includes image termination noise (we generally assume that it does not). Because of these dilemmas, an alternate definition of SSB noise figure has found more common use than the officially sanctioned one. Both definitions, properly applied, lead eventually to the correct noise temperature; therefore, it probably does not matter which is used, as long as consistency is maintained.

SSB Noise Figure (IEEE Definition)

The model for this definition is shown in Figure 5.9(a). The image-frequency termination is assumed to be noiseless, and the mixer noise is assumed to arise only at the RF frequency. The output noise power is, therefore, $G_c(T_{SSB} + T_0)$, where G_c is the conversion gain. The output noise due to the termination is simply G_cT_0. The noise figure is the ratio of these quantities, or

$$F_{SSB1} = \frac{T_{SSB}}{T_0} + 1 \tag{5.25}$$

which is the same as the relationship between noise figure and temperature for two-ports (5.5). Substituting (5.15) gives the relation between noise figure, loss, and effective diode temperature:

$$F_{SSB1} = (L - 2)\frac{T_d}{T_0} + 1 \tag{5.26}$$

which, when $T_d = T_0$, reduces to $F_{SSB1} = L - 1$. This is significantly different from the attenuator case where $F = L$, so it is not valid under this definition to assume that the mixer noise figure is approximately equal to the conversion loss, even if T_d is close to T_0. The ratio T_d/T_0 is sometimes called the mixer's *noise temperature ratio.*

SSB Noise Figure (Alternate Definition)

The previous definition ignores the noise contribution of the termination at the image frequency. This definition includes it, as shown in Figure 5.9(b), and treats it as if it were an intrinsic noise source in the mixer. The total output noise is $G_c(T_{SSB} + 2T_0)$, and the noise due to the input termination at the RF frequency is again G_cT_0. The SSB noise figure, the ratio of these quantities, is

$$F_{SSB2} = \frac{T_{SSB}}{T_0} + 2 \tag{5.27}$$

This differs from the usual relationship between noise figure and temperature in a two-port, expressed in (5.5). Substituting for T_{SSB} as before gives

$$F_{SSB2} = (L - 2)\frac{T_d}{T_0} + 2 \tag{5.28}$$

When $T_d = T_0$, $F_{SSB2} = L$. This relation is the same as that of the attenuator. This result should come as no surprise, because this definition includes all the mixer's power-dissipating elements between the RF input and IF output, and their temperatures are all T_0.

DSB Noise Figure

Paradoxically, the IEEE definition of noise figure does not include the concept of a DSB noise figure, but the latter is the only well-defined noise figure concept for mixers having a significant image response. Figure 5.9(c) illustrates the DSB noise figure. In this case, both the image and RF frequencies are inputs, and they can be used simultaneously.

The output power is $G_c(2T_{DSB} + 2T_0)$ and the output noise due to the input termination is $2G_cT_0$. The DSB noise figure becomes

$$F_{DSB} = \frac{T_{DSB}}{T_0} + 1 \tag{5.29}$$

which is expressed in terms of mixer conversion loss as

$$F_{DSB} = \left(\frac{L}{2} - 1\right)\frac{T_d}{T_0} + 1 \tag{5.30}$$

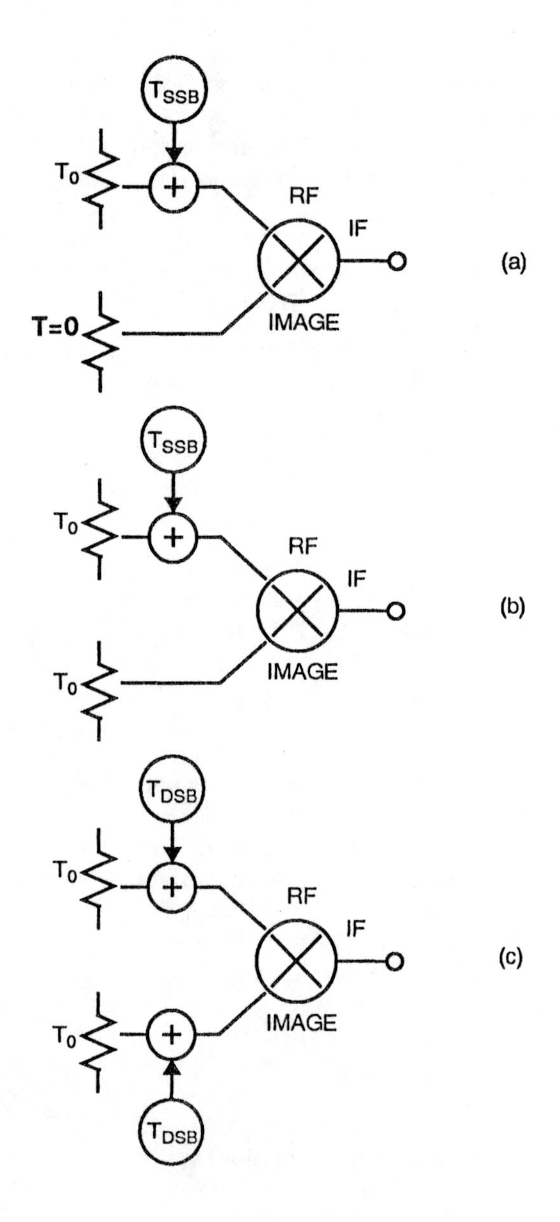

Figure 5.9 Noise models for noise-figure definitions: (a) IEEE SSB noise figure; (b) conventional SSB noise figure; (c) DSB noise figure.

When $T_d = T_0$, $F_{DSB} = L/2$. Thus, the alternate SSB noise figure definition and the DSB noise figure differ precisely by a factor of two. The DSB noise figure and the IEEE SSB noise figure are both 0 dB for a noiseless element, and approach a difference of 3 dB only as the noise figures approach infinity. Comparisons of the various quantities, to complete the reader's confusion, are presented in Table 5.1.

5.2 LO-INDUCED NOISE, INTERMODULATION, AND SPURIOUS SIGNALS

The sensitivity of a microwave receiver is usually limited by its internally generated noise. However, other phenomena sometimes affect the performance of a mixer front end more severely than noise. One of these is the *AM noise* or *amplitude noise* from the LO source, which is injected into the mixer along with the LO signal. This noise may be especially severe in a single-ended mixer (balanced mixers reject AM LO noise to some degree) or when the LO signal is generated at a low level and amplified. Some types of sources are relatively noisy: klystron and backward-wave oscillators used at millimeter-wave frequencies are often very noisy, as are IMPATT sources.

Phase noise is also a concern in systems using mixers. Local-oscillator sources

Table 5.1

Mixer Noise-Figure and Noise-Temperature Relations

L_r: RF-to-IF conversion loss

L_i: Image-to-IF conversion loss

	Noise Temperature	*Noise Figure (IEEE)*	*Noise Figure (Conventional)*
SSB mixer	$T_d\left(L_r - \frac{L_r}{L_i} - 1\right)$	$\frac{T_d}{T_0}\left(L_r - \frac{L_r}{L_i} - 1\right) + 1$	$\frac{T_d}{T_0}\left(L_r - \frac{L_r}{L_i} - 1\right) + \frac{L_r}{L_i} + 1$
DSB mixer	$T_d\left(\frac{L_r L_i}{L_r + L_i} - 1\right)$	Not defined	$\frac{T_d}{T_0}\left(\frac{L_r L_i}{L_r + L_i} - 1\right) + 1$
Imageless mixer	$T_d\,(L_r - 1)$	$\frac{T_d}{T_0}(L_r - 1) + 1$	$\frac{T_d}{T_0}(L_r - 1) + 1$

always have a certain amount of phase jitter, or phase noise, which is transferred degree for degree via the mixer to the received signal. This noise may be very serious in communications systems using either digital or analog phase modulation. Spurious signals may also be present, along with the desired LO signal, especially if a phase-locked-loop frequency synthesizer is used in the LO source. Spurious signals are usually phase-modulation sidebands of the LO signal, and, like phase noise, are transferred to the received signal (occasionally they may also have amplitude components). Finally, the mixer may generate a wide variety of intermodulation products, which allow input signals—even if they are not within the input passband—to generate spurious outputs at the IF frequency. These problems must be circumvented if a successful receiver design is to be achieved.

5.2.1 AM Noise in the LO

A mixer's LO signal can be generated in many ways. The simplest is to use some type of oscillator operating directly at the LO frequency. At the lower microwave frequencies, a FET or bipolar-transistor oscillator may be employed, while at higher frequencies, Gunn oscillators, klystron tubes, or backward-wave oscillators may be used. In some cases, the LO signal is generated initially at a subharmonic of the desired LO frequency and multiplied to the desired frequency. Often, the frequency multiplier requires a relatively high input level, and it is necessary to amplify the original subharmonic signal substantially. Amplifying the LO adds a certain amount of noise; in some cases that noise may have a temperature of several tens of thousands kelvins. This noise is largely *AM noise* or *amplitude noise*; it consists of variations in the amplitude of the LO signal, and can be treated as an additive noise process.

When the noisy LO signal is applied to the mixer, its AM noise components at the RF and image frequencies (Figure 5.10) are downconverted and appear at the IF port just as if they had been applied to the RF input. Therefore, the mixer noise temperature is increased by an amount that, depending on the type of mixer used, may be as high as the LO noise temperature. If no measures are taken to eliminate the LO noise, the increase in the mixer's noise temperature may be very great.

Fortunately, the AM LO noise can usually be reduced by straightforward techniques to a level where it is insignificant. It is most important to pick the IF frequency high enough so that the RF and image frequencies are well separated from the LO, so the noise at these frequencies can be removed effectively by filtering. Typically, the IF frequency should be at least 5% to 10% of the LO frequency if it is expected that AM noise will have to be removed by filtering (this may require a trade-off with IF-amplifier noise temperature, since amplifier noise temperatures generally rise with frequency). A balanced mixer may provide an extra 10 to 20 dB of AM noise rejection. Conversely, the LO-noise spectrum can be minimized through the use of high-Q resonators in the source and narrow amplifier passbands in the LO chain. Operating the amplifier and frequency multiplier stages of the LO chain in gain saturation also helps to minimize noise.

Figure 5.11 shows a well-designed LO amplifier-multiplier chain, consisting of a low-level, high-stability source, a chain of amplifiers, a frequency multiplier, and a filter. The block diagram illustrates how the noise can be raised to a very high level if several stages of amplification are used. The output noise temperatures of the LO amplifiers and the output noise level are as shown in the figure; the latter is calculated via (5.8). The AM noise level applied to the mixer's LO port is nevertheless reduced to an insignificant level by operating the last stage of the amplifier and the frequency multiplier in saturation, and using a filter at the multiplier output.

5.2.2 Phase Noise

The LO signal and its additive noise can be expressed as phasors, as shown in Figure 5.12. The LO voltage, a sinusoid having constant amplitude, frequency, and phase, is a constant phasor, while the noise is a much smaller phasor whose amplitude and phase vary randomly. The resultant phasor representing the waveform applied to the mixer LO port, $V_{LO} + v_n$ in Figure 5.12, varies in phase as well as amplitude. Although these variations may be small compared to the LO level, the

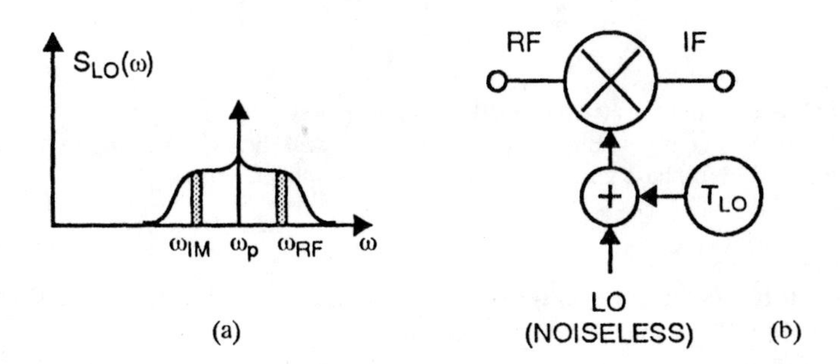

Figure 5.10 AM noise in the local oscillator: (a) AM noise spectrum showing the noise components that can be converted to the IF; (b) AM noise model of the mixer.

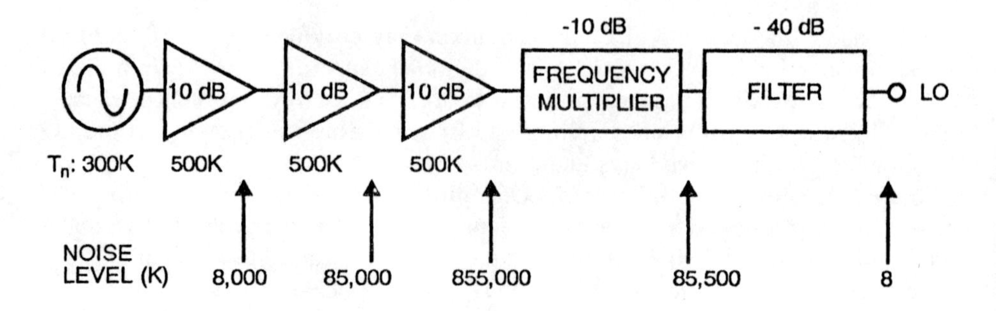

Figure 5.11 A well-designed LO chain. Although the AM noise level is close to 10^6K at the amplifier output, it is reduced to a negligible level by a filter having 40-dB rejection at the RF and image frequencies.

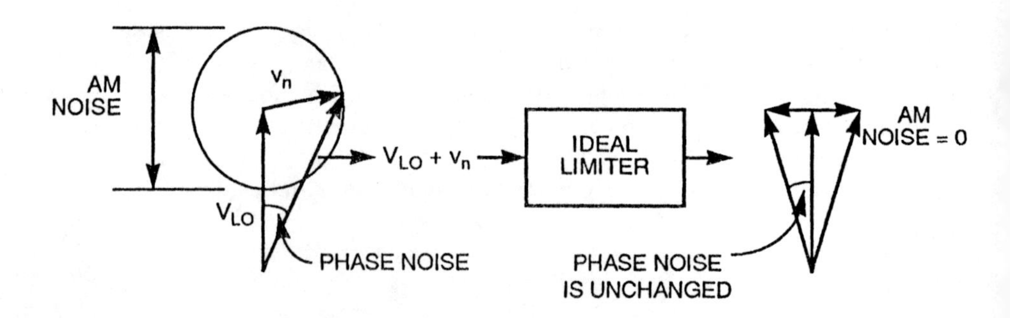

Figure 5.12 LO noise can be treated as a random phasor (shown here as v_n) added to the LO phasor. When the combination passes through an ideal limiter, all amplitude variation is removed, but phase variation—phase noise—remains.

amplitude variation (which is, in fact, the AM noise) may be very large compared to the RF input level. Much of the amplitude variation can be removed by hard limiting; the phase variation, however, is unaffected. Furthermore, because the mixer subtracts not only the LO and RF frequencies, but also their phases, the LO's phase noise is transferred directly to the received signal, radian for radian. It therefore causes the same phase jitter that appears on the LO to exist on the IF signal. In phase- or frequency-shift modulation systems, this phase jitter is demodulated along with the received signal.

The troublesome components of AM noise are those at the RF and image frequencies. However, the troublesome phase-noise components are not those at the image and RF, but those close to the LO frequency, usually within 1 MHz, which correspond to the modulation rates. These noise components arise from low-frequency noise processes within the oscillator, such as $1/f$ noise in solid-state devices, and, unfortunately, these noise components are often very large. Because they are so close to the LO frequency, direct filtering is rarely a practical way to eliminate them.

Little can be done in the mixer to reduce phase noise, but much can be done in the design of the LO source to minimize it. Phase noise in an oscillator is proportional to the inverse square of its resonator Q, so considerable improvement can be attained through the use of a cavity- or dielectric-resonator-stabilized oscillator. Crystal oscillators have low phase noise and are often used with a multiplier chain to generate microwave LO signals. However, frequency multiplication (which is in reality phase multiplication) enhances phase noise spectral density as the square of the multiplication factor, so high-order multiplication of a noisy source should be avoided. Phase-locked loops can also be used as filters to reduce phase noise. More detailed treatment of phase noise in sources is available in [2] through [4].

5.2.3 Internally-Generated Spurious Signals

Another problem closely related to phase and AM noise is the existence of spurious signals in the LO. Like LO noise, these spurious signals are applied to the mixer along with the LO and, depending on their frequencies, may be converted to a frequency within the IF band. The most serious problems arise when the spurious signals are within the RF or image passbands; in this case, the downconverted spurious signals may be much stronger than the RF signals. To illustrate this situation, imagine a balanced mixer having an RF input of –100 dB, and an LO having an RF-frequency spurious signal that is 60 dB below the LO. The IF output level of the desired signal, if the conversion loss is 7 dB, is –107 dBm. If the LO level is 10 dBm, the spurious-signal level is –50 dBm. With 7 dB conversion loss plus 20 dB rejection due to mixer balance, the spurious IF signal may be as great as –77 dBm, or 30 dB above the desired IF output.

The usual source of such "spurs" is a frequency synthesizer used to generate the LO. The spurs are usually phase-modulated sidebands on the LO, rather than simple additive tones, and, in theory, should not produce IF outputs in the absence of an RF input. However, any circuit between the LO source and the mixer having imperfect gain flatness can create an AM component from phase modulation, and this AM component can be downconverted in the manner exemplified above. Avoiding spurious signals requires careful selection of frequencies not only in the receiver front end, but in the LO frequency synthesizer as well. Any frequency used in the synthesizer represents a potential spurious signal in the LO output. A balanced mixer provides up to 20 dB rejection of the AM components of LO spurs, and using a high IF frequency will allow effective LO filtering. It is worthwhile to keep the IF amplifier bandwidth no wider than necessary, so that spurs outside the IF passband are not amplified to the level where they might saturate the IF amplifier or generate intermodulation components.

5.2.4 Two-Tone Intermodulation and Saturation

Mixers, like other components using solid-state devices, are subject to gain saturation and two-tone intermodulation. For a mixer to operate as a linear component (if you missed it, see the footnote on page 156), the RF input level must be kept well below the LO level. As the RF level approaches the LO level, the small-signal assumption used in Chapter 4 is no longer valid, and the RF voltage across the diode is great enough that the junction no longer behaves as a linear, time-varying resistor. The results are gain saturation and distortion.

As with amplifiers, the saturated output level of a mixer is usually defined by its 1-dB compression point, the output power at which its conversion loss increases by 1 dB. In practice, the 1-dB compression point of a diode mixer having an LO level below 13 dBm is rarely greater than 3 to 5 dBm. A workable rough estimate for the saturated output level is $P_{sat} = P_{LO} - L_c$, where P_{LO} is the LO level in decibels above 1 mW and L_c is the small-signal conversion loss in decibels. This estimate often is optimistic, so it should be used with caution.

Because all diodes have fundamentally the same exponential *I*/*V* characteristic, saturated output level depends secondarily on diode parameters and primarily upon LO level. Therefore, increasing the output level of a diode mixer inevitably requires increasing its LO drive. The power that can be applied to a single diode is limited; to increase the LO drive beyond this point it is necessary to create multiple-diode, balanced circuits. Simply increasing the LO level beyond that required to achieve minimum conversion loss does not significantly improve the 1-dB compression point; it may also increase the conversion loss and may damage the diode. At the same LO levels, active FET mixers usually have higher 1-dB compression points than diode mixers. FET resistive mixers, examined in Chapter 9, are capable of significantly greater 1-dB compression points than diode mixers.

Because the Schottky diode is a nonlinear element, diode mixers invariably have high levels of multitone IM*. IM in mixers is similar to IM in static (i.e., time-invariant) circuits such as amplifiers. IM is manifest as the creation of distortion products; when two closely spaced RF input tones (e.g., at frequencies f_1 and f_2) generate other tones at the frequencies

$$f_{\mathrm{IM}} = \pm q f_1 \pm r f_2 \pm s f_{\mathrm{LO}} \tag{5.31}$$

where q, r, and s are positive integers. The worst of these are the infamous third-order ($q + r = 3$) products at $2f_1 - f_2$ and $2f_2 - f_1$ downconverted to the IF.

These distortion products are generated by the nonlinearities in the solid-state device used to perform the mixing. Thus, they are generated in a manner similar to IM in amplifiers or other static circuits; the only difference is that mixers have a large-signal LO, in addition to the small-signal excitations, which converts the intermodulation products to the IF frequency and to sidebands of all the LO harmonics. The spectrum of intermodulation products in a mixer is shown in Figure 5.13. The components near each LO harmonic look like the IM spectrum of any static component.

As long as the mixer is not saturated by the RF tones, the levels of IM at the IF port depend on the levels of the excitations in the same way as IM products in static circuits. A set of two-tone third-order intermodulation curves is shown in Figure 5.14. The level of the intermodulation products at the output of the mixer is given by the well-known relation [5]

$$P_{IMn} = nP_1 - (n-1)IP_n \tag{5.32}$$

where n is the order of the IM product, IP_n is the *nth-order intercept point* of the IM product in question, and P_1 is the output level of the linear responses, assumed to be equal (all power levels in (5.32) are in decibels above 1 mW). Note that IP_n is the extrapolated point, in terms of output power, at which the input-output curves of the linear and IM responses intersect.

Equation (5.32) shows that the levels of the nth-order IM product vary n dB for every 1-dB change in input level of *both* input tones. Closer inspection of the analysis leading to (5.32) would show that the nth-order IM product given by (5.31), where $q + r = n$, varies q dB for every 1-dB change in f_1, and r dB for every 1-dB change in f_2. This principle, in combination with the intercept point and (5.32), can

---

* We use the term *multitone IM* to distinguish IM caused by multiple RF excitations from spurious responses, which are in fact a form of intermodulation distortion. We shall loosely refer to multitone distortion simply as *IM* to distinguish it from spurious responses.

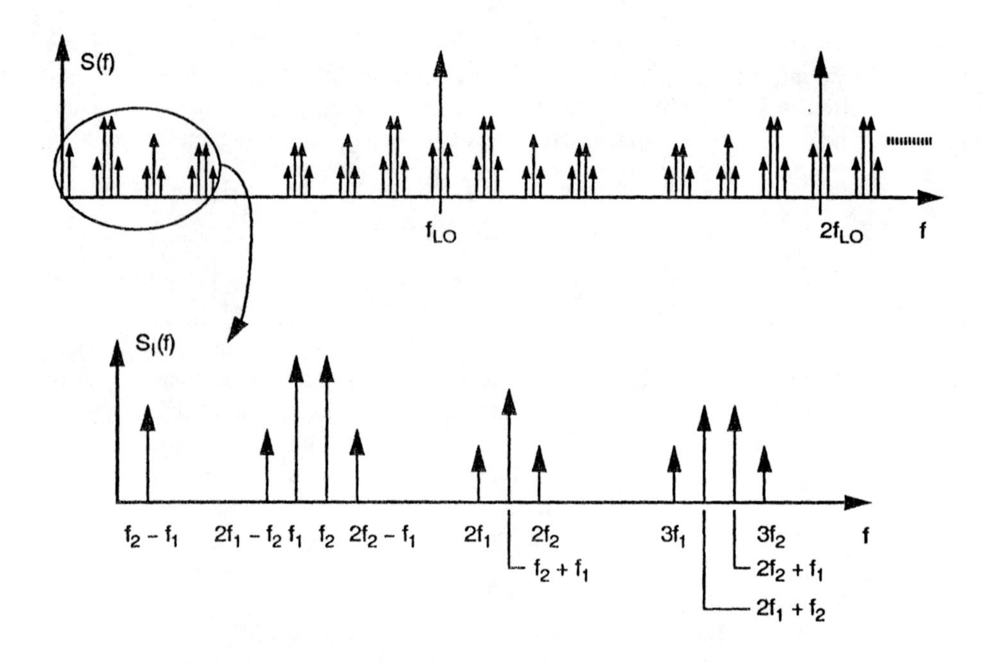

Figure 5.13 Spectrum of the intermodulation products in a pumped diode, resulting from two RF inputs. Intermodulation products up to third order are shown; these are "folded" around each LO harmonic. In the lower curve, f_1 and f_2 are IF frequencies.

be used to find the IM level that results from any set of input levels.

When a mixer is cascaded with either RF preamplifiers or an IF amplifier, the intercept point of the cascade is somewhat problematical. Intermodulation products are generated in the first stage of the cascade. These are amplified or attenuated by the next stage, along with the desired signal, and new IM products are also generated. These new IM products occur at the same frequency as those of the previous stage, but their phase is indeterminate. Thus, the IM products may combine in such a way as to either enhance or reduce their magnitude. This process is repeated at each subsequent stage.

An upper bound on the intercept point can be determined by assuming that the IM products generated in each stage, and those passed along from previous stages, add precisely in phase. This is a reasonable, worst-case assumption. Under this assumption, the worst-case output intercept point of a cascade of stages [5] – [7] is

$$IP_n^{(1-n)/2} = IP_{n,M}^{(1-n)/2} + (G_M IP_{n,M-1})^{(1-n)/2} + (G_M G_{M-1} IP_{n,M-2})^{(1-n)/2} + \ldots + (G_2 \ldots G_M IP_{n,1})^{(1-n)/2} \tag{5.33}$$

where n is the order of the IM product, M is the number of stages, and the subscripts refer to the stage numbers.

In any receiver the mixer is the dominant source of intermodulation distortion. Its contribution can be minimized by using minimal gain in the low-noise RF amplifiers preceding it. Minimizing the noise temperature of the receiver, however, usually requires the use of substantial amplifier gain. These are conflicting trade-

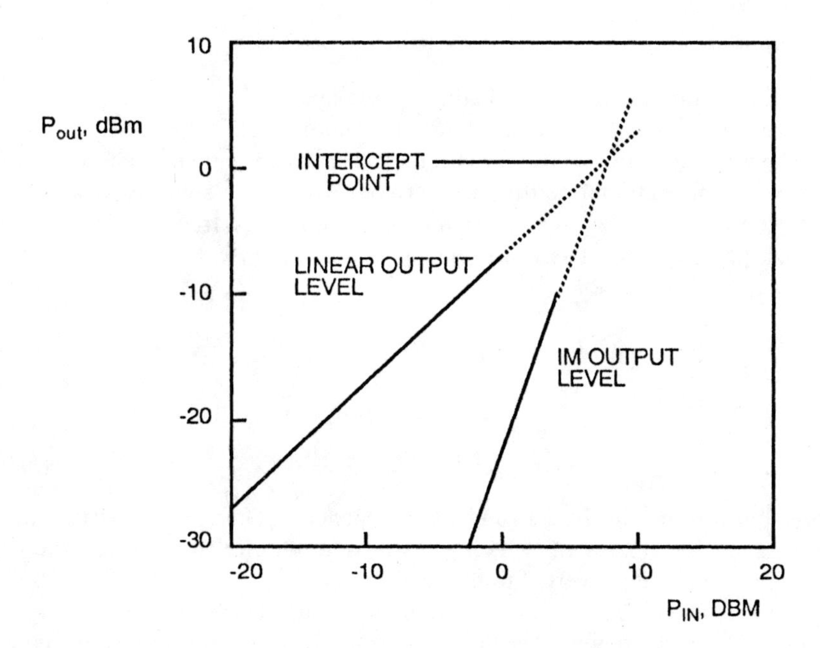

Figure 5.14 The third-order intercept point is the extrapolated point where the IM and linear output levels meet. By convention, P_{in} and P_{out} are the output levels of each tone.

offs, which are resolved only by using no more gain ahead of the mixer than is necessary to meet noise temperature goals. The better the mixer design, the less low-noise amplification is needed, and both noise temperature and intermodulation level will benefit. Minimizing preamplifier gain also improves spurious-response levels (Section 5.2.5) by reducing the input level at the mixer.

One final note is in order: Equation (5.32) is entirely valid when the powers are defined as input levels instead of output levels. The intercept point is then referenced to the input, rather than the output. The input intercept point is simply

$$IP_{\text{in}} = IP_{\text{out}} - G_c \tag{5.34}$$

where all quantities are in decibels or decibels above 1 mW, for all orders of IM products. In diode mixers, $G_c < 0$ and, consequently, $IP_{\text{in}} > IP_{\text{out}}$. Thus, entirely for marketing reasons, diode mixers are often specified in terms of their *input* intercept points, not their *output* intercept points, which are traditionally used to specify other types of components.

5.2.5 Spurious Responses

Although traditionally treated as a distinct phenomenon, spurious responses are nothing more than a type of single-tone intermodulation. The frequencies of spurious responses are given by Equation (5.31) when either $q = 0$ or $r = 0$. However, unlike conventional multitone intermodulation, in a spurious response, a harmonic of the LO other than the one that causes the desired frequency translation often figures prominently. Therefore, in general, the mixer generates output frequencies that satisfy the relation

$$f_{\text{IF}} = mf_{\text{RF}} + nf_{\text{LO}} \tag{5.35}$$

where

$$m, n = 0, \pm 1, \pm 2, \ldots \tag{5.36}$$

and f_{IF}, f_{RF}, and f_{LO} are the IF, RF, and LO frequencies. Thus, there exists a wide range of frequencies, many of which are outside the RF passband, whereby interfering signals can be converted to the IF.

If the input frequencies satisfying (5.35) are well outside the input passband, the interfering signals can be removed easily by filtering. If the spurious response is of a high order (i.e., high values of m and n), the mixer will not convert them efficiently to the IF. The problem becomes serious when the RF frequencies of low-order spurious responses are within the mixer's input passband, or are so close to it that they cannot be filtered effectively. For reasons explained in Chapter 7, certain types of balanced mixers provide extra rejection of spurious responses. However, the best way to avoid spurious responses is to select RF, LO, and IF frequency ranges appropriately.

Figure 5.15 shows a spurious-response plot and Table 5.2 shows a level chart for a 29- to 30-GHz mixer having a 3- to 4-GHz IF passband. The LO is fixed at 26 GHz. The plot in Figure 5.15 is simply a plot of (5.35) for several low-order combinations of m and n. Spurious responses (including the image, which can be treated as the low-order spurious response $(m, n) = (-1, 1)$) are plotted over the range of input frequencies from 20 to 40 GHz and IF frequencies from zero to 8 GHz. Note that we have exceeded the defined RF and IF ranges, because spurious responses outside the defined passband may still be a source of trouble. The response of greatest concern in this example is the (2, –2) response from 27.5 to 28.0 GHz, close to the lower edge of the passband, because it has relatively low order. It is close enough to the RF band that it may be difficult to eliminate by means of a filter.

The spurious-response chart, Table 5.2, shows the spurious-response levels of the mixer. The response levels in the chart were measured with the LO level set at the mixer's standard value, in this case 13 dBm, and at an RF input level of –10 dBm (this input level has become an informal standard for specifying spurious-response levels). The chart includes the effects of filtering and matching circuits that may be used in the mixer. In some cases, the input frequencies that give rise to a particular spurious response are so far removed from the input band that the response cannot be measured; thus, there is no entry in the table for these. For other types of mixers, such as a broadband balanced mixer, the table would be nearly full.

The –10-dBm input level is a compromise: it is not high enough to cause saturation, so the data in the table can be scaled to different input levels, yet it is still high enough to allow accurate determination of some fairly high-order (and therefore weak) responses. These data can be scaled to different RF input levels [6] by recognizing that spurious responses are a type of intermodulation. Like other intermodulation products, the response associated with the mth harmonic of the RF input changes m dB for each dB change in RF input level. Thus, a reduction in RF level of 10 dB reduces the (2, –2) spurious response by 20 dB, or the (–2, 3) by 30 dB. This principle can be used to determine the amount of filtering needed to reduce any spurious response to an acceptable level. Note that this principle does not apply to the LO level, or to RF levels strong enough to saturate the mixer.

We now know enough about intermodulation in mixers to recognize one of the fundamental problems in using LNAs ahead of a mixer to minimize a receiver's noise temperature. Because of the way mth-order spurious responses and

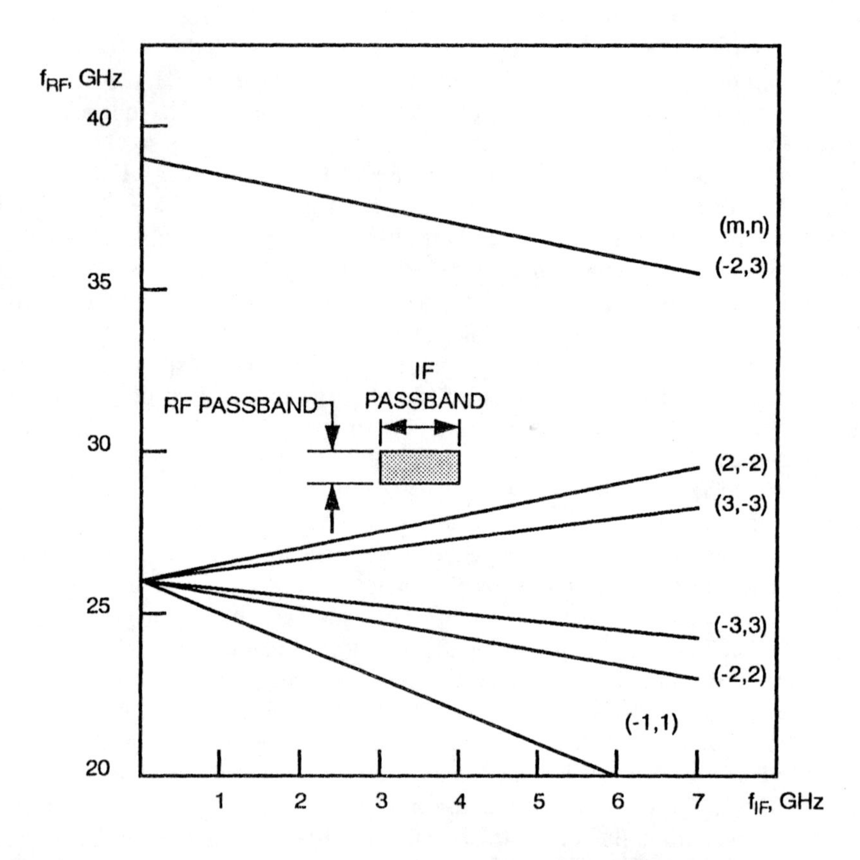

Figure 5.15 Spurious-response plot for a 30-GHz mixer. The curves relate IF and RF frequency ranges that give (m,n) spurious responses. The LO frequency is 26.0 GHz. Because none of these curves intersect the shaded area, this mixer has no in-band spurious responses.

intermodulation products vary with input level, the use of an LNA can seriously degrade the receiver's intermodulation and spurious-response performance. For example, suppose we add an LNA having 20 dB gain to the input of a receiver having a mixer front end. The extra 20 dB of signal level at the input of the mixer will increase the level of a $(2, n)$ spurious-response product by 40 dB, and will degrade the carrier-to-IM level by 20 dB; a $(3, n)$ response will be degraded (in terms of carrier-to-IM ratio) by 40 dB.

Spurious responses can also arise from a combination of intermodulation in the mixer and IF or RF amplifiers. Such *compound spurs* are often difficult to predict. For example, it happens frequently that an (m, n) spurious response has an IF frequency that is far out of the IF passband, and its level may be virtually immeasurable when the mixer is tested. However, a harmonic of this response may mix with LO leakage in the IF amplifier, producing a high-order spurious response that has a surprisingly high level. The cure in this case is to use adequate filtering between the mixer and IF amplifier.

5.3 MISCELLANEOUS CONSIDERATIONS

5.3.1 Sensitivity of Noise Temperature and Conversion Loss to LO Level and DC Bias

An important concern in system applications of diode mixers is that virtually all aspects of their performance—not just conversion loss and noise figure—are quite sensitive to LO level. Minimizing conversion loss requires adequate junction conductance variation, and noise temperature is related strongly to conversion loss.

Table 5.2

Spurious-Response Levels of a 29- to 30-GHz Mixer
× indicates that no spur exists within the 20- to 40-GHz RF range

IF Frequency: 3 to 4 GHz
LO Frequency: 26 GHz
LO Level: +13 dBm
RF Level: −10 dBm

LO Harmonic (n):		*-3*	*-2*	*2*	*3*
	3	−70	×	×	×
	2	×	−45	×	×
	1	×	×	×	×
RF Harmonic (m):	−1	×	×	×	×
	−2	×	×	−46	−80
	−3	×	×	×	−70

Similarly, input and output VSWRs depend strongly upon the junction conductance and capacitance waveforms. We have already seen, in Section 4.6, that nonlinear phenomena such as intermodulation and spurious responses are sensitive to LO level.

Sensitivity of conversion loss to LO level is probably the most fundamental concern in using mixers in systems. A plot of conversion loss of a 10-GHz mixer using silicon beam-lead diodes, as a function of LO level, is shown in Figure 4.13. No external dc bias is used in this mixer. The conversion loss decreases monotonically with LO power until the LO level reaches approximately 6 dBm; it then decreases only slightly–a fraction of 1 dB–as LO level increases from 6 to 10 dBm. If dc bias were used, essentially the same behavior would be observed, except that substantially lower LO power would be required to achieve flat conversion loss, and the increase in loss at low LO levels would not be as steep. Figure 4.15 shows the conversion loss as a function of dc bias, with 0 dBm LO power. At 0.5V bias, the conversion loss is only slightly greater than with 7-dBm LO power and no bias. Clearly, low conversion loss can be achieved with minimal LO power if dc bias is used.

Although many mixers are designed to operate without dc bias, it is generally true that better overall performance can be achieved if bias is used. This improvement results from the use of dc bias and LO level to adjust the diode's conductance waveform, which strongly affects the input and output match, as well as conversion loss. Having the ability to adjust the bias provides an additional degree of freedom in optimizing the mixer.

Input and output impedances of the pumped diode depend strongly on the average value of the diode's junction conductance; the average conductance, in turn, is proportional to the dc component of the diode current. The dc component of the diode current is much more sensitive to LO level when the diode is unbiased than when it is biased. Furthermore, the unbiased mixer usually has a higher IF output impedance than the biased mixer, and is therefore more difficult to match over a wide bandwidth.

The dependence of noise temperature on LO level generally follows that of conversion loss: up to a point, it decreases with LO level. However, as LO level is increased further, the conversion loss remains low, but the noise temperature increases. This increase is relatively gradual in GaAs diodes, but may be much sharper in silicon devices. It occurs because increasing the LO level increases the dc junction current, therefore shot noise, but does not further modify the conductance waveform.

By far the most important parameter in determining a mixer's intermodulation intercept points is LO level. Generally, the second- and third-order intercept points increase monotonically with LO level. High-order intercept points are usually much more sensitive to LO level than low-order ones; unfortunately, second-order intercept points of some troublesome spurious responses (particularly the (2, –2)) often are relatively insensitive to LO level. Nulls–very sharp decreases in IM levels–are observed in some mixers at specific LO levels. It is tempting to try to exploit these nulls to achieve very low distortion; however, such nulls are very sensitive to LO and even RF level, frequency, and diode parameters, as well as to embedding impedances. Furthermore, different IM products do not experience nulls under the same conditions. Thus, it is probably unwise to try to use such nulls to reduce IM distortion.

It is possible to describe qualitatively the reasons for the strong dependence of intermodulation on LO level. To do so, we may consider a single diode, since the results are valid for multiple-diode mixers as well. It happens that IM distortion is generated only while the diode is conducting a moderate current, and its *I/V* characteristic is significantly nonlinear. During the part of the LO cycle when the diode is strongly conducting, the junction is effectively short-circuited, so the junction voltage is nearly zero. During the other half of the LO cycle, the diode is turned off and its total junction current, including the intermodulation current, is also zero. Significant intermodulation current is generated only during the transition between these two extremes, when both the junction voltage and current are nonzero. The higher the LO level, the faster the LO waveform passes through this region, and less intermodulation energy is generated.

Thus, despite its very strong exponential nonlinearity, the Schottky diode generates remarkably low IM distortion. In essence, the diode operates more like a switch than a nonlinear element, and an ideal switching mixer has no intermodulation. Most other mixers do not work this way. Even though an active FET's nonlinearity is much weaker than a diode's, FET mixers generally have intermodulation properties that are, at best, only moderately superior to those of diode mixers. The reason is that the active FET mixer generates intermodulation over the entire positive half-cycle of LO voltage at which the gate voltage is greater than V_t, the FET's turn-on voltage, which is a much longer period than the diode's very brief transition interval.

5.3.2 Port-to-Port Isolation

The Schottky diode is a two-terminal device, and all the important mixing products–the RF, IF, and LO–exist at its terminals simultaneously. Therefore, the embedding circuit must separate these components and present each to its own set of terminals. This separation can be performed by filters, or by using two, four, or more diodes in a balanced structure.

Separating these components by means of filters is clearly impossible when the frequency ranges overlap. In this case, balanced mixers are often used to provide distinct RF and LO ports. In theory, the port-to-port isolations of balanced mixers are infinite, although, in practice, 10 to 30 dB is all that can be attained, depending on a number of factors, such as frequency, bandwidth, and the mixer's structure.

A common problem is that the LO-to-RF or LO-to-IF isolations are not adequate and allow an appreciable amount of LO power to be applied to the preceding or subsequent stages. LO power applied to the output of the RF preamplifier can reduce its dynamic range. If no RF amplifier is employed, LO leakage from the mixer's input may be transmitted by the antenna and interfere with other communications services, or, in military systems, act as a beacon that can be used to locate the receiver. The obvious way to reduce LO leakage is to use an input filter. If the LO and RF bands overlap, an input isolator can be employed.

Although the LO frequency is usually well outside the IF band, LO leakage into the IF can create problems in the IF amplifier. The first potential problem is saturation. IF amplifier chains usually have very high gain, and even an out-of-band signal that is very large, such as LO leakage, can saturate the IF amplifier. The second, and more subtle, potential difficulty is that the LO can generate so-called *compound spurious responses* (i.e., spurious responses that are generated in the IF amplifier when the IF or spurious signals intermodulate with the LO leakage). This problem can often be eliminated through the use of adequate IF filtering and ensuring that the IF amplifier bandwidth is restricted.

Because the FET effectively amplifies the LO, active FET mixers, especially single-gate FET mixers with LO and RF applied to the gate, have relatively poor LO-to-IF isolation. Often, such mixers have greater LO power at the IF port than is applied at the LO port. Other port-to port isolations in FET mixers are very different from those of diode mixers; the FET's nonreciprocal nature causes output-to-input isolation to be very good, while input-to-output isolation is much poorer. Dual-gate FETs usually have better RF-to-LO isolation than single-gate devices.

Because the gate-to-drain capacitance in an unbiased FET is relatively high, much greater than that of a biased device, resistive FET mixers (Chapter 9) have poorer LO-to-RF and LO-to-IF isolation than one might expect. Port-to-port isolation of such mixers is nevertheless no worse than that of comparable diode mixers.

5.3.3 IF-Port VSWR and Reflected Noise

Because a diode mixer's gain is less than one, Equation (5.9) shows that the IF amplifier's noise temperature is a very strong contributor to the system noise temperature. Clearly, other additive noise sources at the IF amplifier input also increase the system noise temperature significantly. One of these other noise sources is the combination of the IF amplifier's input noise and an imperfect mixer IF VSWR. By the term *input noise*, we mean the noise power available from the IF amplifier's input, not the equivalent input noise temperature that we have been using.

A general model of noise in two-ports can be formulated in terms of wave variables. The noise in the two-port consists of two waves at the input, one directed towards the input port and one directed away from it. These noise waves are, in general, correlated [8]. If the mixer's IF VSWR is imperfect, the outgoing noise wave is reflected and returns to the IF amplifier's input, where it combines with the noise represented by the incident wave. The reflected noise is correlated with the incident noise, so the effective input noise temperature of the IF amplifier is changed in some unpredictable way. It is a safe bet, however, that at some frequencies the noise temperature will increase. This problem is fundamentally no different from that of any other mismatched pair of components in a cascade, but it is clearly more critical in a mixer receiver: because the mixer exhibits conversion loss rather than gain, the contribution of IF noise to the receiver's noise temperature is relatively great. Because their IF port VSWRs are often high, this phenomenon is very important in active FET mixers.

If the IF amplifier has an input isolator as shown in Figure 5.16, the noise incident on the mixer's IF port arises from the isolator's termination, not the amplifier, and is therefore uncorrelated. In this case, a meaningful analysis is possible. The noise component reflected from the mixer T_r is

$$T_r = |\Gamma|^2 T_t \tag{5.37}$$

where T_t is the temperature of the isolator termination and Γ is the mixer's IF output reflection coefficient. The IF amplifier's effective noise temperature T_{eff} is

$$T_{\text{eff}} = T_{\text{IF}} + T_r \tag{5.38}$$

where T_{IF} is the IF amplifier's noise temperature with a perfect source. The SSB receiver noise temperature is then found from (5.9). The receiver's SSB noise temperature is increased by $|\Gamma|^2 L T_t$ over the noise temperature that would be achieved with a perfect IF VSWR.

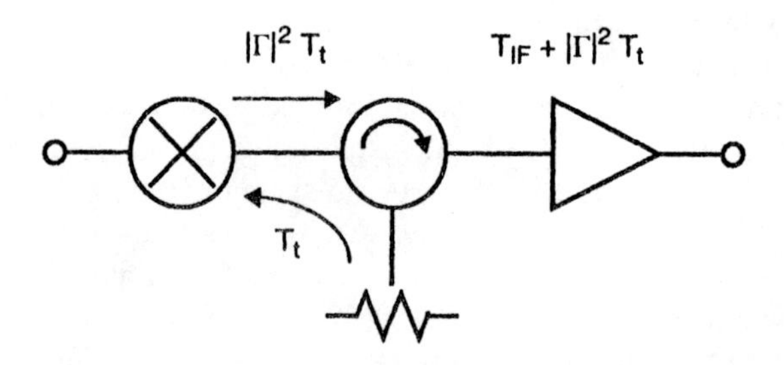

Figure 5.16 Noise-temperature increase caused by IF-port mismatch. The noise from the isolator's termination, at temperature T_t, is reflected into the IF amplifier's input port.

5.3.4 Reactive Terminations in Balanced Mixers

Spurious-response and intermodulation data are invariably measured with broadband resistive terminations at all ports at all mixing frequencies. If any of the ports are terminated reactively at unwanted mixing frequencies, performance may vary significantly. Broadband balanced mixers are especially sensitive to reactive port terminations.

The reason for this sensitivity is that some of the higher-order mixing products are not rejected in the IF, and the diodes "see" the IF termination at these frequencies; that is, the IF termination is part of the diodes' set of embedding impedances. If the IF port is terminated in a reactance at one or more of these frequencies, the result may be a degradation intermodulation performance. The termination of the IF port can increase the mixer's spurious-response and intermodulation levels as much as 20 dB. The effect of reactive terminations at various ports is summarized in Table 5.3. This problem is treated in considerable detail by Will [9].

Often, however, minimizing the effects of inadequate LO-to-IF or LO-to-RF isolation, or LO noise, requires the use of a filter at one or more of the mixer's ports; this filter creates the unwanted reactive termination at frequencies outside the filter's passband. These conflicting requirements can be satisfied by using a broadband isolator between the port and the filter, or by adding a simple diplexing circuit to terminate a troublesome mixing product in a resistive load. Sometimes it is possible to adjust the phase of the reactive termination, by varying the length of transmission line between the filter and mixer, to obtain a reactive termination that has minimal

Table 5.3
Effect of Out-of-Band Terminations in a Broadband, Doubly-Balanced Mixer

Reactive Termination	*Conv. Loss*	*Spurious Resp.*	*Multitone IM*
RF and image	1 to 2 dB	Minor	Minor
LO	Minor	Severe (±10 dB)	Severe (±10 dB)
IF	Normal mismatch loss	Severe (±20 dB)	Severe (±20 dB)

effect on performance. Finally, there exist designs for balanced mixers that reduce sensitivity of intermodulation levels to poor port VSWRs. Figure 5.17 shows one such structure.

5.3.5 Image Noise From an RF Preamplifier

We have seen that, in a receiver having no RF preamplifier or filter ahead of the mixer, the image response can receive additional noise or interference. If an RF

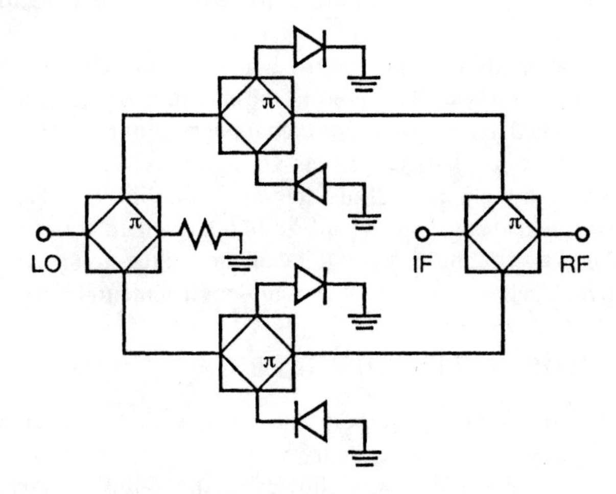

Figure 5.17 A balanced mixer that has low sensitivity to reactive port terminations.

preamplifier is used ahead of the mixer, and its passband excludes the image response, it might be tempting to think that potential problems associated with the mixer's image response are eliminated. However, although the amplifier may have no gain or even substantial loss at the image frequency, it may still have a very high noise output level at the image frequency. If the output noise at the image frequency is as great as the output noise in the RF band, the input noise level at the mixer will double, and the receiver noise temperature will be increased by a factor of two.

To reduce such image noise, it may be necessary to include a filter between the mixer and RF amplifier. If the IF frequency is too low to include such a filter, an image-rejection mixer can often be employed (Chapter 7). The required image rejection is not great; in most cases, 15 to 18 dB is adequate.

5.3.6 LO Power and Bias Leveling

When a mixer has a variable LO frequency, it is often difficult to achieve uniform diode pumping over the entire LO frequency range. Uniform pumping may be difficult to achieve because the LO port may not be well matched, a common problem in single-ended mixers, and the source itself may not have uniform output power. To compound the problem, there may be a narrow range of LO power over which minimum noise temperature can be achieved, especially if the mixer uses a silicon diode. The logical solution is to use an isolator at the LO port and a power-leveling circuit; however, the output power from the LO source still depends upon the frequency response and temperature sensitivity of the detector used in the leveling loop.

A better solution is to use the rectified mixer-diode current as the leveling circuit's indicator of LO power delivered to the mixer. This technique compensates the LO level automatically for temperature changes and uneven frequency response in the mixer's LO circuit. Changes in bias-voltage requirements with temperature can be compensated by using a diode having a similar *I/V* characteristic as a temperature sensor, and designing the mixer bias circuit to track its voltage drop. Implementing these two techniques will keep the mixer noise temperature at its optimum value over a wide range of system and environmental changes.

5.4 CRYOGENIC OPERATION OF MIXERS

Throughout the 1970s, the preferred method for achieving low noise temperatures in such exotic applications as millimeter-wave radio astronomy was to use a cryogenically cooled mixer. Recently, however, the rapid progress in transistor amplifiers using both conventional and heterojunction FETs has made the cooled

Schottky mixer largely antiquated. Even so, a few applications for cooled mixers still exist. The first is at frequencies above 100 GHz, where three-terminal devices are still unavailable. The second is for cooling exotic mixers such as those using superconductors. The recent discovery of high-temperature superconductors may bring about a resurgence of interest in such components.

5.4.1 Methods of Refrigeration

The most common methods of cooling electronic components to low temperatures are liquid cryogens (nitrogen or helium), closed-cycle refrigerators using helium, or, for spacecraft applications, radiation. Liquid cryogens are the least expensive for occasional use, and are probably the easiest to employ. Normal liquid helium has a boiling point of 4.2K at atmospheric pressure, and can be used to achieve temperatures below 4K by vacuum-pumping, but it is expensive and difficult to handle. Liquid nitrogen is less expensive, but its boiling point of 77K is much higher. It is possible to achieve temperatures below 60K by vacuum-pumping liquid nitrogen. Liquid nitrogen is easy to handle; very low mixer noise temperatures can be achieved for laboratory purposes simply by partially immersing a mixer in a foamed plastic container of liquid nitrogen. The nitrogen boil-off gas conveniently purges the container of water vapor, preventing the formation of water frost on the mixer or in the input waveguide.

Closed-cycle helium refrigerators are practical for use in systems or in the laboratory. These refrigerators most commonly use a Gifford-McMahon cooling cycle, and may include a Joule-Thompson expansion, where helium is cooled by the refrigerator to approximately 10K and expanded through an orifice to obtain liquid-helium temperatures. Helium refrigerators are capable of achieving temperatures below 20K with several watts of thermal loading, or 4.2K with less than 1W. These refrigerators are expensive and require several kilowatts of electrical power to run their compressors. They also require considerable maintenance and have *mean times to failure* (MTF) of 8,000 to 10,000 hours; much less if a Joule-Thompson loop is used. Failures are most often related to mechanical failures in the compressor or contamination of the helium gas. In order to achieve such low temperatures, the thermal power dissipation in the cooled stages must be minimized. It is necessary to design the insulating structures very carefully, including a vacuum jacket to minimize conductive heating and radiation shields to minimize radiative heating. Helium refrigerators are most often used for laboratory tests and in satellite or deep-space ground stations, or for radio astronomy. An example of a cooled receiver using a helium refrigerator is presented by Weinreb et al. [10].

Radiative coolers are sometimes used to achieve low temperatures on spacecraft.

It would seem logical that cryogenic temperatures could be achieved simply by attaching a radiator to the components to be cooled and aiming it toward deep space, thereby radiating away all the heat. However, because of two practical limitations, it is rarely possible to achieve temperatures below 150 to 200K. First, as the temperature of the radiator decreases, the amount of radiated heat drops dramatically: the power radiated to cold space is roughly proportional to the fourth power of the radiator temperature. Second, for the same reason, the radiated heat load from other parts of the spacecraft and the earth increases rapidly, as does conducted heat from the wires and cables connecting the electronic components to the rest of the system. Finally, it is very difficult to design a radiator that is never pointed toward the sun, at least briefly, during part of the spacecraft's orbit, or at some orientation that may be required.

5.4.2 Insulation

The purpose of insulation is to prevent heating of the cooled components by their environment. The components can be heated by RF or dc power dissipated within them, by heat arriving via conductive paths, and by radiative heat transfer. The former is rarely dominant and can seldom be reduced very much by the designer. The latter two sources can be effectively minimized.

The best insulation for conductive heat transfer is nothing at all: a vacuum. As well as providing the best possible insulation, a vacuum prevents atmospheric gases, especially water, oxygen, and nitrogen, from condensing on the cooled components. For vacuum insulation to be effective, the pressure in the vacuum vessel must be reduced to 10^{-3} torr or lower, a level achievable with a good mechanical vacuum pump. Once the components are cooled, the remaining gases freeze on the cold surfaces and the pressure drops considerably even without external pumping. A 20K surface of 1 to 2 ft^2 will pump the vacuum chamber at a rate exceeding 1000 liters per second, although the rate may drop if larger quantities of gas freeze out upon the cold surface. The "cryo-pumping" capacity can be increased enormously by adding a trap of activated charcoal to the cold stage. The amount of charcoal required depends on the temperature: a few grams are adequate at 20K, although several kilograms may be necessary at 77K.

When the cryogenic system is warmed, cryo-pumped gasses are released from the cooled surfaces. These can increase the pressure in the vacuum chamber to dangerous levels. Therefore, to prevent the creation of a bomb, the vacuum chamber should include a pressure-relief valve.

Cables, waveguides, and wires used as inputs and outputs and to provide dc bias to the mixer are usually the major source of conducted heat. Thin-wall stainless steel waveguides, with the inside gold- or silver-plated, can be used for high-frequency interconnections. Stainless steel semirigid coaxial cable, with silver-plated current-carrying surfaces, is also available. The dc connections can be made with number 30 gauge or smaller teflon-insulated wire. It is also possible in some cases to locate the feed antenna inside the vacuum chamber, and to mount a half-wavelength-thick window, or even a lens, in the chamber wall. Low-loss, noncontacting waveguide joints using choke flanges are often used. A wide variety of waveguide windows and hermetic-seal coaxial feedthroughs are available for interconnecting components inside and outside the vacuum chamber.

Heat can also be transferred to the cooled components by radiation. Radiative heat loading is potentially very great; therefore, radiative insulation is usually necessary in cryogenic systems. Radiative insulation consists of a thermally conductive, reflective shield surrounding the cooled components. If a multistage helium refrigerator is used, the shield should be mounted on the first stage and cooled to its temperature, usually approximately 50K. By reducing the temperature of the shield surrounding the cold stage, and by polishing it to a high reflectivity, radiative heat transfer is reduced substantially. Heat transfer from the room-temperature walls of the vacuum chamber to the shield can be minimized by wrapping the shield with several layers of aluminized mylar. The mylar should be perforated to allow good evacuation, and should be crinkled to minimize physical contact between layers (and thereby minimize conductive heat transfer between the layers). Aluminized mylar should be used freely on liquid gas cryostats and components cooled by a thermal radiator.

5.4.3 Mechanical Design of Cooled Mixers

The mechanical design of a cooled mixer is critical to its reliability. Failures result most commonly from poorly made whisker contacts or poorly designed diode mounts, where differential contraction, on cooling, creates stresses. The information in Section 2.5.2 on diode-whisker design should be followed religiously, along with some other considerations.

It is important to use materials for the diode mount that have adequate thermal conductivity and low expansion coefficients. It is not always necessary, or even desirable, to use Kovar, Invar, or other low-expansion alloys because their thermal conductivity is poor, they are difficult to electroplate successfully, and difficult to machine. Conversely, aluminum, having excellent thermal conductivity, has a high thermal expansion coefficient and is therefore rarely suitable. Brass or steel are probably the best choices for mount material. Screws should be made of the same material as the mount and should be located to prevent the development of nonuniform stresses. To prevent them from loosening under the vibration of a helium refrigerator, a thread adhesive can be used. Plastics, epoxies, and other materials having large thermal expansion coefficients should not be used in any applications where thermal stability is critical, or where they provide the primary bond between mechanical parts. In particular, these materials should not be used to mount any part of the diode support structure. If a packaged diode is used, one contact should be firmly attached and the other should use some type of contact that can move slightly to prevent the development of stress in the package (a gold-plated beryllium copper bellows or a welded gold ribbon usually works well).

If a whiskered diode is used, the design of the mount is critical. The modified Sharpless mount of Figure 2.22 is a good choice; conventional Sharpless mounts do not have adequate thermal stability for cryogenic use. To prevent diode heating, the diode chip must be attached securely to the IF filter, and the filter must be designed to minimize thermal resistance between the diode and the body of the mixer.

5.4.4 Diodes and Electrical Considerations

For reasons discussed in Section 2.3, the preferred diode for very low temperatures is the GaAs Mott diode. The Mott diode's low doping density minimizes tunneling current, and its relatively flat *C/V* characteristic minimizes LO power requirements. Although Mott diodes do not exhibit exemplary room-temperature performance, they provide outstanding low-noise performance at low temperatures. For cooled operation above 100K, however, the choice of a diode is not as clear. It is probably worthwhile to use a diode having a relatively light epitaxial doping density, although there is probably little benefit in using a true Mott structure. Between 200K and 300K conventional GaAs Schottky diodes are the best choice. For operation at all temperature ranges, silicon diodes are distinctly inferior to GaAs.

Fortunately, room-temperature adjustment of a cooled mixer is usually adequate; it is rarely necessary to readjust a mixer's mechanical tuning after it is cooled. A Schottky diode's *I/V* characteristic does change with temperature, but remains exponential; therefore, it is possible to achieve the same junction conductance waveform at low temperatures as at room temperature by increasing the diode's dc bias and reducing the LO power. The diode's *C/V* characteristic does not change substantially with temperature, although its capacitance waveform may change slightly because of the change in dc bias. This change is usually not enough to mistune the mixer, but it can increase the conversion loss by a few tenths of 1 dB.

5.5 NOISE MEASUREMENTS

5.5.1 Basic Technique

In Section 5.1, we saw that the noise output power from a system or component is $K(T_n + T_t)BG$, where T_n is its noise temperature, T_t is the temperature of its input termination, B is its bandwidth, K is Boltzmann's constant, and G is its gain. In principle, the noise temperature of any receiver system could be found by terminating the input in a load of known temperature; measuring the system's gain, bandwidth and output power; solving for the noise temperature; and finally subtracting the load temperature. This method would not be accurate, however, because the measuring system's bandwidth and gain, especially if it were large, are never known with sufficient accuracy. Therefore, noise measurements are invariably performed in other ways. Two loads having different temperatures can be used to measure noise temperature in a manner that is independent of bandwidth and gain and does not require an absolute power measurement.

The receiver shown in Figure 5.18 has two terminations and a switch connected to its input and a power sensor connected to its output. The terminations have different noise temperatures: T_H is the "hot" termination, the one at the higher temperature. T_C is the "cold" termination; it is at lower temperature. The "power sensor" is used to indicate relative power levels and therefore need not be a power measuring device. It can, for example, be an uncalibrated detector and precision attenuator.

The measurement is made by first switching the cold termination to the receiver input. The output power is

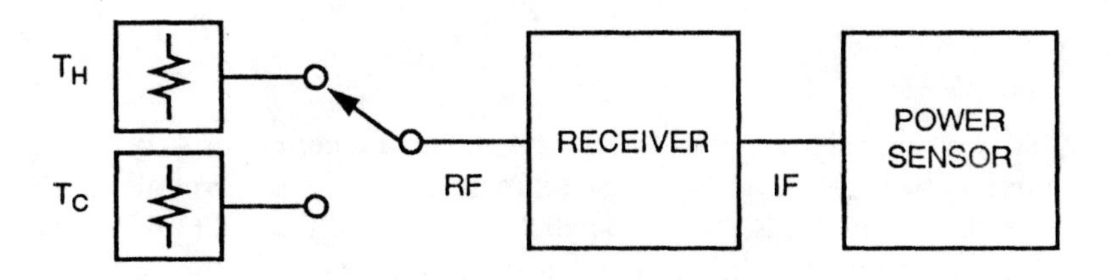

Figure 5.18 Noise-temperature measurement system.

$$P_C = KBG(T_C + T_r) \tag{5.39}$$

where T_r is the receiver noise temperature. The hot termination is then switched to the input, and the noise output power is

$$P_H = KBG(T_H + T_r) \tag{5.40}$$

Dividing (5.40) by (5.39) obtains

$$Y = \frac{P_H}{P_C} = \frac{T_H + T_r}{T_C + T_r} \tag{5.41}$$

which depends only on the hot- and cold-load temperatures and the ratio of the output powers. The noise temperature T_r is found by solving (5.41):

$$T_r = \frac{T_H - YT_C}{Y - 1} \tag{5.42}$$

This is called the *Y-factor method.*

Often the goal is to determine the noise temperature of a single two-port, not that of the entire receiver. If the two-port is an amplifier, it is necessary to measure the receiver noise temperature, connect the amplifier to the receiver's input, and then measure the combination. If the amplifier gain is known (from another measurement), the cascade equation (5.9) can be used to determine the amplifier's noise temperature T_2:

$$T_2 = T_{r2} - \frac{T_r}{G_t} \tag{5.43}$$

where G_t is the amplifier's gain, T_{r2} is the noise temperature of the receiver, including the amplifier, and T_r is the noise temperature of the receiver alone. If T_r is not too high and G_t is more than a few decibels, the second term in (5.43) is usually considerably smaller than the first, and any errors in measuring T_r or G_t do not affect T_2 too severely.

If this technique is applied to a diode mixer, however, several problems appear. First, the gain of the mixer is less than unity. Consequently, the second term on the right side of (5.43) is often much greater than the mixer noise temperature, so small errors in determining the conversion loss create large errors in the measurement of the mixer's noise temperature. Second, the mixer's conversion loss must be determined by a separate measurement. Because the mixer's RF input and IF output are necessarily at different frequencies, it is much more difficult to measure a mixer's conversion loss than an amplifier's gain. Measuring a mixer's conversion loss via absolute power measurements at the input and output frequencies is much less accurate than measuring an amplifier's gain with a detector and precision attenuator, especially if the mixer's input frequency is in the millimeter-wave range. Nevertheless, this method is often used; it requires a very low-noise test receiver, which, if the IF frequency is relatively low, is not difficult to obtain.

Another (and distressingly common) solution to the problem of specifying a mixer's noise temperature is simply to state the noise temperature of the mixer combined with an IF having a stated noise temperature or noise figure. In the latter case, the user is left to wonder whether the mixer has relatively low loss and high noise, or high loss and low noise. This distinction is unimportant only if the noise temperature of the receiver's IF is the same as that of the test system.

5.5.2 Simultaneous Noise and Gain Measurement

Fortunately, there is an elegant solution to the problem of measuring mixer noise temperature. The method presented below can be used to determine not only the mixer noise temperature, but its conversion loss, effective diode noise temperature, and even its output VSWR.

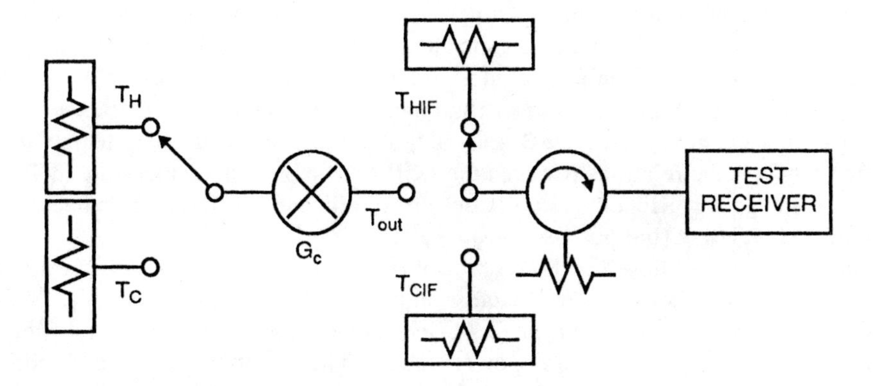

Figure 5.19 System for measuring gain and noise temperature simultaneously.

Figure 5.19 shows the measurement system. A hot or cold termination can be switched to the mixer's input. The test receiver operates at the IF frequency and can be switched to either of two input terminations, a hot or a cold, or to the mixer's IF output. An isolator is connected to the receiver's input to guarantee that the mixer sees the proper load impedance. It will also be useful for determining output VSWR via a minor modification of the system. The purpose of the test receiver is to measure the IF output temperature of the mixer. Therefore, in practice, it must be relatively narrowband and tunable over the entire IF passband.

Measurement of the conversion loss is based on a simple principle: the noise temperatures of the sources represent available power from a standard termination. Furthermore, the receiver measures the power delivered to a standard load. The ratio of these quantities is a transducer gain. Thus, when the noise temperature of the termination at the input of the mixer is changed by a quantity ΔT_{in}, the output power changes $\Delta T_{\text{out}} = G_c \Delta T_{\text{in}}$, and the conversion gain $G_c = \Delta T_{\text{out}} / \Delta T_{\text{in}}$. The input noise temperature of the system of Figure 5.21 can be easily changed by switching between the two terminations at T_H and T_C, so $\Delta T_{in} = T_H - T_C$. The resulting change in output noise temperature, ΔT_{out}, is measured by the test receiver. Thus,

$$G_c = \frac{\Delta T_{\text{out}}}{\Delta T_{\text{in}}} = \frac{T_{OH} - T_{OC}}{T_H - T_C} \tag{5.44}$$

where T_{OH} and T_{OC} are the IF output temperatures that exist when the hot and cold loads, respectively, are switched to the mixer's input. This process can be repeated at frequencies throughout the band to determine the mixer's passband.

When the conversion gain is known, the mixer's noise temperature can be determined: when the cold load is connected to the mixer input, the output noise temperature, which has been measured in order to determine the conversion gain, is $G_c(T_{\text{mxr}} + T_C)$, from which T_{mxr} is easily found. If the mixer has a significant image response, this quantity is the DSB noise temperature. The SSB noise temperature can be found via the relations between the DSB and SSB noise temperature in Table 5.1. If the image and RF conversion losses are significantly different, it is necessary to know their relative (but not absolute) levels.

Finally, we show how the IF test receiver is used to measure the mixer's IF output noise temperature. Two Y-factors must be measured: the first, Y_1, is the conventional Y-factor obtained as in (5.41) by switching between the hot and cold IF terminations and measuring the power ratios. The second, Y_2, is determined similarly, but by switching between the cold termination and the mixer output. The reader should be sufficiently adept at manipulating noise temperatures by now to verify that the output temperature T_{out} is

$$T_{out} = \frac{Y_2 T_{CIF} \, (Y_2 - 1) \, (T_{HIF} - Y_1 T_{CIF})}{Y_1 - 1} \tag{5.45}$$

where T_{HIF} and T_{CIF} are the IF hot- and cold-load temperatures, respectively. Y_2 is measured twice, first with the cold load connected to the mixer input and then with the hot load, and the difference in the two resulting values of T_{out} is $\Delta T_{out} = T_{OH} - T_{OC}$ in (5.44). Of course, it is necessary to measure Y_1 only once at each frequency of interest, provided that the system has good stability and the frequency to which the IF receiver is tuned can be accurately reset. This measurement, although accurate, is clearly laborious and is best implemented as an automatic system with a small computer to operate the switches, collect the data, and perform the calculations. Commercial noise-figure measuring equipment, which provides gain as well as noise figure and de-embeds the IF automatically, works in this manner.

A minor modification of the system of Figure 5.19 allows the mixer IF VSWR to be determined. The isolator termination is replaced by a switch and another pair of noise sources (in practice, a diode noise source is usually used). The noise from the selected source is incident upon the mixer's IF port. As before, the change in output temperature is measured by the test receiver when the switch is operated. This quantity is $\Delta T_{out,r}$, and the change in incident noise when the switch is operated, which is simply the difference between the source noise temperatures, is ΔT_{inc}. The magnitude of the IF port reflection coefficient $|\Gamma_{IF}|$, is

$$|\Gamma_{IF}| = \left(\frac{\Delta T_{out,r}}{\Delta T_{inc}} \right)^{1/2} \tag{5.46}$$

and the VSWR is

$$V_{IF} = \frac{1 + |\Gamma_{IF}|}{1 - |\Gamma_{IF}|} \tag{5.47}$$

Once the IF port VSWR is known, it is possible to correct the measured output noise temperature for reflected noise. If the temperature of the isolator termination is T, the reflected noise level is $|\Gamma_{IF}|^2 T$. This quantity can be subtracted from T_{out}.

As a final note, ΔT_{inc} can be measured by placing a short circuit on the receiver input so that $|\Gamma_{IF}| = 1$, measuring $\Delta T_{out,r}$, and solving (5.46) for T_{inc}.

5.5.3 Noise Sources

A wide variety of noise sources are available for measuring noise temperatures. Although a termination having a known temperature is the basic noise standard, other types of sources have the advantages of convenience or a more suitable output noise level.

For best accuracy, the hot and cold noise sources should be chosen to achieve a Y factor of 3 to 5 dB. If the Y factor is much smaller than this, the measurement will be unduly sensitive to small errors. The fractional error in noise temperature for a given Y-factor error, decreases with increasing Y, but does not improve substantially above 5 dB. Errors may, in fact, increase at higher Y factors, because the error in determining the Y factor itself may rise.

Very few types of noise sources have a variable output noise level. Therefore, in order to adjust the noise output to the desired level, to achieve a Y factor within the desired range, it is often necessary to select a high-temperature hot noise source and to reduce its noise temperature with an attenuator. The same approach that was employed in Section 5.1 to derive the noise temperature of an attenuator shows that the output noise temperature of an attenuated noise source T_{sa} is

$$T_{sa} = T_s G_a + T(1 - G_a) \tag{5.48}$$

where T_s is the source noise temperature, G_a is the attenuator gain (less than 1), and T is its physical temperature.

Another quantity sometimes used for expressing noise-source temperature is the *excess noise ratio* (ENR). This concept is most useful when the cold source is a room temperature load, and the hot source's noise temperature is considerably greater. The ENR is defined as

$$\text{ENR} = \frac{T_H - T_0}{T_0} \tag{5.49}$$

where T_0 is the standard noise temperature of 290K. The ENR is usually expressed in decibels. When the physical temperature of the attenuator is T_0, the ENR of the combination is equal to the ENR of the source, in decibels, minus the attenuation, in decibels.

The choice of a noise source for mixer noise measurements depends on a number of factors, including accuracy, noise temperature, and, not unimportantly, cost and convenience. An ideal noise source has a very low VSWR, an accurately known noise temperature that is constant over a wide bandwidth, a noise level that provides a Y factor of 3 to 5 dB, and is stable over long periods of time. Because a matched termination at a known temperature is a noise standard, one obvious choice for a noise source is a room-temperature termination. Another is a termination that has been heated or cooled to another temperature. Although heated terminations are sometimes encountered, the most commonly used noise standard is a termination cooled to the boiling point of liquid nitrogen. These devices are available commercially and can be home-built quite easily. It is important in building one that the termination remain well matched at the very low temperature of the liquid nitrogen, and that the thermal design be such as to guarantee that the load is effectively cooled to the liquid nitrogen temperature. Even so, losses in the interconnecting coaxial lines or waveguide raise the noise temperature at the output flange to 80K to 85K, rather than the 77K of the nitrogen. These losses must be measured and the noise temperature of the load calculated via (5.48). It is also important that water and ice be prevented from accumulating where they might affect the losses and thereby the noise temperature. It is not unusual to use a dry-nitrogen purge and a heated output flange to prevent condensation.

Cooled loads are adequate for measuring only relatively low noise temperatures; for measuring high noise temperatures (above a few hundred degrees), the combination of a room temperature and liquid-nitrogen-cooled loads may not give an adequately large Y factor. In this case, other combinations are preferable. An argon gas-discharge tube, mounted in a waveguide, is a excellent high-temperature noise source. Because its temperature depends upon the physics of the gas discharge, its noise temperature is quite constant with frequency, and most sources, regardless of frequency or other aspects of their design, have an ENR within 0.5 dB of 15.5 dB. For best accuracy, they are usually calibrated from room-temperature and liquid-nitrogen loads.

The most commonly used noise sources in the microwave industry are avalanche-diode sources. These use shot noise in a reverse-biased diode to achieve ENRs as high as 28 dB. The ENR of such sources is usually reduced to approximately 15 dB by an output attenuator; the attenuator also improves the source VSWR and acts as a room-temperature "cold" load when the diode is turned off. Diode noise sources are simple and convenient to use. Their primary disadvantages are that their noise temperatures are sensitive to a number of factors, such as bias current, diode parameters, and mechanical stability of the diode mount; these can be expected to change with time. Furthermore, the ENR of a diode source typically varies several tenths of a decibel over its useful bandwidth. Thus, a diode source must be calibrated over its frequency range, and must be recalibrated periodically to ensure accuracy.

5.5.4 Noise Measurement Errors

Noise measurements are often suspect (especially if they indicate very low noise temperatures!) because of the large number of subtle errors that can occur. Two sources of inaccuracy have already been discussed: high or low Y factors and large corrections for test-receiver or "second-stage" noise. Uncertainty in the noise source temperatures or Y-factor magnitude are other obvious sources of error. Other, more subtle problems, are as follows:

1. *Spurious outputs:* Because noise measurement fundamentally involves the measurement of power, receiver output power other than noise has the same effect on the measurements as noise. Therefore, the power in any internally generated spurious signals that appear in the receiver's output passband increase the output power and are interpreted in the measurement as noise. These spurious outputs often have greater power levels than the noise, so they can upset the measurement severely. They can also upset noise measurements even when relatively small, at levels where they may be difficult to find with a spectrum analyzer.
2. *Noise source VSWR*: In Section 5.1, we noted that the noise temperature is a function of the source impedance. Thus, for the noise measurement to be valid, the noise source used in testing must have the same impedance as the source that the mixer will have when used in a system. This source is almost always a standard termination. If the noise source's VSWR is poor, the measurement may be invalid. Moreover, because the mixer's gain also depends on source impedance, its gain may change when its input is switched between two imperfect noise sources. This gain change will figure directly into the Y factor, causing either an excessively high or low measurement.
3. *Reflected LO power*: Even if the source VSWR is good at the RF frequency, it may be poor at the LO frequency. If the receiver has a mixer front end and no RF amplifier, there may be appreciable LO leakage from the RF port. This may be reflected back into the mixer by the source, changing the LO level. If the hot and cold sources have different LO reflection coefficients, the LO level may change when the sources are switched, changing the gain and, therefore, the apparent noise temperature of the mixer.

4. *IF VSWR*: If the noise performance of a mixer or a single component is to be determined, the input VSWR of the test receiver must be kept low. Using an isolator at the input of the test receiver will eliminate a number of potential problems. Among these are gain changes due to a poor load VSWR and the possibility of spurious signals caused by LO leakage from the test receiver into the mixer.

5.6 REFERENCES

[1] IRE Subcommittee 7.9 on Noise, "Description of Noise Performance of Receiving Systems," *Proc. IRE*, vol.51, 1963, p. 436.

[2] M. E. Tiuri, "Radio Telescope Receivers," in J. D.Kraus, *Radio Astronomy*, McGraw-Hill, New-York, 1966.

[3] R. M. Gagliardi, *Introduction to Communications Engineering*, John Wiley and Sons, New York, 1978.

[4] D. Scherer, "Today's Lesson: Learn About Low-Noise Design," *Microwaves*, vol 18, no. 4, 1979, p. 116.

[5] S. A. Maas, *Nonlinear Microwave Circuits*, Artech House, Norwood, MA, 1988.

[6] F.C. McVay, "Don't Guess the Spurious Level," *Electronic Design*, Feb. 1, 1967, p. 70.

[7] S. E. Wilson, "Evaluate the Distortion of Modular Cascades," *Microwaves*, vol. 20, no. 3, 1981, p. 67.

[8] P. Penfield, "Wave Representation of Amplifier Noise," *IRE Trans Circuit Theory*, vol. CT-9, 1962, p. 84.

[9] P. Will, "Reactive Loads: the Big Mixer Menace," *Microwaves*, vol. 10, no. 4, 1971, p. 38.

[10] S. Weinreb, M. Balister, S. Maas, and P. Napier, "Multiband Low-Noise Receivers for the Very Large Array," *IEEE Trans. Microwave Theory Tech.*, vol. MTT-25, 1977, p. 243.

Chapter 6
Single-Diode Mixers

Although single-diode mixers are rarely used at frequencies below the millimeter-wave region, they are fundamental to the design of all mixers, especially balanced or multiple-diode mixers. The reason for this mildly paradoxical situation is that any balanced mixer can be reduced to an equivalent single-diode mixer by suitable scaling the RF and IF source and load impedances, the LO level, and by changing the embedding impedances at certain mixing frequencies. Single-diode mixers are also important in their own right, since virtually all high-performance millimeter-wave mixers are single-diode designs.

This chapter is concerned primarily with the design of millimeter-wave mixers using waveguide-mounted chip diodes; these are the most common single-diode mixers. The design of such mixers requires careful selection of the RF, LO, and IF embedding impedances; the diode; and dc bias. We make no apologies for the lower emphasis on low-frequency mixers; the design principles are the same and can be applied in much the same way, regardless of frequency, structure, or transmission medium. The only difference is that, in lower-frequency microwave mixers, a packaged diode may be used, and the package parasitics must be determined and included in the equivalent circuit.

The thrust of this chapter differs from that of Chapter 4 in that it is concerned with the practical design of mixers. The mixer theory developed earlier is fundamentally a method of analysis and does not lead directly to a design methodology. This chapter introduces a design approach that does not rely on heavy and sometimes abstruse theory, replacing it with reasonable approximations. Many of these approximations may at first seem very rough. Nevertheless, their judicious use results in the design of a mixer having good performance. The more serious designer may choose to use this procedure only as a starting point and to optimize the design via a computer program based on the theory in Chapter 4.

6.1 BASIC APPROACH

6.1.1 Design Rationale

The following design approach predates the development of much of the theoretical material in Chapter 4. Although the design process is approximate, it usually results in good performance; however, it must be implemented sensibly and margin must be left in the design to compensate, via tuning, for the inaccuracies inherent in the approximations. The source of the material for this chapter is primarily papers by Maas [1] and Kerr [2].

The design procedure involves matching to the pumped diode at the LO, RF, and IF frequencies. Therefore, the most immediate problem in designing a diode mixer is to determine the diode's input and output impedances. Although the diode is a nonlinear device, and its junction resistance and capacitance vary with time when the LO voltage is applied, it nevertheless has a definable input impedance at each mixing frequency equal to the ratio of the terminal voltage and current components at that mixing frequency.

The time-varying elements in the diode's equivalent circuit are the junction resistance and capacitance. Of these, the junction resistance has the greatest variation, changing from nearly a short to an open circuit over the LO cycle. The capacitance, however, rarely varies over a range greater than two or three to one. It still affects the conversion performance of the mixer, but for the purposes of designing a matching circuit, its variation is small enough that the capacitance can be treated as a constant equal to its time-average value. The average capacitance must, of course, be estimated unless a large-signal analysis of the pumped diode is performed. Once the average capacitance is estimated, it can be absorbed into the matching circuits. The problem then is to match the LO, RF, and IF to the resistive junction.

The diode junction resistance varies widely, from nearly zero to an open circuit. It is not valid to assume that the diode's effective input resistance is simply the time-averaged junction resistance (the average junction resistance is, of course, infinite, because it is infinite over part of the LO cycle). Because the junction resistance converts an RF input signal to an IF output, it can be treated as a two-port, described by an impedance or admittance matrix, where the port currents and voltages are at different frequencies rather than physically separate terminals. In this case, the input and output impedances depend on all four matrix parameters, as well as the input and output terminating impedances themselves. In Chapter 4, we derived these matrices; they consist largely of the Fourier-series coefficients of the conductance waveform, which, in turn, depend on the diode parameters, LO level, and dc bias.

The RF source and IF load impedances that match the diode are the simultaneous-conjugate-match impedances of the two-port. They are generally not known unless the full analysis of Chapter 4, the thing we would like to avoid, is performed. It is possible, however, to estimate those impedances, and to adjust the two-port parameters (via LO level and dc bias) to optimize the input and output match. A good design will also have some provision for empirical adjustment of those impedances.

LO matching is another consideration. The fundamental-frequency LO source impedance has little effect on the conversion loss beyond its obvious effect on power transfer into the diode. Fortunately, the optimum LO source impedance is usually close to the optimum RF source impedance; if the LO frequency is close to the RF, matching the diode at the RF frequency automatically provides an adequate LO match. If the RF and LO frequencies cannot be matched simultaneously, the LO match is usually sacrificed in favor of the RF.

Figure 6.1 illustrates the problem of matching to the diode. The network must match the RF source and IF load impedances to the diode junction's input and output resistances. These resistances are invariably different at the RF and IF frequencies. If image-enhancement is desired (Section 4.6.1), the matching circuit must present a reactive load to the diode at the image frequency. It must also isolate the RF, IF, and LO from each other; clearly, it must provide a very complex filtering function. Although this appears to be a difficult set of requirements to meet simultaneously, the structure of the mixer often provides many of them more or less automatically. For example, in a mixer having a waveguide input and coaxial output, the IF-to-RF isolation is very high as long as the IF frequency is well below the waveguide's cutoff frequency. Because their frequencies are so far apart, the RF and LO are easily separated from the IF. (These points will be clarified in Section 6.2 by design examples.)

The central problem of mixer design is, therefore, to find the diode's effective junction capacitance and input and output resistances. The theoretical work described earlier can be used to make some valid generalization about these. For example, mixer analyses often show that the effective (time-average) junction capacitance is approximately equal to the zero-voltage junction capacitance. This value is remarkably constant with dc-bias voltage and embedding impedances, but not with LO level. The RF input resistance of the junction is usually in the range of 40Ω to 100Ω, and is near the lower end of this range if the diode is biased near its *I*/*V* knee. The IF impedance is sensitive to many of the mixer's operating parameters, but for any set of operating conditions that gives rise to good conversion performance, its real part is nearly twice the real part of the RF resistance. In mixers having a significant image response, it is usually relatively low, near 1.2 to 1.5 times

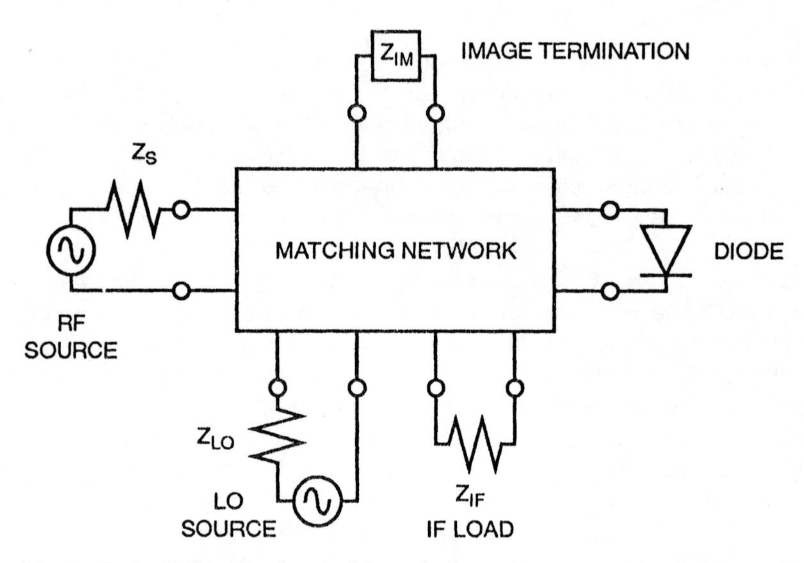

Figure 6.1 A single-diode mixer's matching circuit must separate the important frequency components and present the appropriate impedance to the diode at each mixing frequency.

the RF resistance. In some cases, especially with image-enhanced mixers, it may be greater than twice the RF resistance, but rarely so high that this approximation results in a poor IF match. If the diode is well matched at the RF frequency and the IF frequency is low, the imaginary part of the IF output impedance is usually relatively small.

Some authors have defined a conversion-loss degradation factor, which accounts for the loss in the diode's series resistance. This loss occurs because the diode junction is connected to the embedding network by an RC filter consisting of the diode's series resistance and junction capacitance. The degradation factor accounts for only the loss in R_s at the RF frequency, not the losses at other mixing frequencies.

The conversion-loss degradation factor δ is

$$\delta = 1 + \frac{R_s}{Z_s} + \frac{Z_s f_{\mathrm{RF}}^2}{R_s f_c^2} \tag{6.1}$$

where Z_s is the diode junction's (real) RF input impedance, R_s is the series resistance, and f_c is the cutoff frequency:

$$f_c = \frac{1}{2\pi R_s C_j} \tag{6.2}$$

In contrast to our earlier definition of cutoff frequency, R_s in Equation (6.2) is the series resistance at the RF frequency f_{RF} and C_j is the time-averaged junction capacitance. δ is usually negligible if $f_c << f_{RF}$. By replacing f_{RF} with f_{IF} in (6.1), the loss in R_s at the IF frequency can be accounted for. This IF loss, however, is invariably quite small.

6.1.2 Diode Selection

In Chapter 2 we examined several of the most commonly used types of mixer diodes. The selection of one of these diodes for use in a mixer design involves a trade-off between cost, convenience, ease of manufacture, and performance. The best performance is invariably achieved with some type of dot-matrix diode, usually a Mott structure for cryogenic operation or a conventional structure for room-temperature operation. Next best is a millimeter-wave beam-lead diode, which is usually less difficult to assemble; such diodes are still relatively delicate, however, and require a low-loss substrate, usually fused silica, for mounting. Conventional beam-lead diodes are cheaper and slightly more rugged, but have poorer performance. Finally, packaged mixer diodes are comparable to beam-lead diodes in performance and are simpler to handle, but are somewhat more expensive. At frequencies below 12 GHz, a wide variety of silicon diodes in epoxy packages with ribbon leads are available. Their performance is good enough for most prosaic applications, and they are very inexpensive. Point-contact diodes should be considered obsolete.

The most basic choice in selecting a diode for a mixer design is between silicon and GaAs. GaAs diodes provide superior performance, and they are preferred for high-performance or high-frequency applications. Silicon is usually the choice for moderate-frequency, doubly balanced mixers produced in large numbers, where cost is the primary concern. Silicon diodes are generally available with lower barrier heights than GaAs, and in unbiased applications may require less LO power and achieve a better LO-port VSWR. Single-diode mixers are usually designed for millimeter-wave, high-performance applications, where the extra cost of a GaAs device is easily justifiable.

For applications where low conversion loss is most important, GaAs dot-matrix diodes are the clear choice. A GaAs diode has a higher cutoff frequency than a comparable silicon device, has higher reverse-breakdown voltage, is harder to damage by voltage transients or static-electric discharge, and its noise temperature minimum, as a function of LO power, is much broader. Such diodes are available with a wide variety of junction diameters; the larger diodes have low series resistance and high junction capacitance, and the smaller ones have lower junction capacitance and greater series resistance. Smaller diodes generally have higher cutoff frequencies. There is usually a broad, optimum junction diameter for any application; if the junction is too large, the cutoff frequency will be lower than necessary and C_{j0} will be too large to allow effective matching. However, if the junction is too small, C_{j0} will be small and matching may be easy, but series resistance will be larger than necessary and will cause the conversion loss to be high. For frequencies up to approximately 200 GHz, the optimum diode is usually one that is as large as possible, consistent with matching requisites. Wrixon [3] provides an excellent examination of the problem of diode selection.

Because they are much more convenient than bare chips, packaged diodes are often attractive. However, packaged diodes always represent a trade-off of performance for convenience, because package parasitics always limit the matching possibilities to some degree. Variations in the parasitic inductance and capacitance of the package, caused by normal manufacturing variations, may have a significant effect on the mixer's performance. Packaged diodes usually use ribbon connections to the anode; these ribbons are often long and have high inductance. Furthermore, such diodes must have a diameter of at least 6 to 8 μm in order to make a reliable connection; 12 to 20 μm is more common. These anodes are larger than optimum at frequencies above 15 GHz. In a whisker-contacted diode, the whisker inductance, which is part of the matching circuit, can be optimized by changing the whisker's length and orientation; no such freedom is available in packaged diodes.

A compromise between packaged diodes and chips, which often yields adequate performance, is a low-parasitic, millimeter-wave beam-lead diode. The best of these are almost as good as dot-matrix diodes (Section 2.4.5). They are often delicate, however, and cannot be used in the same way as dot-matrix diodes: they must be mounted on a substrate. This introduces additional complexities in the mixer design.

A final consideration is the anode-junction metallization. If the diode will be subjected to high temperatures in the bonding process, it is wise to avoid platinum anodes, which degrade rapidly at high temperatures. For gold-tin or gold-germanium eutectic bonding, titanium anodes are best. If diodes having platinum anodes must be used, they can be attached with indium-alloy solders or silver-filled epoxies. Because silver-filled epoxies can be lossy at high frequencies, a minimal amount should be used, and the material should not be allowed to cover the sides of the chip or the mounting surface.

6.1.3 Design Methodology

Figure 6.2 shows the equivalent circuit of a single-diode mixer. The RF and LO use the same input port; in practice, the RF and LO are applied via a filter diplexer. Because they use the same input, the RF and LO share the same matching circuit. The IF matching circuit is physically separate from the RF/LO circuit and matches the diode's output impedance to the IF load. An image filter, designed to reject the image and pass the RF and LO frequencies, may be used at the input to provide image enhancement. The diode's image-frequency termination is adjusted by varying the distance from the filter to the diode; this can be accomplished easily in waveguide by the use of metal spacers. For wide image-enhancement bandwidths, the filter's short-circuit plane must be as close as possible to the diode. The filters should be noninteracting; the RF/LO matching circuit must not affect the IF matching, and the IF circuit should not affect the RF matching.

The design process is as follows:

1. Estimate the effective value of the diode's junction capacitance. C_{j0} is a good guess, although the effective capacitance can be as low as 0.5 C_{j0}. Determine values of the diode's package reactances, whisker inductance, or other parasitic inductances and capacitances.
2. Design the RF/LO matching circuit to match the diode's junction resistance, 40Ω to 100Ω, to the RF source impedance. The LO matching is of secondary importance. The RF circuit must not affect the IF matching (i.e., it must present an open circuit to the diode at the IF frequency for the equivalent circuit in Figure 6.2). The diode's package parasitics must be absorbed into the matching circuit.

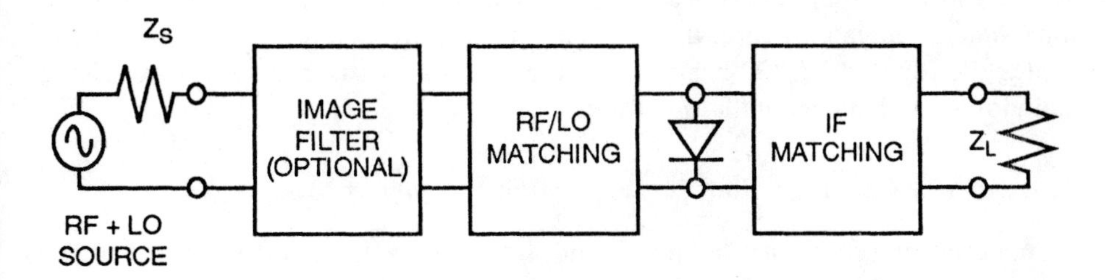

Figure 6.2 Equivalent circuit of a single-diode mixer.

3. If a match to the expected RF junction impedance cannot be achieved, it may be necessary to select a different diode, with a larger or smaller junction area, and thus different C_{j0}.
4. Design the IF matching circuit to match the diode's IF impedance, approximately twice the RF resistance, to the IF load. The junction capacitance and package parasitics are rarely significant at the IF frequency. It is important, however, that the circuit have good RF and LO rejection. The IF circuit must not upset the RF matching.
5. Once the mixer is built, adjust the dc bias and LO level to achieve a suitable match to the diode and good conversion performance. It may be necessary to perform some tuning of the matching circuits to optimize the performance.
6. If image enhancement is desired, an image filter must be produced. Rejection across the image band of 15 dB is adequate for enhancement purposes. The image filter must have very low RF insertion loss, or the improvement in conversion loss due to image enhancement (at most 1 dB) will be offset by RF losses. Image enhancement is optimized empirically, after the mixer is built, by varying the distance from the diode to the filter.
7. It is wise to check the design by performing an analysis; the techniques of Chapter 4 should be used. If the embedding impedances at the higher-order mixing frequencies are not known and cannot be calculated accurately, they may be set to zero. The results in this case are necessarily approximate, but they are likely to be surprisingly accurate.

The detailed design of the matching circuits depends on the type of mixer to be produced. In a microstrip mixer, which may be used as one half of a balanced mixer, the RF matching circuit may consist of nothing more than a judicious choice of the diode junction capacitance and bond-wire inductance. For a millimeter-wave mixer, a careful analysis of the diode mounting structure is necessary, because it and the diode parasitics comprise most of the matching network. In a quasi-optical submillimeter-wave structure, the RF matching circuit includes the antenna, which couples the focused RF/LO beam to the diode. In most cases, however, the design methodology is fundamentally the same.

6.1.4 Special Considerations for Very-High-Frequency Mixers

The material presented in Chapters 2 and 4 and up to this point in this chapter is valid at least to 200 GHz. Between 200 and 300 GHz, the material is valid if some minor modifications are made; for example, the frequency sensitivity of the diode's series resistance, caused by skin effect, must be included. Above 300 GHz, and especially approaching 1 THz (1,000 GHz), new phenomena arise [4], [5], and these may dominate the mixer's performance. They may even, in some cases, completely reverse our earlier conclusions about mixer operation, which were based on lower-frequency considerations. They also show that diode selection is critical for such high-frequency applications; indeed, the diode in many cases must be designed specifically for the mixer.

An intuitive approach to optimizing the diode for high frequencies is to reduce the area of the junction. Modern fabrication techniques allow diode diameters smaller than 0.5 μm to be fabricated. Unfortunately, however, the diode's conductance is proportional to the total junction current, not current density. Therefore, to maintain the same conductance waveform, the junction's current density must increase as the diode diameter is reduced. The ability of the diode to sustain high junction-current densities is limited, so eventually the peak diode current, and hence the peak conductance, must decrease as junction diameter is reduced. As a result, the peak conductance no longer is nearly infinite, and conversion loss is increased accordingly. This phenomenon was first examined in a classic paper by McColl [6].

Another problem with small diodes is that they must be biased at a relatively high voltage, because the knee of the I/V characteristic moves to a higher voltage as anode area is reduced. Consequently, the peak forward LO voltage moves closer to the built-in voltage, and the junction-capacitance variation increases. As a result, the time-averaged junction capacitance does not decrease in proportion to the junction area; this increases both the conversion loss and the LO-power requirements. Finally, the high electric fields in the junction, necessary to achieve the high current densities, cause hot-electron noise to become significant. In small diodes, this may be the dominant noise source. In Section 6.2.5, we describe a mixer that uses an unusually large diode to avoid this problem.

For mixers operating below 300 GHz, the so-called dc cutoff frequency (based on dc series resistance and low-frequency (megahertz) measurements of the capacitance) was deemed to be a valid figure of merit, and the dominant effect in determining the series resistance was the substrate's bulk resistivity. We noted that skin effect in the diode could increase the series resistance at high frequencies by as much as a factor of two. In the terahertz range, however, several changes occur [4]. First, as might be expected, skin effect increases the series resistance significantly. Second, above the dielectric relaxation frequency of GaAs, much of the diode current in the epitaxial layer becomes displacement current. The dielectric relaxation frequency, ω_d, is

$$\omega_d = \frac{\sigma}{\varepsilon} \tag{6.3}$$

where σ is the conductivity of the GaAs and ε is its dielectric permittivity. Third, carrier inertia is significant above the scattering frequency,

$$\omega_s = \frac{q}{\mu m^*} \tag{6.4}$$

where q is the electron charge, m^* is the effective mass, and μ is the mobility. These combined effects lead to plasma oscillations, which increase the diode's series resistance and add a significant complex part at frequencies near the plasma frequency,

$$\omega_{pl} = \sqrt{\omega_s \omega_d} \tag{6.5}$$

In GaAs, ω_{pl} is approximately 3 THz. The dc series resistance is often insignificant compared to the increase in series resistance brought about by these phenomena; the real part of the series resistance of a 2-μm diode with a measured dc series resistance of 5Ω may be 80Ω at 4,000 GHz. The resulting conversion loss degradation factor δ may then be several decibels. It happens that, above 1,000 GHz, δ is very sensitive to diode diameter and is minimized for a specific diameter. In a 1,500 GHz mixer, this diameter is approximately 1 μm.

At terahertz frequencies, transit-time effects may become significant. At such high frequencies, the electron's transit time through the depletion layer is no longer small compared to the inverse of the RF frequency [5]. The forward-current degradation factor, caused by transit-time limitations, is given by van der Zeil [7]:

$$|D(j\omega)| = \frac{1 - 2\left(\frac{\omega}{\omega_{pl}}\right)\sin\left(\frac{\pi}{2}\frac{\omega}{\omega_{pl}}\right) + \left(\frac{\omega}{\omega_{pl}}\right)^2}{\left[1 - \left(\frac{\omega}{\omega_{pl}}\right)^2\right]^2} \tag{6.6}$$

The conversion loss varies as $|D(j\omega)|^4$; this factor may be very great at frequencies above 1,000 GHz.

High epitaxial doping concentrations reduce the complex series resistance at frequencies above 1 THz, and also reduce the transit-time degradation. The result is that heavy doping provides the best conversion loss. However, at low temperatures, heavily doped diodes have substantial tunneling current, so low-temperature noise performance may be poor. If one is tempted to improve the low-temperature performance by reducing the epilayer's doping density, hot-electron noise, which is not reduced by cooling, may become dominant. Consequently, at frequencies of several terahertz, it may happen that no advantage is achieved by cryogenic cooling.

Below 2 to 3 THz, there may be some advantage to cryogenic cooling if the epilayer doping density is judiciously chosen. It is necessary to trade off hot-electron noise effects, which dominate at low doping densities, with tunneling noise that occurs at high doping densities. These factors also affect the conversion-loss degradation factor through the diode diameter and series resistance. Furthermore, these considerations may dictate the use of a very small diode, which suffers from the current-density limitations described earlier.

6.2 SINGLE-DIODE MIXER DESIGNS

6.2.1 Untuned Mixer

The most commonly used mixers at microwave frequencies are balanced mixers having either two or four diodes. These circuits, although seemingly complicated, can always be reduced, for the purposes of analysis, to an equivalent single-diode circuit; thus, the analytical techniques we have examined so far are directly applicable to balanced mixers. The use of multiple diodes has many benefits, such as RF-to-LO isolation and rejection of LO noise. The RF and LO are generally applied through some type of microwave hybrid coupler, and because of the complexity of the circuits, it is often impossible to use any type of RF, LO, or IF matching network. Hence, the mixer is completely untuned, and the only real means to optimize the mixer's performance is via diode selection. Even then, that choice is usually strongly dictated by cost.

There is another—probably the best—reason for using an untuned diode in a balanced mixer, even in one where a matching circuit is practical. The most important property of a balanced mixer is usually its rejection of even-order spurious responses and intermodulation products. In order to maximize this rejection, the balance must be very good. If a matching circuit is used ahead of each diode, differences between the circuits caused by manufacturing tolerances or manual tuning may unbalance the mixer. Although conversion efficiency may be improved by a decibel or two, the reduction in IM rejection may be several decibels.

This is often an unacceptable trade-off.

In an untuned balanced mixer it must be made certain that the good balance obtained by dispensing with matching circuits is not lost through the use of dissimilar diodes. If the diodes in a balanced mixer are not identical, some degree of imbalance will result. Fortunately, it is possible to obtain diodes that have several devices fabricated on the same chip; these are likely to be very well matched.

Therefore, it is worthwhile to know what level of performance can be attained at microwave frequencies in an unmatched mixer having a prosaic silicon Schottky diode. We can calculate this through the use of the theory in Chapter 4. A single-diode mixer was analyzed at an LO frequency of 10 GHz and an IF frequency of 40 MHz. The RF, LO, and IF terminating impedances were all 50Ω, real. The diode had 10Ω series resistance, 0.1 pF C_{j0}, and an ideality factor of 1.25. The current parameter I_0 was 2.8×10^{-9}A, which gives 1 mA of junction current at 0.4V forward bias. The dc bias is zero.

The calculated performance is as follows:

- LO power: 14 mW
- DC junction current: 13.7 mA
- LO input impedance: 49 – j19
- RF input impedance: 53 – j15
- IF output impedance: 59 – j0.1
- Conversion loss: 5.8 dB

Although these results are strictly valid only at 10 GHz, it is fair to expect that they would be virtually the same at lower frequencies; only the LO and RF input impedances would be reduced. It is remarkable to note how close all the impedances are to 50Ω. The highest port VSWR of this mixer is below 1.5, and the conversion loss of 5.8 dB (an acceptable value for many purposes) is consistent with that of commercial mixers.

6.2.2 Waveguide Mixer

Figure 6.3 shows a cross section of a millimeter-wave waveguide mixer. In this structure, the unpackaged diode is mounted transversely in a reduced-height waveguide and is contacted by a whisker. A movable waveguide short, or *backshort*, is placed behind the diode to help match it to the waveguide. The IF circuit is realized in coaxial transmission line, and the diode chip is mounted on the end of the coaxial structure. The diode is biased through the IF port. An image-enhancement filter is not shown, but can be included between the waveguide taper and the diode mounting structure.

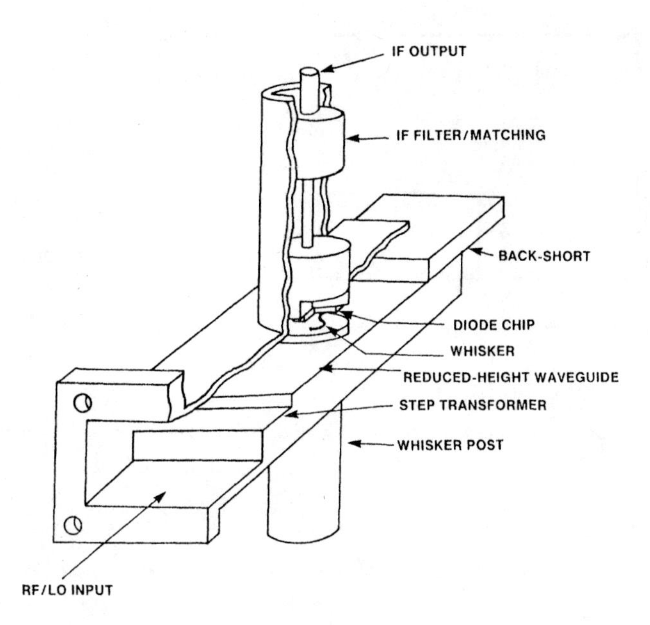

Figure 6.3 Cutaway view of a single-diode waveguide mixer. The diode and IF filter are partially removed to show the whisker.

The diode mounting structure is shown in detail in Figure 6.4, which is a view down the axis of the waveguide. The diode is mounted on the end of the coaxial IF filter, which protrudes slightly into the waveguide. The whisker post is press-fit into the waveguide mount after the whisker is attached and formed to the correct shape. The waveguide height is reduced by a factor of approximately four compared to the standard waveguide. The reduced height and the protrusion of the mounting and whisker posts into the waveguide allow the whisker to be made very short, and its inductance to be minimized.

In order to design the RF matching circuit, which in this case consists of the backshort, whisker inductance, diode junction capacitance, and diode mounting waveguide, it is necessary to have a circuit model for the mounting structure. A workable (if approximate) model is shown in Figure 6.5. Z_g is the characteristic impedance of the waveguide, according to the power-voltage definition:

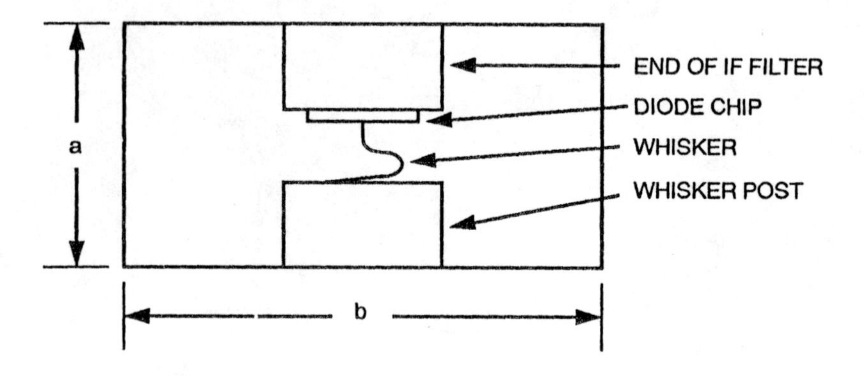

Figure 6.4 Diode-mount geometry of the waveguide mixer. The view is down the axis of the waveguide.

$$Z_g = 754 \frac{b}{a} \frac{\lambda_g}{\lambda_0} \tag{6.7}$$

where b and a are the height and width of the waveguide, respectively, and λ_g/λ_0, the ratio of the guide wavelength to the free-space wavelength, is

$$\frac{\lambda_g}{\lambda_0} = \frac{1}{\sqrt{1 - (f_c/f)^2}} \tag{6.8}$$

where f_c is the waveguide cutoff frequency. C_p is the capacitance of the post ends; for small insertions into the waveguide, it is approximately equal to 1.5 times the parallel plate capacitance of the ends. The capacitance is

$$C_p = \frac{1.5\,\varepsilon_0 A}{h} \tag{6.9}$$

where $\varepsilon_0 = 8.854 \times 10^{-14}$ F cm^{-1}, A is the area of the ends of the posts in centimeters squared, and h is their separation in centimeters. If the insertion is greater than about 25% of the waveguide height, the effect of the posts is much more difficult to model.

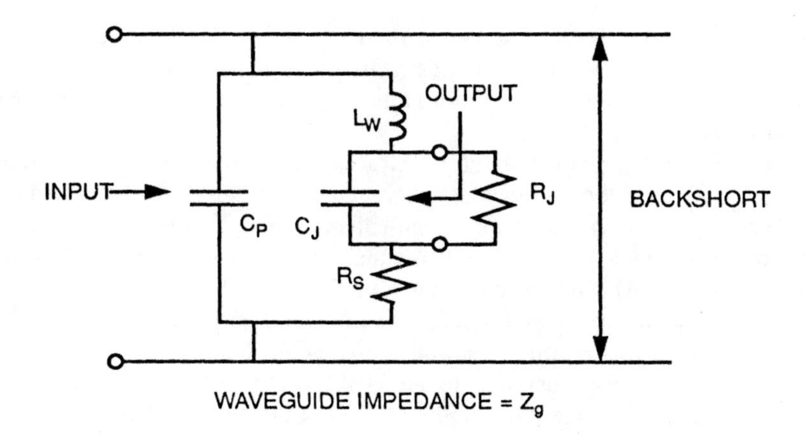

Figure 6.5 Equivalent circuit of the diode mount. In a waveguide mixer, the diode mounting structure comprises most of the RF/LO matching circuit.

The work of Eisenhart and Khan [8] can be used for this purpose. The whisker inductance L_w is given by the expression

$$L_w = 2 \times 10^{-9}\, l\, \ln\left(\frac{2a}{\pi r}\right) \tag{6.10}$$

where l is the length of the whisker wire in centimeters, a is the waveguide width, and r is its radius. Equation (6.10) is a modification of the well-known Schelkunoff formula for the inductance of a straight wire across a waveguide. Generally, the whisker is not straight, but is curved to absorb the stress of the contact. Therefore, it is wise to increase the value of L_w in (6.10) by 20% to 40% depending on the length and shape of the curve. Other quantities in Figure 6.5 are C_j, the time-average junction capacitance, and R_s, the series resistance. R_j is the diode's junction resistance at the RF frequency. The input impedance of the IF filter at the RF frequency, which is in series with the diode, is not shown; in a good design, it is low enough to be negligible.

An analysis of the circuit in Figure 6.5 shows that it is not always possible to match the structure to any values of C_j and L_w simply by adjusting the backshort reactance, even if R_j is not constrained to have a specific value. When a match is possible, two values of backshort reactance will match the structure, and these result in different values of R_j. Successful matching of a waveguide mixer generally requires that the normalized susceptance of the junction capacitance, $\omega C_j Z_g$, be less than unity, as well as the inductive whisker reactance $\omega L_w/Z_g$. Acceptable matching conditions are best found through the use of a general-purpose, microwave circuit-analysis program.

Successful matching of the waveguide structure depends strongly on the correct choice of Z_g. Z_g can be adjusted to achieve the desired R_j by varying the waveguide's height and width. If an image-enhanced mixer is to be designed, it is important to keep the waveguide cutoff frequency low enough so that is well below the image frequency. The image can then propagate with minimal dispersion, and the change in phase of the image reflection coefficient over the image band, seen by the diode junction, will be minimized, and image-enhancement bandwidth will be maximized. It is generally desirable to select the waveguide dimensions to allow only a single mode to propagate throughout the RF, LO, and image frequency range.

The IF matching circuit consists of a filter to reject the LO, RF, and image frequencies, and some additional matching components to realize the required IF load impedance. The filter must be designed to have adequate rejection at these frequencies. Perhaps even more important, it must have a near-zero input impedance (measured at the diode end) so that the diode is electrically connected to the waveguide's top wall; this low input impedance prevents the RF and LO matching from being upset by the IF filter. Usually, the IF is well below the waveguide cutoff frequency, so the waveguide matching elements have no effect upon the IF match; however, on occasion a high IF frequency can generate evanescent modes in the waveguide that are affected by the backshort position or other tuning elements. Similarly, the junction capacitance and whisker inductance are usually negligible at the IF frequency.

It is important that the first two or three sections of the IF filter be cut off at the RF and LO frequencies to all but TEM modes; the effect of the IF filter on non-TEM modes is likely to be very different from the effect on TEM. The cutoff frequency to the lowest non-TEM mode in coaxial line is given approximately by

$$f_c = \frac{7.51}{d + D} \tag{6.11}$$

where D and d are the outer and inner diameters of the coaxial line, respectively in inches, and f_c is in gigahertz. If the filter is realized in microstrip or some other medium, it is usually adequate to design the size of the cavity enclosing it to be cut off to all waveguide modes (all cross-sectional dimensions must be less than one-half wavelength at the RF, LO, and image frequencies).

Design Example of a Waveguide Mixer

A 60-GHz image-enhanced mixer will be designed. The IF frequency is relatively high, 11 GHz (if it were less than approximately 10% of the RF frequency, the image filter might be impossible to realize). The required bandwidth is 1 GHz, which is uncomfortably broad for an image-enhanced mixer. The fixed LO frequency is, therefore, 49.0 GHz and the center of the image band is 38 GHz. A high-quality 3-μm GaAs Schottky diode is available and has the parameters $C_{j0} = 0.02$ pF, $R_s = 4\Omega$, $\eta = 1.15$. The chip dimensions are 0.02 inch square by 0.004 inch thick. It is desirable to use this diode if possible.

The design process begins with the diode mount. In order to pass the image frequency, the waveguide cutoff frequency must be as low as possible. However, the waveguide should be designed to be single-moded at the highest RF frequency, 60.5 GHz. This means that the waveguide cutoff must be above 30.25 GHz. We choose a cutoff frequency of 32 GHz to allow for manufacturing tolerances. This corresponds to a waveguide width of 0.185 inch ($\lambda_0/2$ at 32 GHz).

As a starting point, a waveguide height of 0.030 inch is chosen. The resulting waveguide impedance Z_g is, from (6.7), 144Ω, usually a good value. The mounting-post and whisker-post protrusion into the waveguide are each set at 0.01 inch. With the chip thickness of 0.004 inch, the gap between the surface of the whisker post and the chip is 0.006 inch. The mount dimensions are shown in Figure 6.6. Equation (6.10) gives 0.17 nH for the whisker, although to account for the curve in the whisker and to allow some room for tuning, the inductance will be increased by an

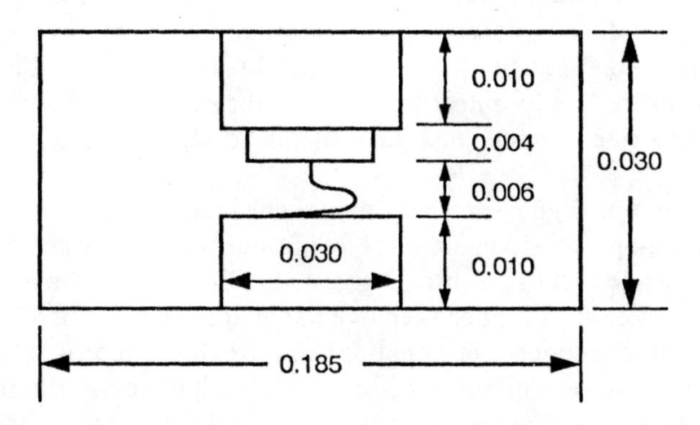

Figure 6.6 Dimensions of the diode mount in the 60-GHz mixer example.

estimated 40% to 0.24 nH. The whisker reactance at the RF frequency X_L is 91Ω and, normalized to the waveguide impedance, $X_L/Z_g = 0.63$. Similarly, we estimate 70% of C_{j0} for the effective junction capacitance; its normalized susceptance $Y_c Z_g$ is 0.76. With some help from a computer circuit-analysis program, we find that a normalized backshort reactance of −0.77 gives $R_j = 60\Omega$, a good starting value.

The parallel-plate capacitance between the end of the whisker post and the surface of the diode chip is 0.026 pF. Including 50% for fringing fields, we find that this capacitance is approximately 0.04 pF. The susceptance of the combined backshort and capacitance must present a normalized susceptance of −0.87, or an inductive reactance of 1.15. This corresponds to an electrical distance from the diode to the short of 49 deg.

The circuit-analysis program used to design the diode mount verified that the circuit has adequate bandwidth at the RF frequency, and a reasonable match at the LO frequency. The source impedance Z_g was then replaced by a length of waveguide and a short circuit, representing the shorting plane of the image filter, and, after R_j was removed, the image-embedding impedance at its terminals was calculated. The program verified that the phase of the image-frequency reflection coefficient varied only moderately across the image band (we do not know at this point whether the diode requires a short- or open-circuit image; both cases were examined).

The diode mount is now designed and matched for $R_j = 60\Omega$. If the results were not acceptable, the process could have been repeated with other waveguide and mounting-post dimensions. If acceptable results still could not be obtained, it might have been necessary to use a different diode.

The rest of the input circuit is relatively simple. Because the standard WR-15 waveguide (0.148 × 0.074-inch interior dimensions) has a cutoff frequency of 39.9 GHz, above the highest image frequency of 38.5 GHz, the input waveguide itself is the image filter. All that is necessary is to design a quarter-wave transformer to match the input waveguide to the reduced-height waveguide at the RF and LO frequencies. To ensure that its cutoff frequency is the same as that of the waveguide, the sections of the transformer must have the same width as the standard WR-15 waveguide.

It is occasionally suggested that image enhancement can be achieved by adjusting the diode-mount waveguide's cutoff frequency to a value between the LO and image frequencies, thus terminating the diode in a reactance at the image frequency. Although this will indeed remove the image response, it also removes the freedom to adjust the image termination empirically. Empirical adjustment is invariably necessary for optimizing an image-enhanced mixer. Furthermore, we saw in Section 4.6.1 that the optimum termination is usually closer to a short circuit than an open circuit; however, a cutoff waveguide behaves more like a large inductive reactance. Thus, attempting to achieve image enhancement in this manner is usually unsuccessful.

The next task is to design the IF filter. Because the RF embedding impedance is 60Ω, the IF load will be approximately 120Ω. It will be found that this is too high to be realized in practice in coaxial line, but 80Ω to 100Ω is probably achievable. There are many possible approaches to the design of the structure; one of the author's favorites was chosen. The filter is a series of high-impedance and low-impedance coaxial sections, which are each one-quarter wavelength long at the center of the rejection band. These present the requisite short circuit at the diode end of the filter, and rejection can be made arbitrarily great by increasing the number of sections. At low frequencies, where the sections are only a few degrees long, the filter operates as a transmission line having a characteristic impedance approximately equal to the geometric mean of the section impedances:

$$Z_c \approx \sqrt{Z_h Z_l} \tag{6.12}$$

If the overall filter is an odd number of quarter wavelengths long at the IF frequency, the entire structure can be used as a section in a multisection quarter-wave IF transformer. It is necessary only to design the remaining sections of the transformer and to optimize the entire structure on the computer.

It is usually a bad idea to try to realize this filter as a conventional stepped-impedance coaxial low-pass filter. The first difficulty is that such filters are not guaranteed to have a low rejection-band input impedance, and can therefore upset the RF/LO matching. The second is that creating an IF termination greater than 50Ω invariably requires the use of high-impedance sections of the filter that are physically unrealizable.

For the initial design of the IF filter, Z_h and Z_l are chosen as high and low, respectively, as possible, consistent with practical limits in realizing the transmission lines. Z_h is chosen to be 100Ω and Z_l is 25Ω, giving, from (6.12), $Z_c = 50\Omega$. The sections of the filter are made 0.25 wavelength long at the center of the rejection band (which is, of course, equal to the LO frequency), 49 GHz. These sections are 20 deg long at the 11-GHz IF center frequency, so four sections plus a 10-deg length of 50Ω line are a total of 90 deg. If the diode is to see 90Ω at the IF frequency, the opposite end of the filter must be terminated in $Z_c^2/90 = 27\Omega$. The second section of the transformer, which matches this impedance to the standard 50Ω output, must have a characteristic impedance of 37Ω.

The initial IF design is shown in Figure 6.7(a). Because of the approximations involved, the structure must be computer-optimized for best performance. The optimized IF circuit is shown in Figure 6.7(b). The IF load presented to the diode is

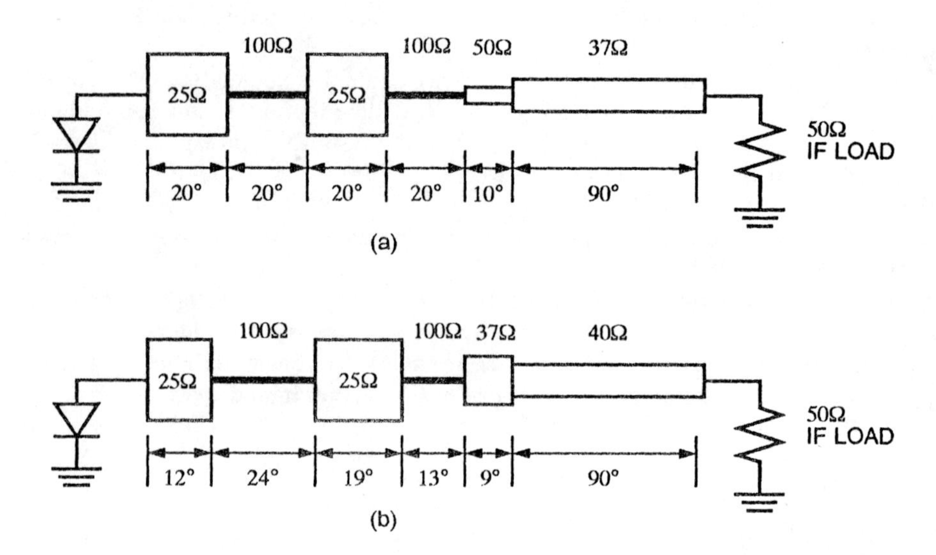

Figure 6.7 (a) Initial and (b) optimized designs of the IF filter for the 60-GHz waveguide mixer.

between 83 and 85Ω resistive, with a small, negligible reactive part. Over the RF, LO, and image bands, the real part of the input impedance is less than 1Ω; the reactive part is very small (less than 27Ω), and has negligible effect on the RF/LO matching.

This mixer was fabricated and optimized with very little difficulty. Conversion loss was slightly greater than 5 dB when the mixer was first turned on. The position of the diode in the vertical plane of the waveguide and the whisker length were optimized empirically, and the image termination was optimized by inserting spacers between the diode mount and the waveguide transformer/image filter. For test purposes, the LO was injected via a coupler; in practice, a filter diplexer would be used. The optimized mixer exhibited conversion loss between 3.8 and 4.2 dB across the 1-GHz passband, remarkably low loss for a 60-GHz mixer.

The crudeness of some of the approximations in this design may be disturbing to some. These approximations are nevertheless acceptable, because the goal of this procedure is not to produce an exact design (probably a hopeless goal in millimeter-wave technology), but to produce an approximate design that can be tuned empirically to obtain acceptable performance. In this regard, it is thoroughly successful. Furthermore, one is still free to use this as the beginning of a more scientific design process: this can be treated as an initial design, tested analytically by means of the techniques in Chapter 4, and redesigned to obtain more appropriate embedding impedances and thus better performance. The use of this approximate design approach, combined with analysis, has been very successful for the development of high-performance millimeter-wave mixers.

Design Example of a Microstrip Mixer

Essentially the same procedure can be used to design a very different structure, the microstrip mixer shown in Figure 6.8. Single-diode microstrip mixers are not used frequently in microwave systems. They are found more often in singly balanced mixers, where two such mixers are connected to a pair of mutually isolated ports of a 90- or 180-deg hybrid. The RF and LO are applied to the other pair of ports, and as a result both the RF and LO are applied to each single-diode mixer. The low-frequency IF is separated from the RF and LO by an L-C or microstrip low-pass filter. It is necessary also to include dc and IF returns and a dc/IF block in each mixer's RF/LO microstrip. In the mixer shown in Figure 6.8, the return is realized by a simple stub. If necessary, this stub can be used as a tuning element; if it is not needed for tuning, it can be made one-quarter wavelength long at the RF and LO frequencies. The block is realized by a parallel-line dc block; this has a narrower bandwidth than a simple chip capacitor, and will therefore have a greater impedance at the IF frequency.

The figure also shows provision for dc bias; the advantages of using dc bias are described in Section 4.6.4. It is important that the diode not be open-circuited at dc; thus, if dc bias is not used, the bias terminal must be connected to ground.

The mixer is designed to operate at an LO frequency of 20 GHz with a 1-GHz IF. The diode is a beam-lead device having a junction capacitance of 0.05 pF and 10Ω series resistance. It also has 0.10 pF overlay capacitance (Section 2.4.4). As long as the diode is mounted properly, it has no significant series inductance.

The combined junction and overlay capacitances have a reactance of –53Ω at 20 GHz; this is easily resonated by selecting the IF return to have a characteristic impedance of 100Ω and a length of 28 deg. Thus, unless further tuning is employed, the diode's RF and LO embedding impedances are approximately 50Ω. If the structure is used as a single-diode mixer, the IF load impedance is also 50Ω; however, in a balanced mixer, the IF outputs of the individual mixers are connected in parallel (often with a common IF filter), and each diode is effectively terminated in a 100Ω load. Figure 6.9 shows the equivalent circuit of the mixer.

The simplicity and small size of this mixer, even when a hybrid is included,

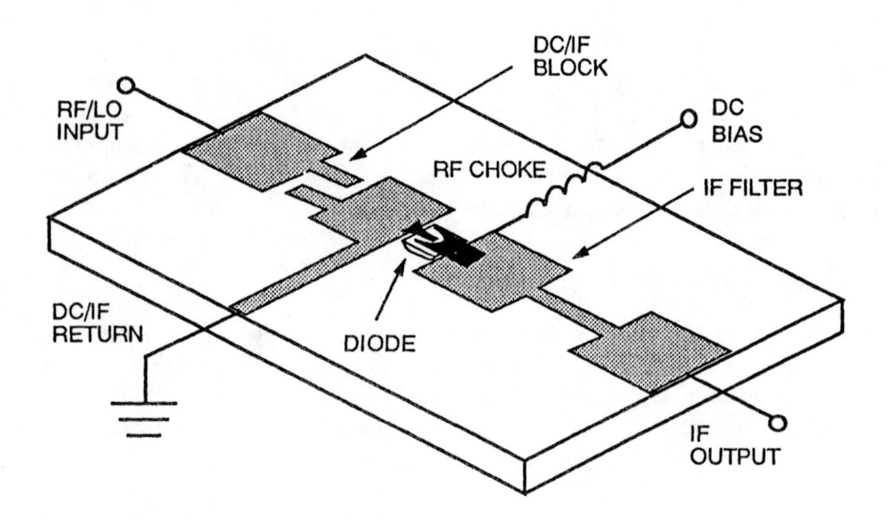

Figure 6.8 Single-diode microstrip mixer.

makes it attractive for use as a low-cost mixer in integrated microwave receivers. The low Q of the diode makes it relatively broadband; the bandwidth of such mixers is usually limited by the hybrid or parts of the circuit other than the mixer itself. Balanced mixers similar to this have conversion losses around 7 dB. They use fewer than six low-cost parts and usually do not require tuning.

6.2.3 Accurate Determination of the Embedding Impedances

In order to model a waveguide mixer accurately, the set of impedances presented to the diode junction by the mount and matching structure must be known. In Chapter 4 we called these the *embedding impedances* of the mixer. The model described in the previous section, illustrated in Figure 6.5, uses a number of relatively crude approximations, and is therefore clearly inadequate for this purpose.

One venerable technique for finding the mixer's embedding impedances is to build a scale model of the mixer and to measure them at a low frequency. This technique was used successfully by Held and Kerr [9]. Scale models have been used successfully to design other diode circuits as well, particularly frequency multipliers [10]. Scale models have obvious disadvantages: they are difficult to modify, expensive to construct, and require tedious measurements. More fundamental, however, is the fact that the mixer must be designed before the data to analyze it are

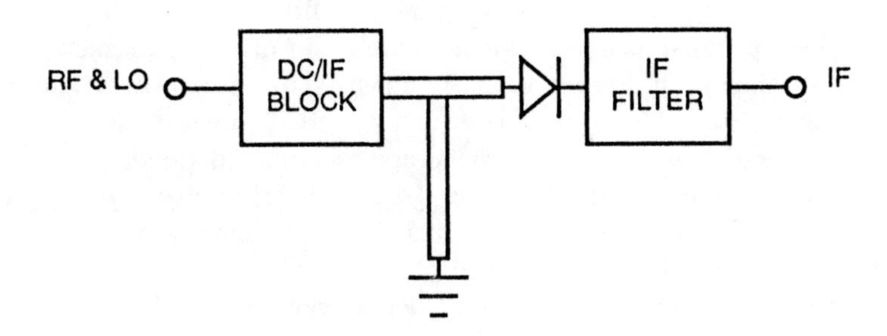

Figure 6.9 Equivalent circuit of the single-diode microstrip mixer.

available. If the scale model shows that the design is inadequate, the model must be redesigned and rebuilt, with no guarantee that the new design will be successful.

A more satisfying method of determining embedding impedances is to model them with an electromagnetic-field simulator. Such simulators have recently become available. However, they are expensive, as is the computer necessary to run them, and they often do not analyze certain types of circuit structures efficiently. Especially difficult are circuits having complex structures, especially circular ones.

A satisfactory approach is to use the theory of Eisenhart and Khan to model the waveguide mount and treat the rest of the circuit as an ideal, perhaps multimoded, waveguide. This method does not require expensive computing facilities; an ordinary desktop computer should be adequate. The greatest uncertainty in this approach is in the inductance of the whisker.

Microstrip mixers can be analyzed with surprisingly good accuracy by commercial harmonic-balance programs. The limited accuracy of the models in such programs at high mixing frequencies and LO harmonics does not seem to hurt the accuracy as much as might be expected, probably because the diode's junction capacitance is virtually a short circuit at these high frequencies and becomes the dominant termination. In such mixers, it is imperative to include an adequate number of LO harmonics and mixing frequencies, even if the circuit models are not accurate at those frequencies.

6.2.4 Single-Diode Millimeter-Wave Mixers Using Beam-Lead Diodes

If a millimeter-wave beam-lead diode is to be used instead of a whisker-contacted chip, it cannot be mounted directly in the waveguide; instead, it must be mounted in a strip transmission-line circuit. Microstrip and suspended-substrate stripline are favored media. Inevitably, some parts of the mixer's filters and matching circuits are realized in the strip transmission medium as well. An example of such a mixer in common use is shown in Figure 6.10. It consists of three filters and has two waveguide inputs. The filters are realized in suspended-substrate stripline, usually fabricated on a single substrate, and the diode is mounted on this substrate, not directly in the waveguide as with the mixer described earlier. For high frequencies a fused silica substrate is preferred, although sometimes alumina, or even fiberglass-reinforced polytetrafluoroethylene, is used.

The diode is located in the center of the RF waveguide, as shown in the figure; the mounting structure can be designed in much the same was as the chip diode mount. The filters F_1 and F_2 must present a short circuit to the diode at the RF frequency so the diode is effectively connected between the top and bottom walls of the waveguide. Because the beam-lead diode has relatively low series inductance, matching this diode is somewhat easier and broader bandwidth can be obtained than in the waveguide mixer of Section 6.2.2.

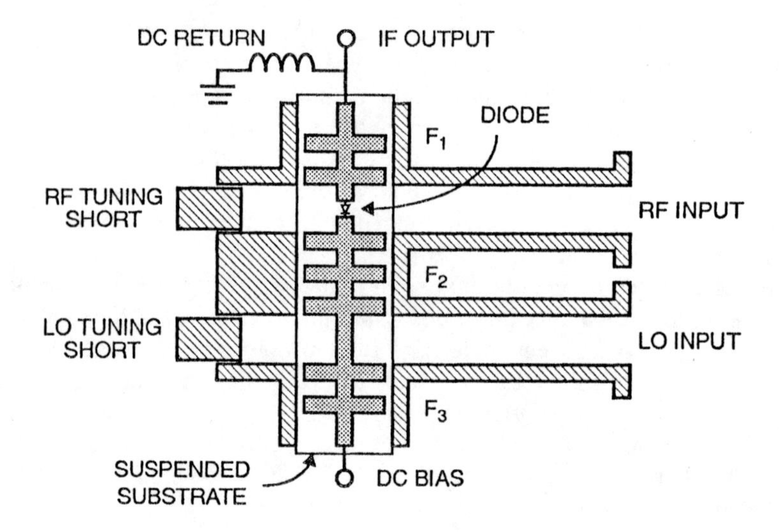

Figure 6.10 A millimeter-wave mixer using a beam-lead diode. The RF and LO are applied through separate waveguide inputs and are coupled to a suspended-substrate circuit. F_1 is the IF filter, designed according to the same criteria as in the mixer using a waveguide-mounted chip diode. F_2 is the LO filter; it passes the LO and rejects the RF. F_3, the bias-injection filter, is virtually identical to F_1.

The three filters are designed for different purposes. F_1 is the IF filter, designed in a manner identical to that of the chip-diode mixer; the range of realizable characteristic impedances for the high- and low-impedance sections will, of course, differ from those of a coaxial line. F_2 is designed to pass the LO so that the diode can be pumped effectively, but to reject the RF frequency so that no RF loss occurs through the LO waveguide. F_3 grounds the lower end of the structure to the bottom wall of the LO waveguide. This cannot be accomplished simply by physically connecting it to the bottom wall, because low-impedance mechanical connections may be unreliable at millimeter wavelengths. It also creates a convenient connection point for dc bias.

Local-oscillator leakage from the RF input can be reduced by the use of a filter, or by the selection of an RF waveguide having a cutoff frequency above the LO frequency. A cutoff waveguide or other type of image filter can be used to obtain image enhancement. Millimeter-wave beam-lead diodes are particularly well suited for balanced mixers and subharmonically pumped mixers; these applications will be addressed in later chapters.

6.2.5 Single-Diode Mixers Above 100 GHz

The single-diode mixer design techniques described in Section 6.2.2 can be used at least to 100 GHz, and perhaps to 250 GHz, with some modifications. Image enhancement is generally impractical above 100 GHz; above this frequency, filter and waveguide losses become very great, and because (at a fixed IF) the image frequency becomes proportionately closer to the RF passband, image-filter requirements become more severe. The mixer is also inherently more lossy, and consequently the improvement achievable through image enhancement is less. Because of these effects, it is rarely possible to achieve any reduction in conversion loss through image enhancement.

Because of high waveguide losses at frequencies above 100 GHz, it is often advantageous at high frequencies to use quasi-optical components instead of metal transmission media. In quasi-optical systems the millimeter-wave signals are treated much like optical beams and are transmitted from the antenna and LO source to the mixer via free space. Reflectors are used to direct the beam, and lenses collimate it. Couplers, resonators, and other passive components are realized by their optical analogs. Although the use of free-space transmission introduces problems of alignment and stability of the optical system, as well as diffraction loss, the loss in quasi-optical systems can be controlled much more effectively at frequencies of a few hundred (or even thousand) gigahertz than in waveguide.

The matching considerations described in Section 6.1 are valid even for the very-high-frequency mixers described here. However, in many cases it is impossible to design and test a filter or matching structure directly at the operating frequency. In this instance, a scale model of the structure can be fabricated and measured at a lower frequency. In modeling mixer structures in this manner, one must remember that diode parasitics and circuit losses cannot be scaled, and unique submillimeter-wave phenomena such as transit-time limitations also cannot be included.

The simplest quasi-optical mixers are more or less conventional waveguide mixers having a feed antenna connected directly to the input and no intervening waveguide. The rest of the LO- and antenna-coupling circuit is quasi-optical. Figure 6.11 shows one type of mixer useful in quasi-optical systems, the biconical mixer [11]. It consists of two cones with small flats machined on the ends, with the diode chip mounted on one, and the whisker on the other. The two cones are, in effect, a biconical antenna, and the received signal and LO, combined in a quasi-optical diplexer, are focused on them. Groves one-quarter wavelength in depth are cut in the cylindrical part of the structure to prevent RF currents from flowing along it. The RF and LO impedance presented to the diode is a function of the angle of the cones, and for this mixer is approximately 50Ω.

A curved backshort surrounds the biconical structure. Part of it is movable to tune the mount. An aperture is left on one side to allow the bicone to be illuminated; its beam pattern is nearly circular in cross section. The IF is connected to the cone on which the diode is mounted. No IF filter is needed; the choke groves serve its function. The cone with the whisker is press-fit into the mixer body. This mixer has achieved DSB noise temperatures of 4,000K at 361 GHz and 15,000K at 671 GHz.

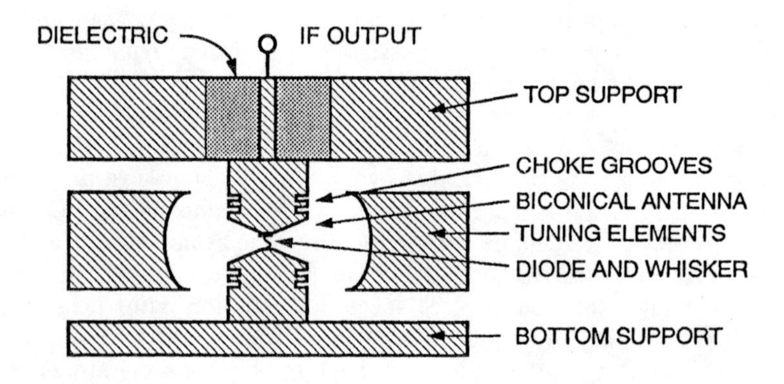

Figure 6.11 Biconical mixer.

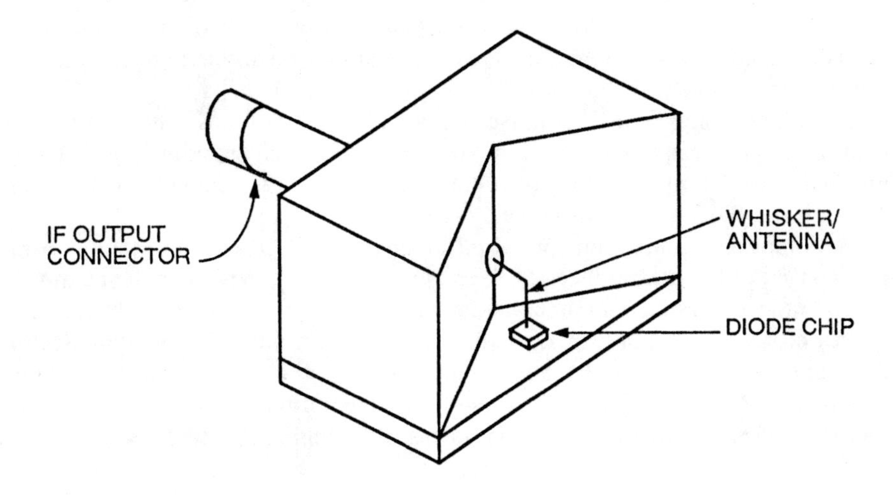

Figure 6.12 Corner-cube mixer of Fetterman et al. [13].

Another ingenious design is the corner reflector or corner-cube mixer [12], [13], a realization of which is shown in Figure 6.12. In this mixer, the diode chip is mounted on a post pressed into the bottom of the mixer body. Its anode diameter is 1.0 μm, and it has a junction capacitance of 1.5 fF and series resistance of 45Ω. The whisker acts as an antenna, and is approximately four wavelengths long. The antenna was optimized via a 100× scale model. Its beam is approximately circular, 14 by 15 deg at the half-power points, and makes a 25-deg angle with the antenna wire. This mixer has good performance and wide bandwidth, and has therefore been widely accepted. It is one of the few submillimeter-wave mixers sold commercially. It has been used at frequencies from 300 to 2,500 GHz, and has achieved DSB noise temperatures of 4,000K at 338 GHz and 7,600K at 762 GHz.

A more conventional structure for a 200- to 270 GHz mixer is shown in Figure 6.13 [14]. This mixer includes a corrugated horn antenna followed by a combined circular-to-rectangular transition and a rectangular waveguide transformer. The diode is mounted in a structure similar to that used in the design example, although the IF filter is realized in microstrip on a fused silica substrate, and the backshort is adjusted by the selection of end sections having different waveguide depths. This mixer is intended for cryogenic operation, and has achieved outstanding noise performance largely because of an astute choice of diode parameters. The diode's junction diameter is relatively large by conventional standards, 2 μm, with 7 fF C_{j0}

and 12Ω series resistance. This large diode exhibits low levels of hot-electron noise when cooled, and its large junction capacitance allows the diode to be matched by a small, low-inductance whisker, resulting in broad bandwidth.

This mixer achieves a SSB noise temperature below 600K when adjusted for operation at any frequency between 200 and 270 GHz, including the effects of diplexer losses. The minimum noise temperature is 511K. Conversion loss varies between 8 and 9 dB over this frequency range.

One important application for submillimeter-wave mixers is in imaging arrays. These consist of a large number of mixers, each integrated with an antenna mounted on a common substrate. The antenna can assume one of many possible forms, but the most common are dipoles, "bow-tie" antennas, and slots. For such applications, the mixer must be very simple (often consisting of a single diode) and must be easy to reproduce in quantity, ideally by monolithic integration. In such arrays, matching circuitry is often impractical; the sole means of matching the diode is to adjust the

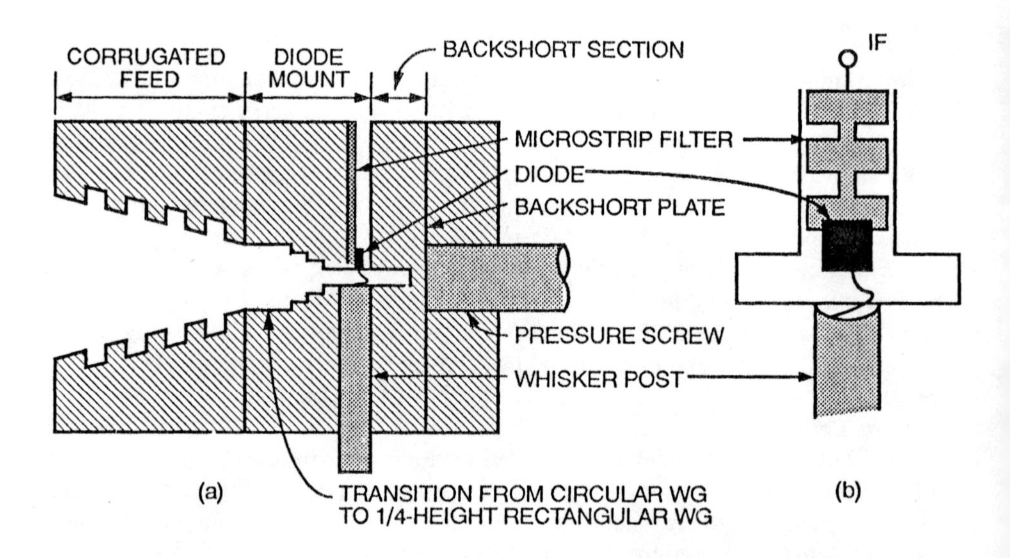

Figure 6.13 The 200-GHz mixer of Archer and Faber [14]: (a) cross section of the mount; (b) the diode and IF filter. The dimensions of the rectangular diode-mount waveguide are 0.0384 × 0.0048 inch; the diode protrudes 0.0015 inch into the waveguide.

antenna impedance. If a quasi-optical diplexer similar to those described below is used, it is also necessary that the antenna have a nearly conical beam.

Figure 6.14(a) shows a bow-tie mixer. Such mixers are usually fabricated on low-permittivity substrates to prevent the generation of surface-wave modes that might distort the antenna's radiation pattern. IF connections (and the necessary dc return for the diode) are made to the outer part of the antenna, because the fields are relatively weak at this point. Such mixers have been fabricated as subharmonically pumped mixers [15] and on GaAs substrates; in the latter, surface-wave modes are eliminated by the use of a hemispherical lens. Figure 6.14(b) shows a dipole mixer; the IF connection is made directly to the diode through a simple low-pass filter. Unfortunately, the radiation pattern of such structures is not as symmetrical as that of the bow tie.

Mixers such as those described above require a quasi-optical diplexer to combine the RF and LO beams while providing adequate RF-to-LO isolation. Two commonly used designs for quasi-optical LO diplexers are shown in Figure 6.15 and Figure 6.16. The diplexer in Figure 6.15 is best used for subharmonically pumped mixers. It could also be used for mixers having high IF frequencies and, thus, wide separation between RF and LO frequencies. It uses a Fabry-Perot mesh diplexer, tuned to the RF frequency, to pass the RF input and to reflect the LO. Lenses are used to collimate the LO beam and to focus the combined beam onto the mixer's horn antenna.

The diplexer in Figure 6.16 is more appropriate for fundamentally pumped mixers, and is widely used. It employs a polarizing grid to isolate the RF and LO

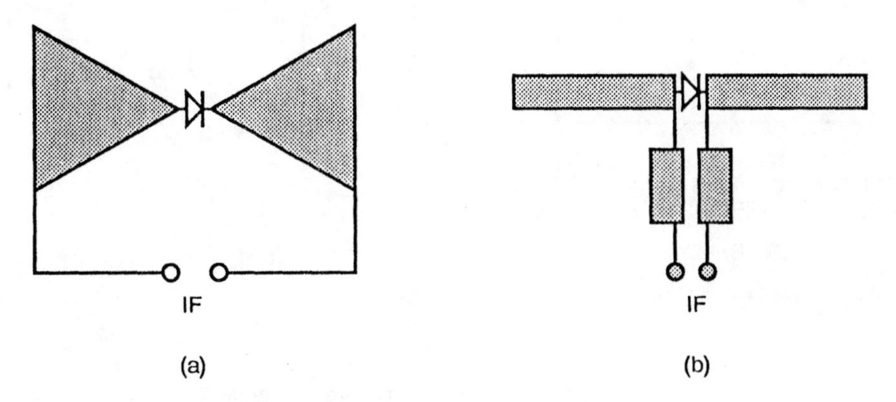

Figure 6.14 Two mixer elements suitable for use in quasi-optical imaging arrays: (a) bow-tie mixer; (b) dipole.

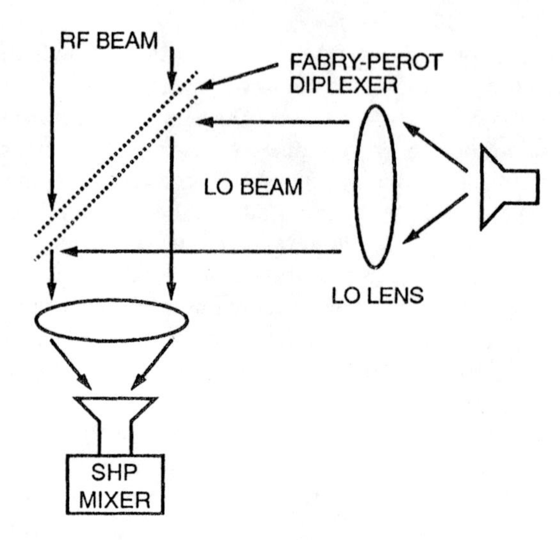

Figure 6.15 A simple quasi-optical LO diplexer suitable for subharmonically pumped mixers.

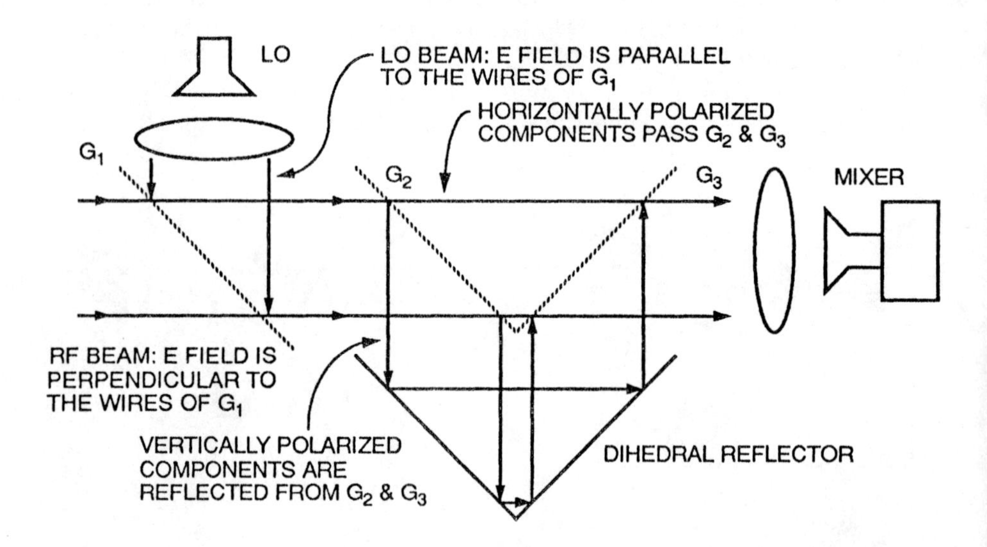

Figure 6.16 A quasi-optical diplexer. G_1, G_2, and G_3 are polarizing grids. The polarizations of the RF and LO beams are orthogonal, 45 deg to the plane of the figure. The position of the dihedral reflector is adjusted to form linearly polarized RF and LO beams at the mixer.

beams, and a set of grids and a reflector to split and recombine them with identical polarizations. At entry, the RF beam's polarization is 45 degrees to the plane of the figure, and the LO beam's polarization is perpendicular to the RF. The grid G_1 is oriented to reflect the LO beam and to pass the RF. This grid provides LO-to-RF isolation, but because the RF and LO components at its output are orthogonally polarized, they cannot be applied to the mixer. The grids G_2 and G_3, plus the dihedral reflector, rotate the polarizations. The wires of G_2 and G_3 are oriented perpendicularly to the plane of the figure and act as a beam splitter, directing half the LO/RF beam's energy toward the dihedral reflector. The reflector's position is adjusted until the beams, which combine at the mixer's input, have the same polarizations. Dielectric lenses, usually of polystyrene plastic or polytetrafluoroethylene, are used to collimate and focus the beams.

The beams emerging from the grids and the reflector will have identical, linear polarizations only if two conditions are satisfied:

1. The phase shifts in the RF and LO beams, caused by the path difference, are an integral number of wavelengths.
2. The higher-frequency (RF) beam undergoes 180 deg more phase delay than the LO.

The second condition dictates that the path difference must be an odd number of half wavelengths at the IF frequency; it may not be possible, in theory, to achieve this condition precisely and still meet the first. However, the error in meeting this requirement is rarely more than a few degrees; this error is negligible in practice.

6.3 REFERENCES

[1] Maas, S. A., "Design EHF Mixers with Minimal Guesswork," *Microwaves*, Vol. 18, No. 8, 1979, p. 66.

[2] Kerr, A. R., "Low Noise Room Temperature and Cryogenic Mixers for 80-120 GHz," *IEEE Trans. Microwave Theory Tech.*, MTT-23, 1975, p. 781

[3] Wrixon, G. T., "Select the Best Diode for Millimeter Mixers," *Microwaves*, Vol. 15, No. 9, 1976, p. 56.

[4] Kelly, W. M., and G. T. Wrixon, "Conversion Losses in Schottky-Barrier Diode Mixers in the Submillimeter Region," *IEEE Trans. Microwave Theory Tech*, MTT-27, 1979, p. 665.

[5] Mattauch, R. J., T. W. Crowe, and W. L. Bishop "Frequency and Noise Limits of Schottky-Barrier Mixer Diodes," *Microwave Journal*, No. 3, 1985, p. 101.

[6] McColl, M., "Conversion-Loss Limitations on Schottky-Barrier Mixers," *IEEE Trans. Microwave Theory Tech*, MTT-25, 1977, p. 54.

[7] van der Zeil, A., "Infrared Detection and Mixing in Heavily-Doped Schottky-Barrier Diode," *J. Appl. Phys.*, Vol. 47, 1976, p. 2509.

[8] Eisenhart, R. L., and P. J. Khan, "Theoretical and Experimental Analysis of a Waveguide Mounting Structure," *IEEE Trans. Microwave Theory Tech.*, MTT-19, 1971, p. 706.

[9] Held, D. N., and A. R. Kerr, "Conversion Loss and Noise of Microwave and Millimeter-Wave Mixers," *IEEE Trans. Microwave Theory Tech.*, MTT-26, 1978, p.49.

[10] Tolmunen, T. J., and A. V. Räisänen, "An Efficient Schottky-Varactor Frequency Multiplier at Millimeter Waves. Part I: Doubler," *International J. Infrared and Millimeter Waves*, Vol. 8, 1987, p. 1313.

[11] Gustincic, J. J., "A Quasi-Optical Receiver Design," *IEEE MTT-S International Microwave Symposium Digest of Technical Papers*, 1977, p. 99.

[12] Kräutle, H., E. Sauter, and G. V. Schultz, "Antenna Characteristics of Whisker Diodes Used as Submillimeter Receivers," *Infrared Physics*, Vol. 17, 1977, p. 477.

[13] Fetterman, H. R., et al., "Far-IR Heterodyne Radiometric Measurements with Quasi-Optical Schottky Diode Mixers," *J. Appl. Phys. Lett.*, Vol. 33, 1978, p. 151.

[14] Archer, J. W., and M. T. Faber, "Low-Noise, Fixed-Tuned, Broadband Mixer for 200-270 GHz," *Microwave Journal*, Vol. 27, No. 7, 1984, p. 135.

[15] Stephan, K. D., and T. Itoh, "A Planar Quasioptical Subharmonically Pumped Mixer Characterized by Isotropic Conversion Loss," *IEEE Trans. Microwave Theory Tech.*, MTT-32, 1984, p. 97.

Chapter 7
Balanced Mixers

Although single-diode mixers are entirely practical (and for millimeter-wave receiver applications often preferred), they have some undeniable faults. The most obvious difficulty is the need for a filter diplexer or other device to allow LO injection. Balanced mixers overcome these problems and have additional advantages as well. Balanced mixers generally have better power-handling capabilities and reject certain spurious responses, LO noise, and spurious signals. Their only disadvantages are generally poorer conversion performance (because it is difficult to design a balanced mixer having dc bias or a carefully optimized embedding network) and the need for greater LO power to drive the increased number of diodes. For most applications in the microwave range, where the lowest possible conversion loss is not required but small size and low cost are, balanced mixers are used almost exclusively.

There are a relatively limited number of basic designs for balanced mixers. These are *singly balanced* mixers based on 90-deg (quadrature) and 180-deg hybrids, and *doubly balanced* mixers, which use diodes in a ring or star configuration. Virtually all balanced mixer designs are variations on these themes; the only real difference between them is in the design of the hybrids. A fifth type of mixer, which is justifiably considered to be a balanced mixer, is the *subharmonically pumped* mixer, which uses an antiparallel diode pair. The most commonly encountered balanced mixer at microwave frequencies is the ring mixer. The least commonly used mixer (for good reasons) is probably a singly balanced mixer employing a 90-deg hybrid.

To our great fortune, all balanced mixer circuits can be reduced to an equivalent single-diode circuit. Reducing a complex balanced mixer circuit to a single-diode circuit considerably simplifies the process of designing and analyzing the mixer, and allows the single-diode mixer analysis described in Chapter 4 to be applied to virtually any mixer. The process of generating the single-diode equivalent circuit and using it to design a balanced mixer is a recurring theme in this chapter.

7.1 MICROWAVE HYBRIDS

7.1.1 Fundamentals of Hybrid Couplers

Microwave hybrid couplers are fundamental building blocks of balanced mixers. It is therefore important to examine them in some detail. Microwave hybrids are four-port devices having a special set of characteristics: (1) all ports are matched*, (2) RF power applied to any one port is split equally between two of the other ports, and (3) the remaining port is *isolated* (i.e., no output is obtained from it). In a 180-deg hybrid the output voltages differ in phase by 0 deg or 180 deg, depending on which port is excited. In a 90-deg hybrid, the phases differ by 90 deg at the two outputs, regardless of which port is excited.

The properties of the two hybrids are illustrated in Figure 7.1. We assume initially that the hybrids are ideal. The lines between the ports show the phase shift between them. For example, if a signal is applied to port 1 of the ideal 180-deg hybrid in Figure 7.1(a), it appears at ports 4 and 3 with identical phases and at a level 3 dB below the input. No output appears at port 2. Similarly, if port 2 is excited, the outputs are ports 4 and 3, and port 1 has no output. The outputs at ports 4 and 3 are 180 deg out of phase, and, of course, are 3 dB lower in level than the input. In this hybrid, ports 3 and 4 are often called the *sum (sigma)* and *difference (delta)* ports, because the output voltage at port 3 is proportional to the sum of the input voltages at ports 1 and 2, and the voltage at port 4 is proportional to their difference. Similarly, ports 1 and 2 can be considered the sum and difference ports, respectively, with ports 4 and 3 the inputs.

Figure 7.1(b) shows a quadrature hybrid. This component has a 90-deg phase delay between ports 2 and 4 and between ports 1 and 3. Ports 1 and 2 and ports 3 and 4, are mutually isolated pairs.

The two types of hybrids can be described by their S matrices. An ideal 90-deg hybrid, with the ports defined as in Figure 7.1, has the S matrix

---

* In the sense that $S_{nn} = 0$ at each port. This is not necessarily a *conjugate* match to the source.

$$S_{90} = \begin{bmatrix} 0 & 0 & -j & 1 \\ 0 & 0 & 1 & -j \\ -j & 1 & 0 & 0 \\ 1 & -j & 0 & 0 \end{bmatrix} \tag{7.1}$$

and the 180-deg hybrid has the S matrix

$$S_{180} = \begin{bmatrix} 0 & 0 & 1 & 1 \\ 0 & 0 & 1 & -1 \\ 1 & 1 & 0 & 0 \\ 1 & -1 & 0 & 0 \end{bmatrix} \tag{7.2}$$

Equations (7.1) and (7.2) imply specific phase shifts between the input and output ports. In real hybrids, the input-to-output phase shift is rarely important, but the phase difference between the two output ports, as we shall see, is critical. The

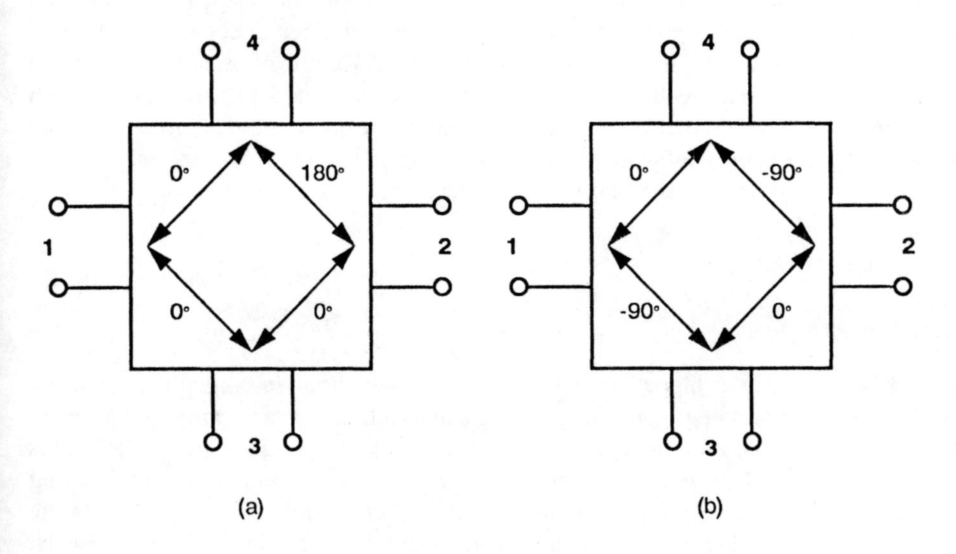

Figure 7.1 Ideal (a) 180-deg and (b) 90-deg hybrids.

input-to-output phase shifts implied by these equations and by Figure 7.1 are important only insofar as they define the phase difference between the outputs.

One may wonder whether other types of hybrids are possible. It can be shown from the properties of the S matrix, without regard for the physical structure of the hybrid, that these are the only ones possible, as long as the circuits are lossless, have four ports, and are matched at each port. A fuller explanation of this point may be found in Collin [1].

Real hybrids differ from the ideal hybrids described above in several ways: the most important nonidealities are phase and amplitude balance, loss, and VSWR. *Balance* refers to the matching of phase and power levels at any two output ports. *Phase balance* is the deviation in phase from the ideal phase difference between any pair of outputs, and *amplitude balance* is the difference in output amplitudes, usually expressed in decibels. Phase and power match between the outputs is critical; the phase shift through the hybrid, between the input and outputs, usually is not. *Isolation* is the ratio of power at the isolated port to that applied at the input, usually expressed in decibels. Isolation and balance are usually frequency-dependent and may be different for different pairs of ports. They are generally the same only if the hybrid has a symmetrical structure. Like all other real components, microwave hybrids introduce some dissipative power loss, which is usually specified as loss above the unavoidable 3 dB of coupling loss.

A hybrid's port VSWR may also be imperfect. VSWR often depends strongly on the discontinuities in the connection of input-output striplines or connectors, as well as the properties of the intrinsic hybrid. Port VSWRs are of special concern in balanced mixers because the single-diode mixers connected to those ports often have poor VSWRs as well. The resulting combination of poor source and load VSWR degrades the flatness of the mixer's passband, which is usually not spectacular in even the best situations, and increases its conversion loss.

7.1.2 Hybrid Couplers

Transformer Hybrid

The ubiquitous center-tapped transformer commonly used in ring mixers can be used as a type of 180-deg hybrid. Figure 7.2 shows a transformer configured as such. The ports are numbered in a manner that corresponds to those in Figure 7.1(a). This hybrid has unequal port impedances; to be matched at all ports, the terminating resistances at ports 3 and 4 must be half those at ports 1 and 2. We assume that the three transformer windings are ideal, and have the same number of turns in each winding. We shall refer to the winding connected to port 4 as the primary winding,

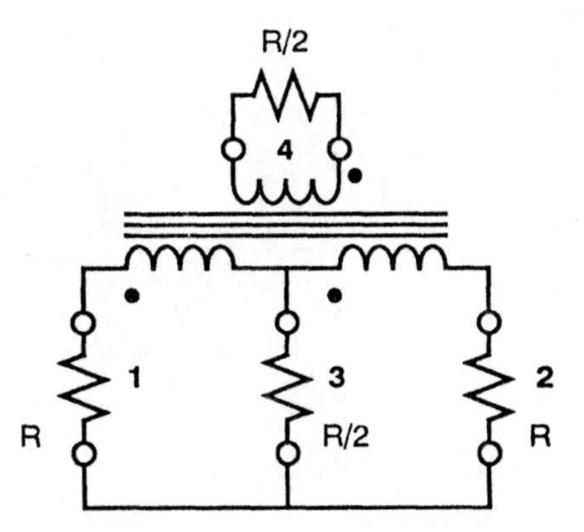

Figure 7.2 The transformer hybrid. The port numbering follows Figure 7.1(a). All windings are ideal and have the same number of turns.

and the two others as the secondary windings.

The voltages and currents that result when the ports are excited are shown in Figure 7.3. In Figure 7.3(a) port 4 is excited. If a voltage V appears across the primary, V must also appear across each secondary, and by conservation of power the current in each secondary must be $I/2$, half the primary current. The net current in port 3 is clearly zero, and the power dissipated in each of the terminating resistors of ports 1 and 2 is $VI/2$, or half the input power. The polarities of the voltages at these ports are opposite. Because there is no voltage across port 3's load, the center tap of the secondary has the same voltage as the lower terminals of ports 1 to 3; if this is a ground point, the center tap is a virtual ground.

Figure 7.3(b) shows port 3 excited. Because of the hybrid's symmetry, the current I in port 3 must be split equally between the two secondary branches. These oppositely directed currents in the two secondary windings induce no currents in the primary, giving zero primary voltage. Because the voltages across each of the secondary windings must be equal to that of the primary, the secondary windings also have zero voltage. Hence, V appears across ports 1 and 2 in phase, with port currents $I/2$. The power dissipated in each of these ports is again $VI/2$.

Figure 7.3(c) shows port 2 excited. The port current I must exist in the secondary winding closest to port 2, so an equal current must exist in the primary. The primary voltage is $IR/2 = V/2$, so all winding voltages must be $V/2$. By Kirchoff's laws, port 1's voltage and current must be zero. Thus, the phase characteristics are as shown in

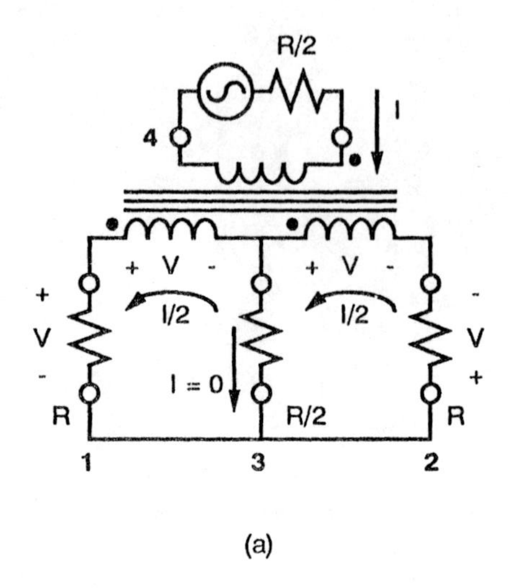

(a)

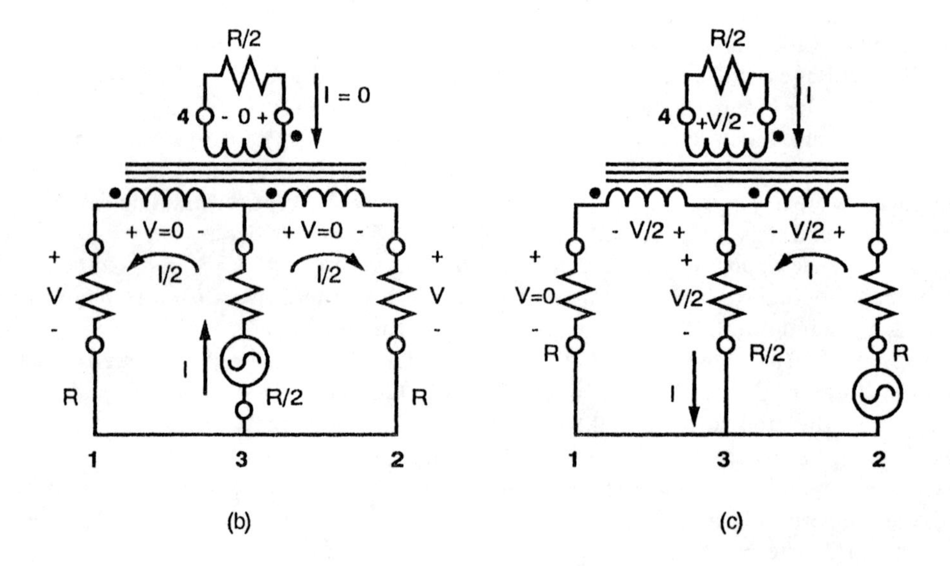

(b) (c)

Figure 7.3 Currents and voltages in the transformer hybrid with (a) port 4 excited, (b) port 3 excited, and (c) port 2 excited.

Figure 7.1(a) and the power split is equal. Although we did not demonstrate explicitly that $S_{nn} = 0$, this could have been demonstrated in much the same manner.

Transformer hybrids often have very wide bandwidths, sometimes exceeding several decades. Many designs for such transformers have been proposed [2]. One straightforward approach is to use a trifilar winding (i.e., three wires in a single winding) on a toroidal ferrite or powered iron core (Figure 7.4). Because the magnetic fields are well contained by the high-permeability toroid, the magnetic coupling between the windings is nearly ideal. This transformer is capable of achieving bandwidths from approximately 10 MHz to as much as 2 GHz, although 0.5 to 1 GHz is more common. The remarkably broad bandwidth of such transformers occurs because they operate as conventional transformers at low frequencies, but as transmission line structures at high frequencies. In conventional transformers, the winding capacitances and inductance create resonances that limit high-frequency performance. In the toroidal transformer, however, the bifilar secondary windings realize a transmission line wherein the interwinding capacitances are distributed capacitances. The high-frequency resonances are

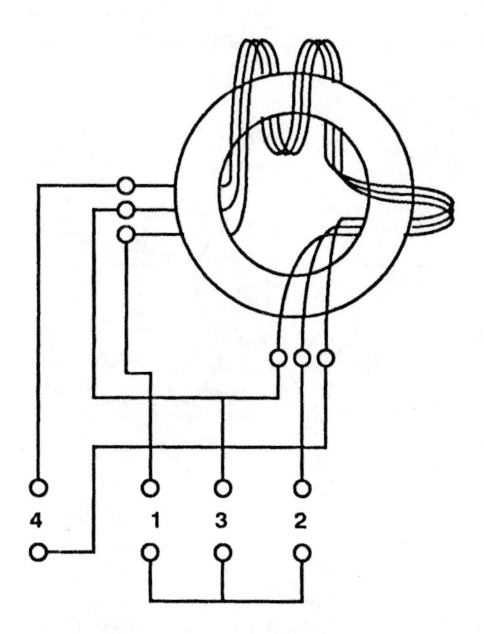

Figure 7.4 The transformer hybrid realized as a trifilar toroidal transformer.

thereby avoided, and the high-frequency performance depends upon the length of the equivalent transmission line. The bandwidth of a balanced mixer using a transformer hybrid is usually limited primarily by the transformer, and consequently extremely broadband mixers can be realized by using them.

Ring (Rat-Race) Hybrid

One of the oldest types of 180-deg couplers, and one of the simplest to design, is the so-called "rat-race" or ring hybrid. A microstrip realization of the ring hybrid is shown in Figure 7.5. It consists of a ring of transmission line (although sometimes waveguide is used) 1.5 wavelengths in circumference. Its characteristic impedance is $\sqrt{2}$ times the port impedances. Transmission lines for the four ports are connected to the ring in such a way that two are 0.75 wavelength apart, and the spacings between the rest are 0.25 wavelength. The port numbering in Figure 7.5 corresponds to that in Figure 7.1(a).

The operation of the ring hybrid is illustrated in Figure 7.6. To describe its operation, we make note of the symmetries of the structure. For simplicity, we shall assume that the port impedances are all 50Ω, and the ring characteristic impedance is 70.7Ω. In Figure 7.6, port 1 is excited, and waves travel around the ring toward ports 3 and 4. Because the path from port 4 to port 2 is one-half wavelength longer

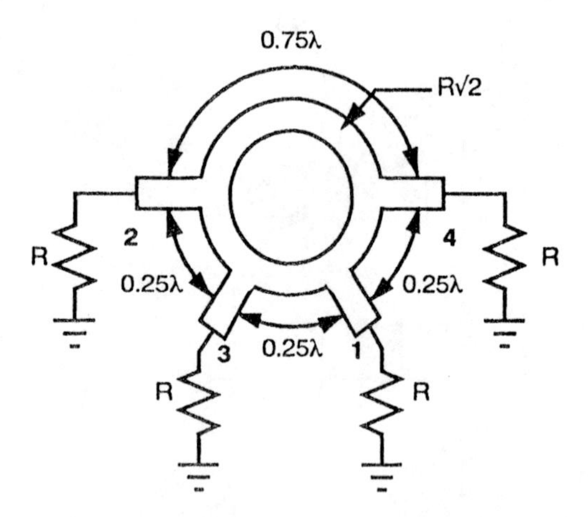

Figure 7.5 The ring hybrid. The figure shows a top view of a microstrip realization. The straight lines between the ring and terminations are interconnections and may have arbitrary length.

than the path from port 3 to port 2, the waves meeting at port 2 are 180 deg out of phase, so the voltage at this point must be zero. Port 2 is therefore a virtual ground, and the parts of the ring from port 2 to port 3 and from port 2 to port 4 are, in effect, quarter-wave shorted stubs. The terminal impedances of these stubs are infinite, so they have no effect on ports 3 and 4. The remaining parts of the 70.7Ω ring act as quarter-wave transformers, which transform the 50Ω loads on ports 3 and 4 to 100Ω each at port 1. The parallel combination of these is 50Ω, so the port is matched, and the power from the source is split evenly between them.

The same reasoning can be applied when any of the other ports is excited. For example, when port 2 is excited, the waves reaching port 1 are 180 deg out of phase, so port 1 is a virtual ground. The power is split between ports 3 and 4, but because the path to port 4 is 0.5 wavelength longer than the path to port 3, the outputs are 180 deg out of phase.

Coupling loss, input reflection coefficient, and phase balance of an ideal ring hybrid are shown in Figure 7.7. The bandwidth of this hybrid is approximately 10% to 15%, although when the effects of junction discontinuities and transmission line dispersion are included, it may be slightly lower. The hybrid is relatively narrowband because its dimensions are frequency-dependent; its bandwidth is limited most strongly by the requirement that the difference in length between the port 2-3 and port 2-4 paths must be 0.5 wavelength, a requirement that can be

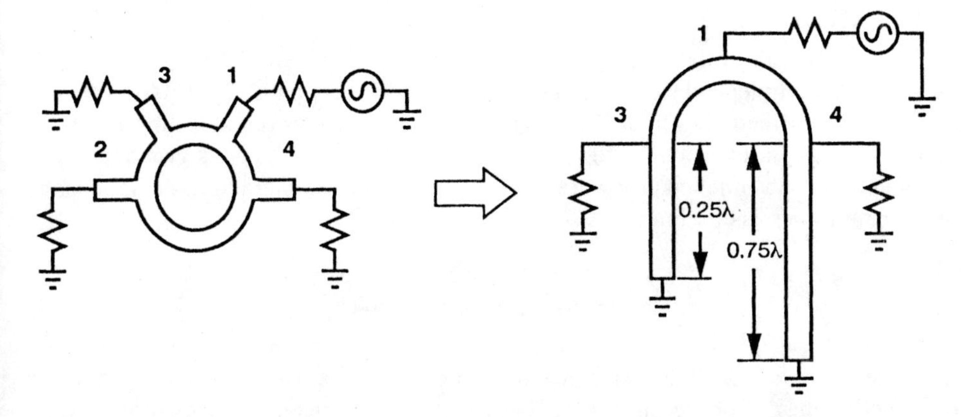

Figure 7.6 Equivalent circuit of the ring hybrid when port 1 is excited. Port 2 is a virtual ground, and the input power is divided between ports 3 and 4. Essentially the same equivalent circuit is valid if port 3 is excited. If port 2 or 4 is excited, the equivalent circuit is also similar; the only difference is that both stubs are 0.25λ and one of the paths from the excitation point to a load is 0.75λ.

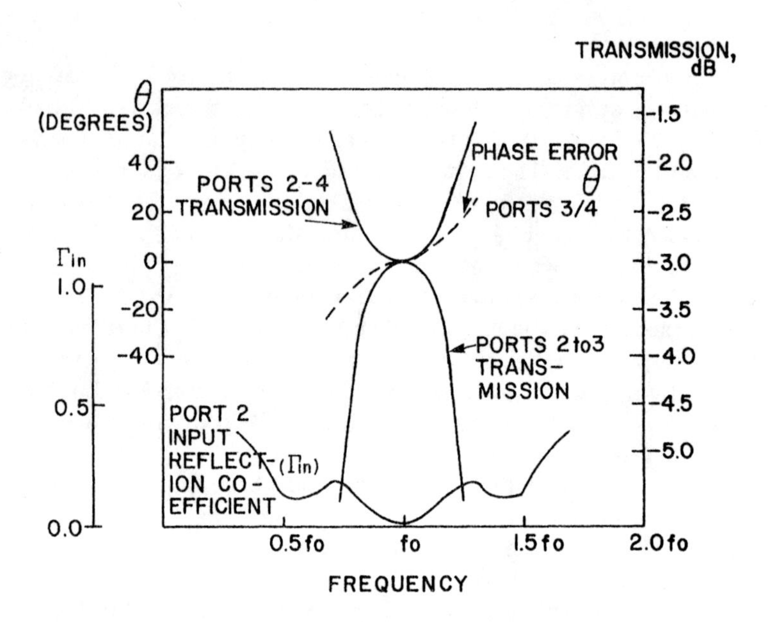

Figure 7.7 Input reflection coefficient, phase error, and transmission in the ring hybrid when port 2 is excited; f_0 is the design frequency.

satisfied exactly at only one frequency.

The bandwidth of the ring hybrid can be improved considerably through the use of a frequency-independent phase reversal in the port 2-4 path [3]. This is the idea behind the modified ring hybrid in Figure 7.8. The 0.75-wavelength line has been replaced by a quarter-wavelength coupled-line section with its ends grounded. The coupled line is designed as if it were a 3-dB coupler for the characteristic impedance of the ring transmission line; that is,

$$Z_{0e} = 3.414R \qquad Z_{0o} = 0.585R \tag{7.3}$$

where R is the terminating resistance at each of the ports. It can be shown that this line section has a phase shift of 270 deg at the frequency at which it is 0.25 wavelength long. Additionally, its phase shift over a wide frequency range is very close to that of a quarter-wave transmission line having an additional, frequency-independent phase shift of 180 deg. Performance of this hybrid is shown in Figure 7.9; it is capable of bandwidth approaching an octave. Its disadvantages are the need

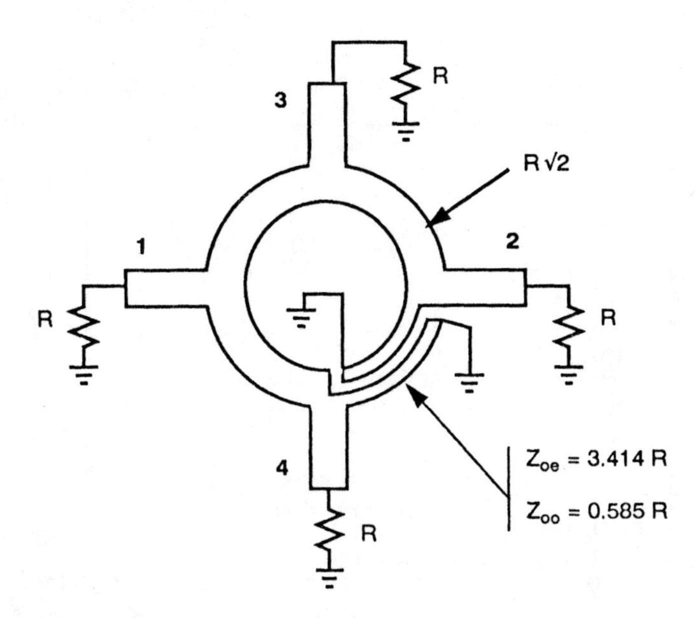

Figure 7.8 Modified ring hybrid.

for ground connections, which may require holes in the substrate and thus may be difficult to fabricate on hard substrates such as alumina, and the difficulty in achieving the required even- and odd-mode impedances in the coupled section. The coupled lines must often be realized by multiple conductors, as in a Lange hybrid. The bandwidth may also be limited somewhat by differences in even- and odd-mode phase velocities in the coupled sections; these increase its phase dispersion when the structure is realized in microstrip.

Four-Port Waveguide Tee

One of the most commonly encountered 180-deg waveguide hybrids is the four-port waveguide tee, or "magic tee." Its power split, isolation, and phase reversal occur because of the symmetry of the waveguide junction, and do not require wavelength-related dimensions. However, it is not inherently matched, and it requires additional tuning to remove mismatches caused by its junction discontinuities. Its bandwidth is limited to 5% to 10% by that tuning if good port VSWRs are necessary. If not, an untuned hybrid exhibits good isolation and balance (but poor VSWR) over a full

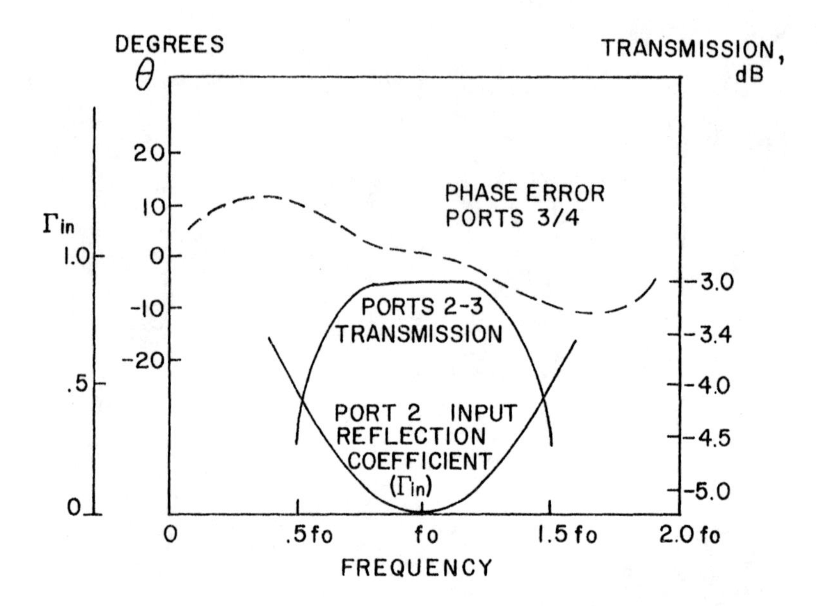

Figure 7.9 Input reflection coefficient, transmission, and phase error of the modified ring hybrid of Figure 7.8.

waveguide bandwidth. The waveguide tee is shown in Figure 7.10.

The waveguide tee must be tuned to eliminate reflections from its junction discontinuities. A traditional (and completely empirical) method of matching is to insert a post (a finely threaded screw is used in practice) into the bottom of the junction so that it points toward port 4, and to locate an inductive iris close to the junction in the waveguide from port 4. These elements are usually sufficient to tune the entire structure; no additional tuning is needed in ports 1 and 2.

For mixer use, ports 3 and 4 are usually the RF and LO inputs, and single-diode mixers are mounted on ports 1 and 2. This arrangement separates the IF outputs and makes it difficult to combine them with good phase match. In order to move the IF ports closer together and to reduce the overall size of the mixer, waveguide-tee hybrids are often designed with sharp mitred bends close to the junction, so the waveguides from ports 1 and 2 are parallel and close together. These are called *folded hybrid tees*.

Disk and Patch Hybrids

It is possible to use a planar microstrip resonator as a hybrid if the coupling points and line widths are carefully chosen [4] - [6]. Both 90- and 180-deg hybrids are possible, and, if desired, unequal couplings also can be achieved. Such devices are, unfortunately, quite narrowband, rarely exceeding a few percent in bandwidth, and have therefore not been used extensively in mixer circuits. Nevertheless, the ease of fabricating such hybrids, even when they are quite small, may make them useful for monolithic and hybrid millimeter-wave circuits.

The design of these hybrids is a fairly complicated problem in electromagnetics, and is thus beyond the scope of this book. The interested reader should consult the references.

Coupled-Line Hybrid

A coupled-line hybrid using strip transmission lines is shown in Figure 7.11; the port numbering is consistent with Figure 7.1(b). Under the right conditions, the pair of transmission lines can become a 90-deg hybrid coupler. It is necessary to invoke even-odd mode analysis to describe the operation of the coupled-line hybrid [7] because it is impossible to describe it qualitatively in any satisfying manner. Therefore, only the design equations will be given here.

This coupler is designed according to its even- and odd-mode impedances, Z_{0e} and Z_{0o}. For any specified coupling, the design equations are

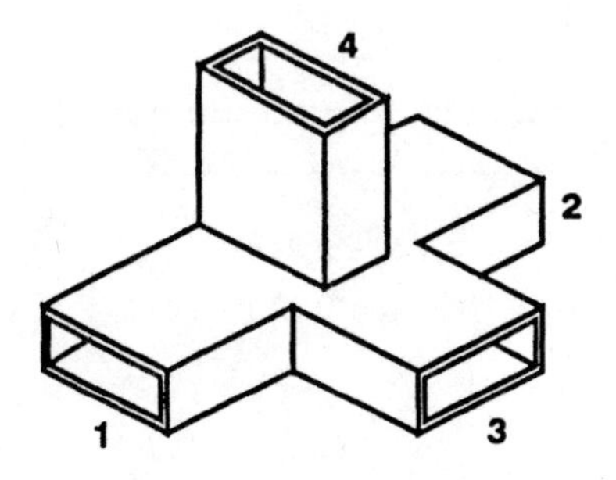

Figure 7.10 Waveguide tee hybrid.

$$Z_{0e} = R\sqrt{\frac{1+c}{1-c}}$$
$$Z_{0o} = R\sqrt{\frac{1-c}{1+c}} \qquad (7.4)$$

where R is the port impedance and c is the voltage coupling factor, the square root of the power coupling (e.g., for a 3-dB coupler, $c = 0.707$). Note that (7.4) implies that $R = \sqrt{Z_{0e}Z_{0o}}$. Once the values of Z_{0e} and Z_{0o} are obtained, the dimensions of the coupled transmission lines that realize these impedances must be found. In general, numerical techniques must be employed [8], [9].

The required even- and odd-mode characteristic impedances of a coupler having 50Ω port impedances are 120.7Ω and 20.7Ω, respectively. It is virtually impossible to achieve these impedances, even on high-permittivity substrates, because the required spacing between the lines is too small. Another problem with coupled-line hybrids is that the two outputs emerge on opposite sides of the isolated port, so a symmetrical circuit layout is difficult to achieve.

The Lange coupler is a solution to both problems. It is derived from the simple coupled-line hybrid in the manner shown in Figure 7.12. In order to increase the

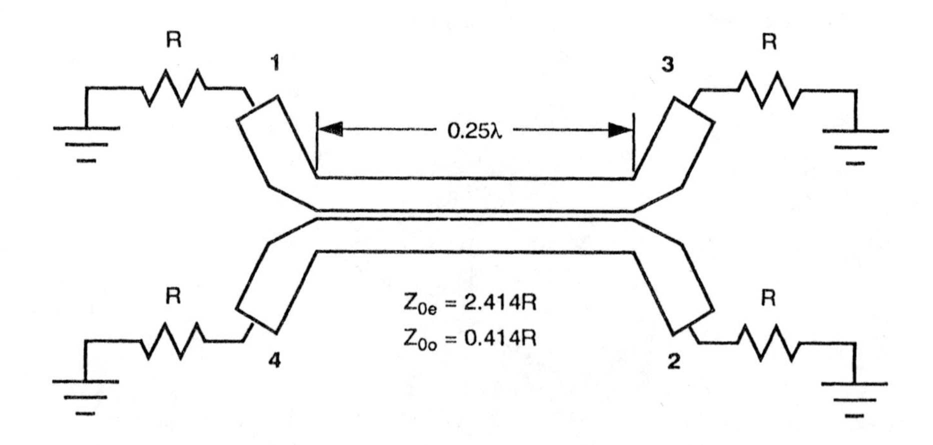

Figure 7.11 Microstrip hybrid coupler. A quadrature hybrid cannot usually be realized with only a single pair of coupled strips. The even- and odd-mode impedances specified in the figure realize a 3-dB coupler.

coupling, the two coupled lines are split into four. This change allows three pairs of edges to be adjacent instead of just one, increasing the capacitance between them. The four-strip geometry has dimensions that are readily realized, although on low-permittivity substrates, it may be necessary to use six strips instead of four. The structure is made symmetrical by cutting the lower-edge strip in half and moving one half of it to the opposite side of the structure. It is necessary to connect the pieces with a wire.

The calculation of the even- and odd-mode impedances of the multiple coupled lines is, as might be expected, more difficult than that of the single pair of coupled lines. The most straightforward approach is to use moment methods [10]. This approach is very mature, and is embodied in commercial software [11]. Simpler methods have also been proposed [12] – [14]. One simplification, which is reasonably accurate, is to ignore the capacitance between nonadjacent lines [12].

Performance of the ideal coupled-line hybrid is shown in Figure 7.13. The coupling bandwidth is much wider than that of the ring hybrid, and approximately the same as the modified ring hybrid. However, the bandwidths of the phase and port VSWR are theoretically infinite; its input VSWR is 1.0, and its phase is precisely 90 deg, at all frequencies. In practical hybrids, phase errors occur and bandwidth is limited because of asymmetries in the structure, inductance of the connecting wires,

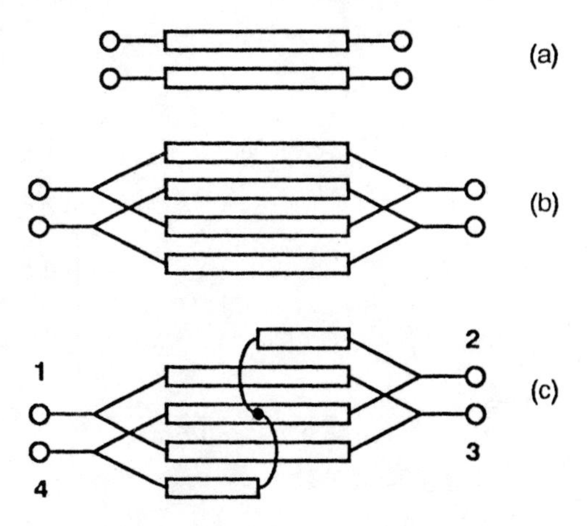

Figure 7.12 Steps to the realization of the Lange hybrid: (a) the microstrip parallel-line coupler; (b) each line is split in two to increase the capacitance between them; (c) half of the lower strip is placed on the opposite side of the hybrid so that the outputs are on the same side.

discontinuities where the coupled section connects to the input or output lines, and, most importantly, differences in the even- and odd-mode phase velocities along the coupled lines. The latter may be especially great in microstrip couplers on high-permittivity substrates.

Branch-Line Hybrid

The branch-line hybrid has historically been more favored for mixer applications than the coupled-line hybrid because the dimensional tolerances in its manufacture are much greater, and because it is suitable for fabrication on inexpensive, soft substrates. The branch-line hybrid is generally larger than a coupled-line hybrid, however, and is limited in microstrip realizations to lower frequencies.

In the branch-line hybrid, coupling is achieved between two transmission lines by periodic interconnecting branches. The interconnecting lines are one-quarter wavelength long and are spaced one-quarter wavelength apart. Very wide bandwidth can be achieved through the use of multiple branches. There are many approaches to

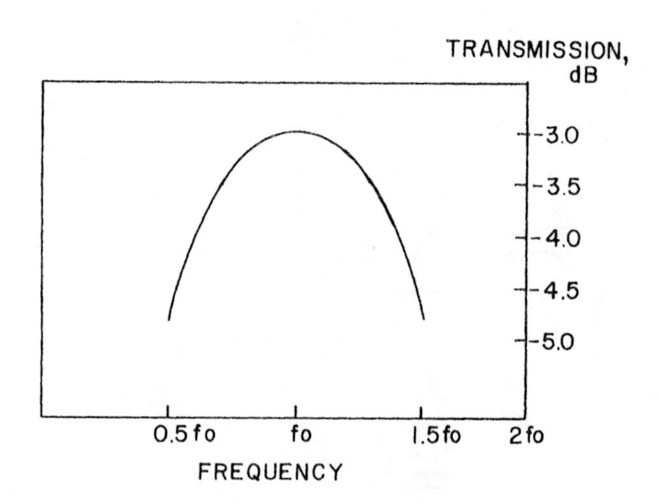

Figure 7.13 Transmission of the ideal coupled-line hybrid. The ideal hybrid has perfect phase balance and zero input reflection coefficient at all frequencies.

the design of branch-line hybrids [7], including some wherein either the series lines or shunt branches have equal impedances. For the simple single-section 3-dB quadrature hybrid shown in Figure 7.14, however, the characteristic impedance of the series lines between the branches is $R\sqrt{2}$ and the branch impedance is simply R, the terminating impedance of the four ports. Performance of this single-section branch-line hybrid is shown in Figure 7.15. It is noteworthy that the transmission bandwidth between ports 1 and 3 is much greater than that of ports 1 and 4. This characteristic limits this simple hybrid to narrowband applications.

Unfortunately, the line impedances of the branch-line hybrid are often not realizable, especially in multisection designs. Also, as frequency increases, the line lengths decrease until they are comparable to the line widths, the junction discontinuities become dominant, and performance suffers accordingly. The upper frequency limit for the branch-line hybrid's operation is generally lower than that of the other hybrids, since the branches must be long compared to the width of the low-impedance lines.

Lumped-Element Hybrids

It is possible to fabricate hybrids from lumped-element circuits. Such hybrids are inherently smaller than those using distributed elements and are thus useful in monolithic mixers and low-frequency circuits. Many designs for such couplers have been described; a few are listed in [15] – [17]. Some designs include transformers as

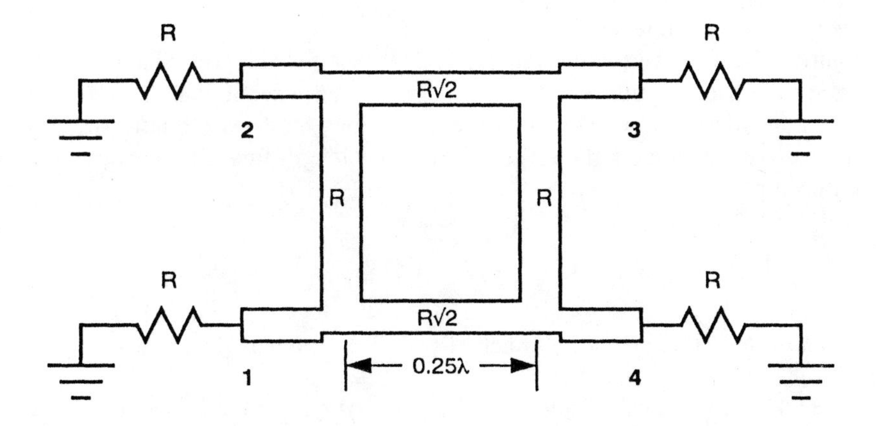

Figure 7.14 Single-section branch-line hybrid.

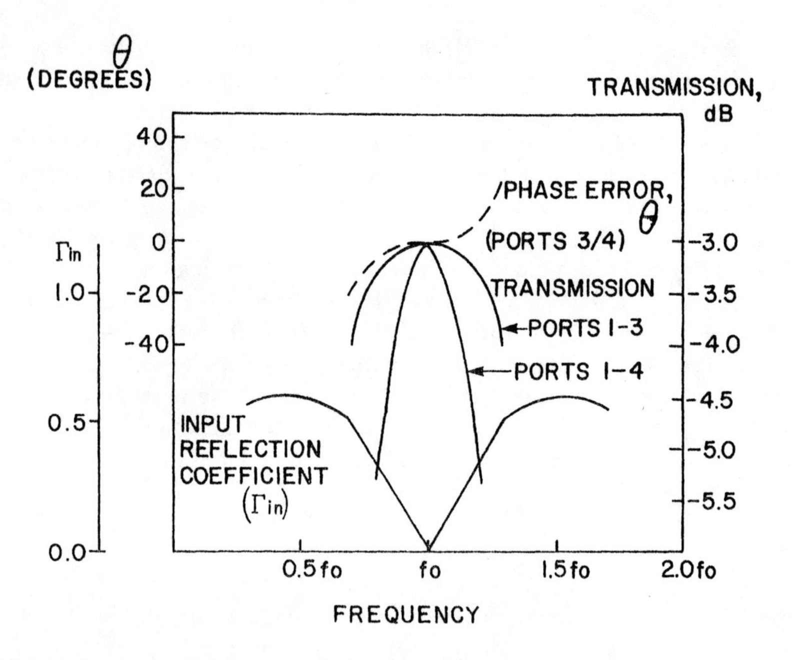

Figure 7.15 Reflection coefficient, transmission, and phase of a branch-line hybrid. port number 1 is excited.

well as transmission-line sections.

Figure 7.16(a) shows a design for a 180-deg hybrid [15]. This structure is analogous to a ring hybrid. It consists of four tee and pi sections that model the ring hybrid's transmission-line segments; the pi sections model the quarter-wave lines, and the tee section models the three-quarter wavelength line. The design equations are quite simple:

$$\omega L = \frac{1}{\omega C} = 1.414R \tag{7.5}$$

The bandwidth of this structure is actually slightly broader than that of the simple ring hybrid.

Figure 7.16(b) shows a lumped quadrature hybrid [16]. Its design principle is similar to that of the lumped 180-deg hybrid: the pi sections (C_0 and L) approximate transmission lines, and the capacitors C_1 provide the coupling between the "lines."

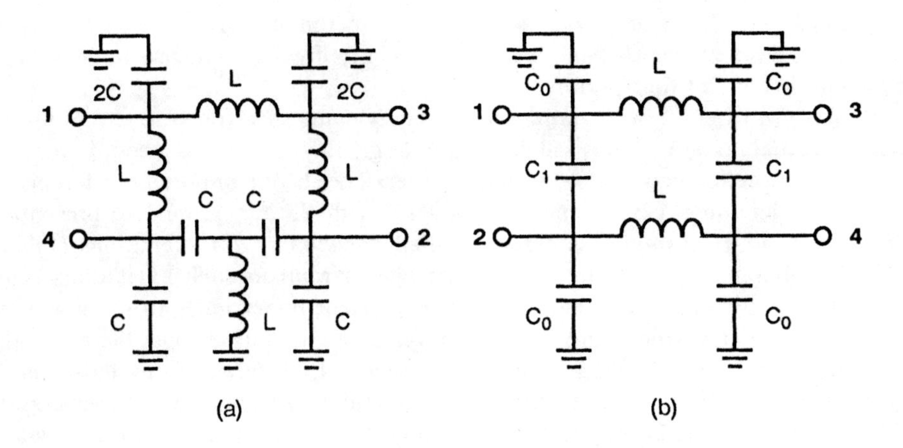

Figure 7.16 Lumped-element hybrids: (a) 180 deg; (b) 90 deg.

The design equations are

$$C_1 = \frac{1}{\omega R\sqrt{K}}$$

$$L = \frac{R}{\omega\sqrt{1 + \omega C_1 R}} \tag{7.6}$$

$$C_0 = \frac{1}{\omega^2 L} - C_1$$

where K is the ratio of powers at the output and coupled ports; that is, for a 3-dB hybrid, $K = 1$. This design procedure results in a relatively narrowband coupler; the coupling is within 0.2 dB of the bandcenter value over only a 2% bandwidth, although the VSWR bandwidth is much broader. Reference [17] includes design charts and tables that can be used to design a much more broadband coupler.

7.1.3 Baluns

Microwave baluns are structures that couple a balanced transmission line to an unbalanced line (the word *balun* is, in fact, an acronym derived from the terms *bal*anced and *un*balanced). Many mixers, especially doubly balanced diode mixers, use baluns instead of four-port hybrids.

Baluns are often viewed as microwave equivalents of a transformer. Although microwave baluns are often used in mixers (e.g., the doubly balanced diode ring mixer) as a structure equivalent to a transformer, there is one important difference: a transformer having a center-tapped secondary, with the tap grounded, presents a short circuit to even-mode (sometimes called *common-mode*) excitations (Figure 7.17). The balun, on the other hand, is an open circuit to such excitations. This means that transformers and baluns present different impedances to the diodes at mixing frequencies where the diodes' currents are in phase, and therefore the performance of mixers using baluns is, in general, different from those using transformers. Unfortunately, much of the conventional wisdom about the operation of balanced mixers is based on the assumption that transformer baluns are used at the RF and LO ports.

Parallel-Line Balun

The most common baluns used in mixers are parallel-line and Marchand baluns. Figure 7.18(a) shows a parallel-line balun. It consists of a section of parallel-strip transmission line one-quarter wavelength long; this section can be used as a transformer to match a value of Z_L other than Z_s. The simplest analysis of this structure is to view the parallel-strip line as a transformer between the microstrip and the load; then

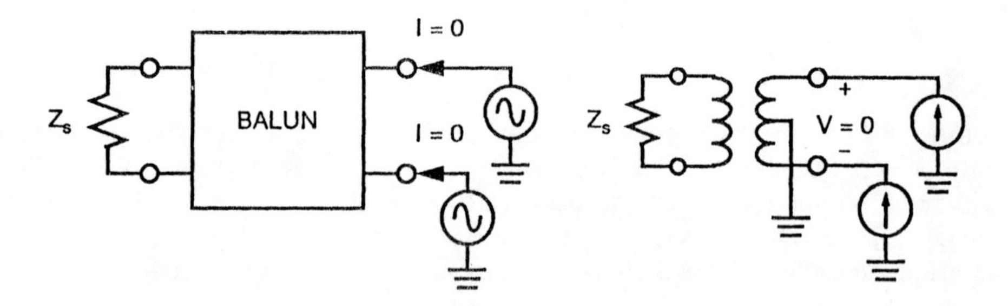

Figure 7.17 A fundamental difference between baluns and a transformer: when the outputs are excited by even-mode signals (i.e., signals having the same magnitude and phase), the balun is an open circuit, but the transformer is a short circuit.

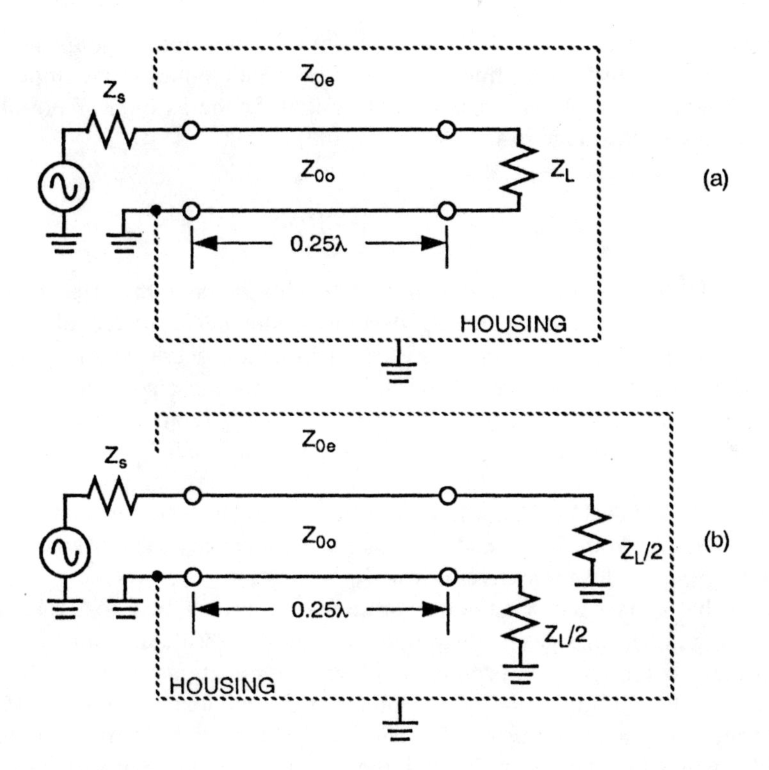

Figure 7.18 Parallel-line balun. Z_{0o} is the odd-mode impedance of the coupled lines; Z_{0e} is the even-mode impedance. Note that Z_{0o} is half the parallel-plate line's characteristic impedance. The balun can be used (a) with a "floating" (ungrounded) load, or (b) as a phase-splitter driving two grounded loads. In (b), the ports are not all matched.

$$Z_{0p} = \sqrt{Z_s Z_L} \tag{7.7}$$

Although (7.7) gives the correct characteristic impedance, Z_{0p}, of the parallel-strip line, a model of the structure as a simple transmission line is not adequate to describe all aspects of the balun's performance. The capacitance between the strips and the housing (which is the ground surface) has a significant effect on the balun's operation, and the transmission-line model does not include that capacitance. It is best to treat the parallel-strip section as a pair of coupled lines; this way, the strip-to-housing capacitance is modeled by the even-mode characteristic impedance, Z_{oe}. The parallel-strip transmission-line impedance is twice the odd-mode impedance of the coupled lines, Z_{oo}. Clearly, it is necessary that Z_{oe} be as high as possible (the strip-to-housing capacitance is as low as possible) and

$$Z_{0o} = 0.5 Z_{0p} = 0.5\sqrt{Z_s Z_L} \tag{7.8}$$

If Z_{0e} is high enough, this balun has very wide bandwidth, often one to two octaves in practical circuits. As Z_{0e} decreases, the performance of this balun degrades in several ways. Not only does the bandwidth decrease and port VSWR increase, but the balun develops a notch in its transmission characteristic precisely at center frequency*. Unfortunately, this type of balun is relatively sensitive to low values of Z_{0e}; as a general rule, a parallel-strip balun must have $Z_{0e} > 10\, Z_{0o}$ to avoid these difficulties. Such a high value of Z_{0e} is virtually impossible to achieve unless the balun is fabricated on a low-permittivity substrate and is located far (i.e., tens of line widths) from the housing or other ground surface. The Marchand balun, as we shall see, is far less sensitive to low Z_{0e}.

Often a balun is used as a phase-splitter and drives two loads having the impedance $0.5\, Z_L$. In this case, illustrated by Figure 7.18(b), the end of each load not connected to the balun is grounded. In such applications, a low value of Z_{0e} degrades both the amplitude and phase balance at the two loads. In mixers, the balun is often used this way, driving a pair of diodes. These diodes may or may not be grounded. Even if they are not grounded, the mixer may operate properly only if the connecting node between the diode is a "virtual ground," causing the voltage to be the same at both diodes. In such circuits, low Z_{0e} may degrade performance much as if the diodes were physically connected to ground.

Marchand Balun

The Marchand balun [18] is a bit more complicated, but has greater bandwidth and less sensitivity to low even-mode impedance. Through the use of multiple quarter-wave sections, it is theoretically possible to achieve a Chebyshev response up to six octaves wide [19], [20] (although three to four octaves is rarely exceeded in

* This notch can be modeled by a circuit-analysis program only if a "floating" load, as shown in Figure 7.18(a), is used. If the load consists of two resistors, each connected from one output terminal to ground, the notch is not observed. the author has seen experimental evidence of this phenomenon.

practical baluns). Figure 7.19 shows a Marchand balun. It consists of two coupled sections, each one quarter wavelength long; each section has one terminal grounded, and the load is connected between the terminals opposite the grounded ones (it can also be used, like the parallel-line balun, as a phase splitter driving two loads). The structure is surrounded by a ground plane, usually the housing in which the balun is mounted.

The rightmost pair of lines in Figure 7.19 is open-circuited at its right end. It is therefore an open-circuit quarter-wave stub and has zero input impedance. Because of this low impedance, the points b - b' in Figure 7.19 are electrically connected together, and the right side of Z_L, point b', is effectively connected to point a on the upper strip. Initially, the Marchand balun appears to be little more than a glorified parallel-line balun.

This balun, however, is more than just a parallel-plate balun having a stub in series with the load. The extra coupled-line section provides two benefits. First, the stub provides a second resonance that increases the balun's bandwidth. Second, the lower strip of each coupled-line section and the housing realize a coaxial line. The impedance of these coaxial lines is Z_{0e}, their lengths are one-quarter wavelength, and they are shorted. Thus, even-mode currents in Z_L excite the input of a pair of high-impedance, quarter-wave stubs, a very high impedance. Because these stubs' inputs are effectively in series, the even-mode input impedance of the balun is very high, much higher than that of the parallel-line balun.

As with the parallel-strip balun, the even-mode impedance of the coupled lines is

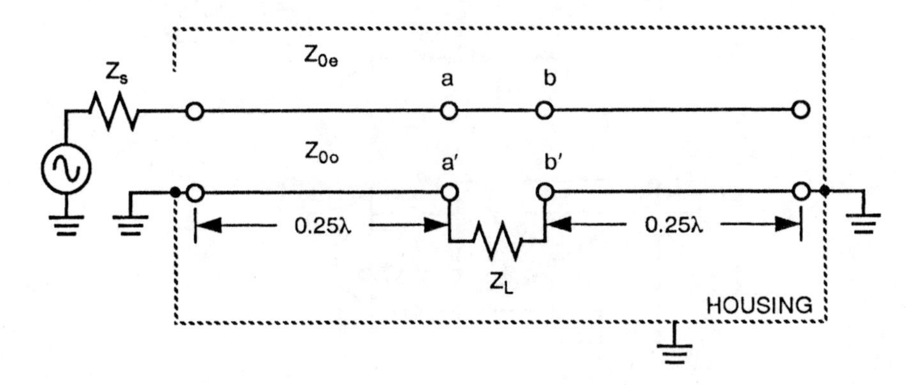

Figure 7.19 Marchand balun.

ideally infinite, and the odd-mode impedance is given by (7.8). The minimum Z_{0e} in the Marchand balun is much lower than in the parallel-strip balun; usually $Z_{0e} \cong 3Z_{0o}$ is adequate to achieve good performance in mixer applications.

7.2 SINGLY BALANCED MIXERS

7.2.1 General Concepts

Figure 7.20 shows the two fundamental types of singly balanced mixers: those using 180-deg and 90-deg (quadrature) hybrids. In both cases the balanced mixer consists of two complete single-diode mixers connected to two mutually isolated ports of the hybrid. These "mixers" can be complete, single-diode mixers; more often, however, they are just single diodes. The other two ports, also mutually isolated, are used for the RF and LO inputs. Typically, the IF outputs from the two mixers are connected in parallel, although they sometimes are combined by another hybrid.

Virtually all singly balanced diode mixers are one of these two types. Therefore, the fundamental properties of all mixers of each type are at least qualitatively the same. Differences in performance between individual designs are due primarily to details: the type of hybrid used, the use or absence of matching circuits, or the use of

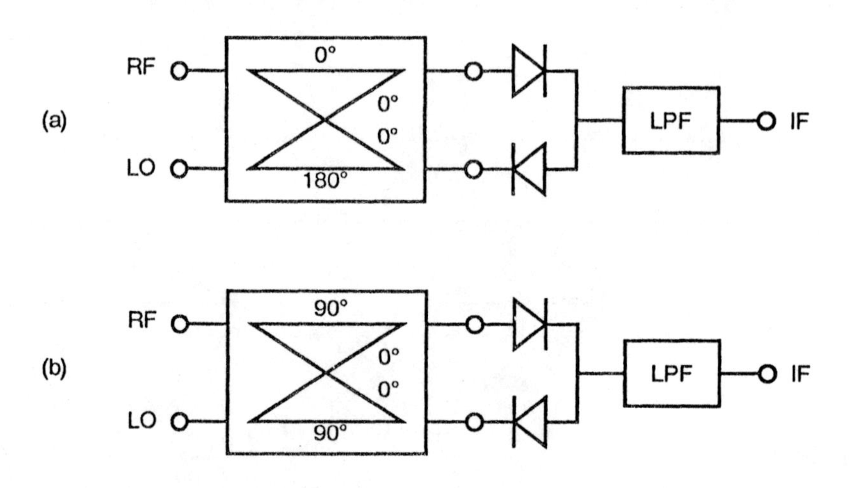

Figure 7.20 Singly balanced mixers using (a) 180-deg and (b) 90-deg hybrids.

dc bias.

We must examine several properties of balanced mixers. These properties are the following:

1. *Diode orientation and RF/LO phases.* Balanced mixers have special properties, such as the rejection of spurious responses and LO noise, because of the phases of the RF and LO at the two diodes. These, in turn, dictate a specific diode orientation; otherwise, the mixer will reject the desired signal instead of the undesired responses. Unfortunately, the necessary diode orientation is often inconvenient; for example, it may preclude the use of dc bias. Ways to circumvent this problem exist, but care must be taken in implementing them so that the spurious-rejection properties of the mixer are not changed or eliminated. True balanced mixers are not simply two mixers connected together via a hybrid coupler. It is possible to propose many mixers that look like balanced mixers, but are not; they are simply two single-diode mixers in parallel.
2. *AM noise-rejection properties.* In Chapter 5 we noted that balanced mixers have the ability to reject AM LO noise. In an ideal balanced mixer, this rejection is perfect; in real balanced mixers, the rejection is limited by the balance of the hybrid and the conversion-loss match between the single-diode mixers. In millimeter-wave mixers, this is an important concern: millimeter-wave LO sources are often so noisy that AM noise, rather than conversion loss and diode noise, is sometimes the dominant factor in establishing the receiver's noise temperature.
3. *Spurious-response rejection.* Balanced mixers have the delightful property of rejecting certain spurious responses associated with the even harmonics of either the LO or the RF, or both. Not all such responses are rejected by every type of balanced mixer, however, so it is important to select the proper type of mixer for any system where rejection of spurious responses is important. The need for "spur"' rejection is often the pivotal reason for using a balanced mixer instead of a single-diode mixer.
4. *LO-to-RF isolation.* The LO-to-RF isolation of the 90-deg and 180-deg hybrid mixers are different and have important implications in systems.
5. *Equivalent single-diode mixer.* It is possible to reduce any singly balanced mixer to an equivalent single-diode mixer having the same conversion loss and noise temperature. Hence, the theory presented in Chapter 4 can be used to analyze balanced mixers.

We shall address these matters as we describe each type of balanced mixer.

7.2.2 180-Deg Hybrid Mixers

A 180-deg hybrid mixer is shown in Figure 7.20(a). It consists of two individual mixers connected to two mutually isolated ports of the hybrid, with the other ports used for the LO and RF inputs. These ports are also mutually isolated, so the LO-to-RF isolation of the mixer is usually about as good as that of the hybrid itself, even if the individual mixers' input VSWRs are relatively poor. If the hybrid is ideal, the input VSWRs at the LO and RF ports are the same as those of the individual mixers.

The theory necessary to calculate the conversion loss and noise temperature of a diode mixer was presented in Chapter 4. Fortunately, it is not necessary to invoke such involved theory to describe qualitatively the spurious-response and noise-rejection properties of balanced mixers. It is possible to describe these special properties by considering only the RF, LO, and IF frequency components of the diode's voltage and current, and by approximating the *I*/*V* characteristic by a power series. It is necessary to consider only the *I*/*V* characteristic of the diode; including the effects of its nonlinear junction capacitance would complicate the description but would not change the results significantly.

One might rightly object to the use of a polynomial series to describe the *I*/*V* characteristic of a Schottky diode, because a series representation must at best be very long and at worst may not converge. Such objections, however, miss the point: we use the power series to show how the even and odd symmetries of the I/V characteristic of the diode (or, for that matter, of any nonlinear element) affect the mixer's properties. These are qualitatively the same for both the strong nonlinearity of the diode or the weaker nonlinearity expressed by the polynomial.

Figure 7.21(a) illustrates schematically the operation of the balanced mixer. In this case, the LO is applied to the delta port of the hybrid, so the LO voltage has 180 deg phase difference at the two diodes. The RF is applied to the sigma port, so the RF voltage is in phase at the diodes. Because the diodes are reversed, the junction-conductance waveforms of the two diodes are in phase. The small-signal current in the time-varying conductance is

$$i(t) = g(t)v(t) \tag{7.9}$$

where $v(t)$ is the total small-signal voltage across the diode, consisting of voltage components at all the mixing frequencies. The applied RF voltage is in phase at the diodes, and the conductance waveforms are in phase, so all the voltage and current components in the diodes, including those at the IF frequency, must also be in phase. Hence, the IF currents combine at the node joining the diodes.

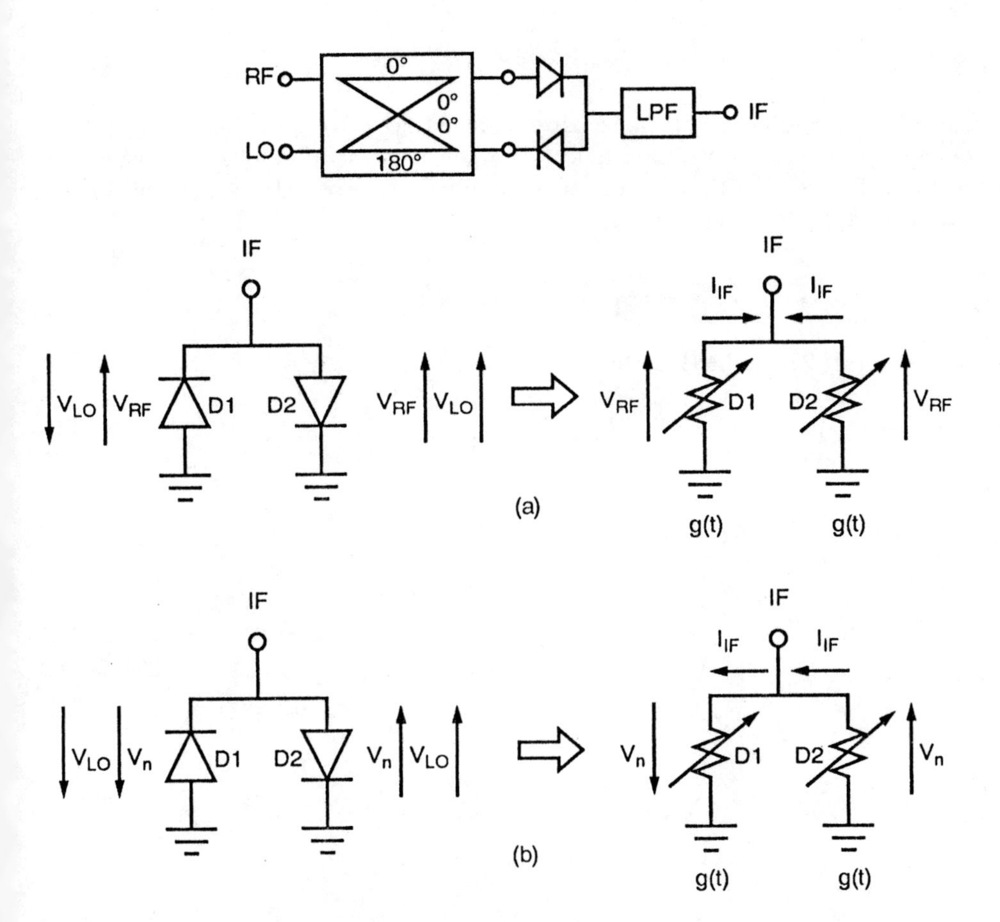

Figure 7.21 LO and RF phases in the balanced mixer: (a) normal operation; (b) AM noise from the LO (V_n).

The situation is not the same for the LO noise-voltage components shown in Figure 7.21(b). These enter the mixer at the LO port, are 180 deg out of phase at the diodes, and cancel at the IF output. Thus, AM noise and spurious signals on the LO (but not phase-modulated spurious signals) are canceled in the ideal, 180-deg mixer.

We approximate the diode junction's I/V characteristic, as shown in Figure 7.22(a), by the power series

$$I_1 = aV_1 + bV_1^2 + cV_1^3 + dV_1^4 + \dots \tag{7.10}$$

where V_1 is the total ac voltage across the diode (the RF plus the LO voltage), I_1 is the current, and the lower-case letters rare constants. If the diode is reversed, as shown in Figure 7.22(b), only the applied voltage is reversed, so the signs of the odd-power terms become negative:

$$I_2 = -aV_2 + bV_2^2 - cV_2^3 + dV_2^4 + \dots \tag{7.11}$$

From Figure 7.22(c), the IF current is

$$I_{IF} = I_1 - I_2 \tag{7.12}$$

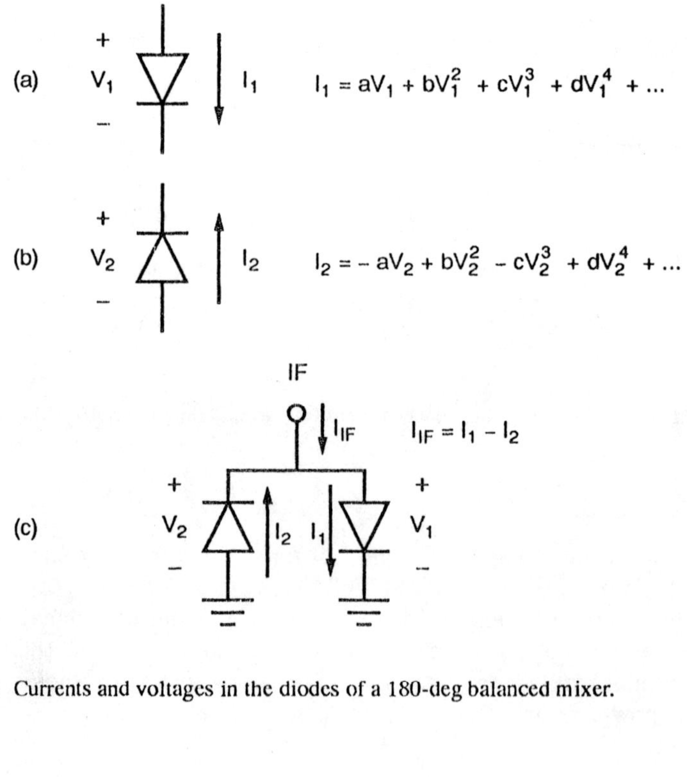

Figure 7.22 Currents and voltages in the diodes of a 180-deg balanced mixer.

With the LO applied to the delta port of the hybrid,

$$V_1 = -V_L \cos(\omega_p t) + V_{RF} \cos(\omega_p t) \tag{7.13}$$

$$V_2 = V_L \cos(\omega_p t) + V_{RF} \cos(\omega_p t) \tag{7.14}$$

Equations (7.13) and (7.14) are substituted into (7.10) and (7.11). These are, in turn, substituted into (7.12), and trigonometric identities are used to find the current components at the various mixing frequencies. The following results are obtained:

1. kth-order responses (those arising from mixing between $mf_{RF} + nf_{LO}$ where $m + n = k$) arise only from the terms of the kth power in (7.10) and (7.11).
2. All (m, n) spurious responses, where m and n are even, are eliminated.
3. The (m, n) spurious response is eliminated if m is even and n is odd, but not if m is odd and n is even.

Interchanging the LO and RF ports does not change the conversion loss and noise temperature, but it does change the spurious-response rejection. In this case, the diodes' conductance waveforms are out of phase, but the applied RF voltage is also out of phase, so the diodes' IF currents combine in phase. AM LO noise is still rejected, because the LO noise is applied in phase to the diodes. The spurious response properties can be derived by replacing (7.13) with

$$V_1 = V_L \cos(\omega_p t) - V_{RF} \cos(\omega_p t) \tag{7.15}$$

and repeating the process. The result again is that all (m, n) spurious responses are eliminated when both m and n are even. However, in this case, the (m, n) response with m even and n odd is not eliminated, but the (m, n) response with m odd and n even is rejected. Therefore, although the selection of ports for the LO and RF does not affect conversion loss, it very significantly affects the spurious-response rejection properties.

It is possible to create a balanced mixer wherein the diodes have the same orientation; that is, both anodes or cathodes are connected to the input hybrid. This configuration has some practical advantages: it may be easier to include dc bias and is useful for certain devices that cannot be reversed (e.g., FETs). In this configuration, the IF currents from the diodes are out of phase; therefore, there must be a 180-deg IF hybrid, as shown in Figure 7.23, to subtract the individual diodes' output currents. This circuit has the same conversion performance as the

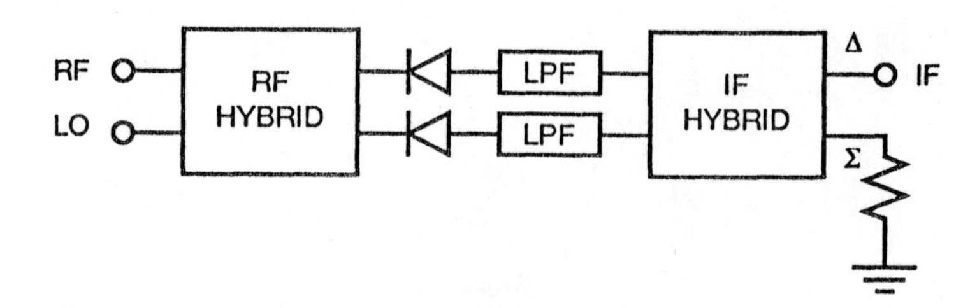

Figure 7.23 Balanced mixer having the diodes oriented in the same direction. This mixer requires an IF hybrid.

configuration having reversed diodes, but the spurious-response characteristics are the opposite.

One may wonder if it is possible to create a balanced mixer having identically oriented diodes and no output hybrid. The possibilities usually suggested are shown in Figure 7.24. We have already seen (Section 7.1.1) that it is impossible for a hybrid to have mutually isolated RF and LO ports and no phase difference; such a hybrid would be required for these mixers. Even if such a hybrid existed, the structure in Figure 7.24(a) suffers from another flaw: it is simply two single-diode mixers in parallel, not a true balanced mixer. It does not reject LO noise or most spurious responses. Paradoxically, the structure in Figure 7.24(b) does reject (m, n) spurious responses when either m or n is even, but not both. The fundamental requirement for a true singly balanced mixer is that one signal–either the RF or the LO–must be applied in phase at the diodes, and the other must be applied out of phase.

The single-diode equivalent circuit of the singly balanced mixer is relatively simple to generate. Since two separate mixers are connected to the output ports of the hybrid, the single-diode equivalent circuit is almost exactly that of one of the individual mixers. The only difference is in the IF impedance. If the IFs are connected together directly, as in Figure 7.20, the IF load for each diode is twice the mixer's IF load impedance Z_{IF}. This point is illustrated in Figure 7.25. The IF load can be separated into two loads, each of which has the impedance $2Z_{IF}$, twice the actual IF load impedance. Because the diode currents are equal, no current exists in the interconnection and breaking it has no effect. The mixer is then reduced to two entirely separate mixers having separate loads, each of which is twice the IF load impedance of the original balanced mixer. For the mixer configuration in Figure 7.23, the IF load impedance of the single-diode equivalent circuit is the port impedance of the IF hybrid. In all cases, the balanced mixer requires twice the LO

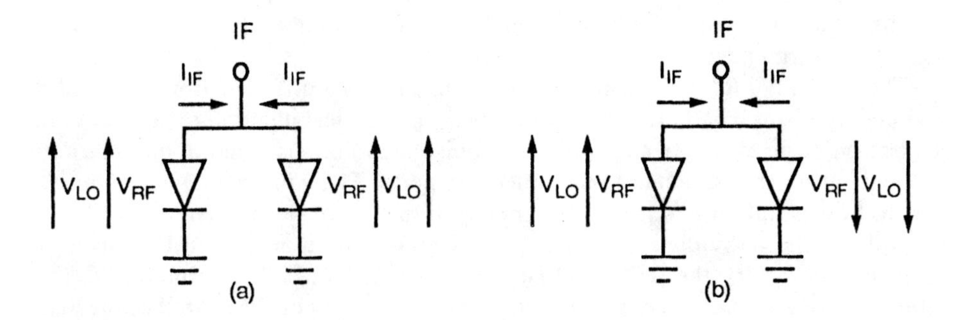

Figure 7.24 Two circuits that work, but are not true balanced mixers: (a) is simply two mixers in parallel; (b) rejects certain spurious responses, but not LO noise; furthermore, it has no inherent RF-to-LO isolation.

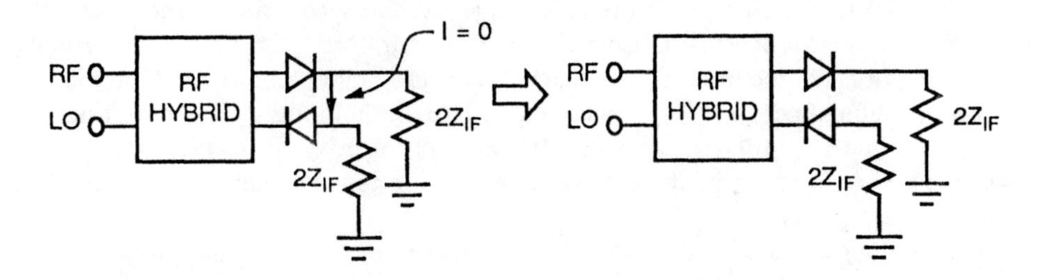

Figure 7.25 Single-diode equivalent circuit of the singly balanced mixer. The IF load can be split into two separate loads, dividing the diodes and IF network into two separate mixers.

power of the single-diode equivalent, and provides twice the maximum output power. The conversion loss is, however, the same.

7.2.3 Quadrature-Hybrid Mixers

Figure 7.20(b) shows a balanced mixer using a quadrature hybrid. As with the 180-deg hybrid mixer, the diodes are connected to two mutually isolated ports, and the RF and LO are connected to the other pair of mutually isolated ports. The mixer is symmetrical, so interchanging the RF and LO ports does not affect any aspect of the mixer's performance.

The port VSWRs and RF-to-LO isolations are very different from those of the 180-deg hybrid mixer. One important property of an ideal quadrature coupler is that the transmission loss between two mutually isolated ports is equal to the return loss of identical loads connected to the other two ports. Thus, in quadrature mixers, the port-to-port isolation is equal to the input return loss of the individual mixers; for example, if the individual mixers are identical and the input return loss at the RF frequency is 10 dB, the RF-to-LO isolation is also 10 dB. Furthermore, the input return loss of any port is the sum of the output return loss of the source connected to the other port, plus twice the return loss of the identical loads. The input VSWR at either port, therefore, depends on the input VSWR of the individual mixers *and the source* VSWR *of the other port.* For example, if the LO source's return loss is 6 dB at the RF frequency, and the mixers' input return loss is 10 dB as before, the RF input return loss is 26 dB.

Usually, the input return losses of the individual single-diode mixers used to realize a balanced mixer is relatively poor, and it is to be expected that the RF source's VSWR is also poor at the LO frequency and vice versa. Hence, both the input VSWR and the RF-to-LO isolation of the quadrature-hybrid mixer are usually relatively poor. Worse is the fact that a poor LO-source return loss at the RF frequency unbalances the RF power applied to both diodes, thus upsetting the mixer's balance. Similarly, a poor RF-port return loss at the LO frequency unbalances the pumping of the diodes. These limitations account partially for this mixer's unpopularity.

The same procedure used for the 180-deg hybrid mixer can be used to determine the noise-rejection and spurious-response properties of the quadrature-hybrid mixer. These are the following:

1. The mixer is a true balanced mixer, with LO noise rejection properties identical to those of the 180-deg mixer.
2. It has no appreciable (m, n) spurious-response rejection, however, if either m or n is even but the other is odd.
3. It has the same (m, n) spurious-response rejection as the 180-deg mixer if both m and n are even.

As with the 180-deg mixer, orienting the diodes in the same direction requires the use of an output 180-deg hybrid to subtract the IF output currents of the individual diodes. The single-diode equivalent circuit of the quadrature mixer is the same as that of the 180-deg hybrid mixer. Like the 180-deg mixer, if no output hybrid is used, the effective IF load impedance presented to each of the individual diodes is twice that of the actual IF load.

Quadrature-hybrid mixers have one other surprising (and little-recognized) property: configured as shown in Figure 7.20, they do not work as upconverters! Specifically, if a low-frequency signal is applied to the IF port, there is no RF output. To make the mixer operate as an upconverter, one of the diodes must be reversed; however, in this case it will not work as a downconverter. This unilateral property of the quadrature-hybrid mixer may be advantageous in some applications.

7.3 DOUBLY BALANCED MIXERS

7.3.1 Ring Mixers

The most commonly used mixers between 1 and 18 GHz are doubly balanced mixers of the ring or star configuration. Their advantages over the mixers considered earlier are inherent isolation between all ports, rejection of LO noise and spurious signals, rejection of spurious responses and certain intermodulation products, and extremely broadband operation. Their disadvantages are the need for at least four diodes and two hybrids, greater LO power requirements, and generally higher conversion loss than single-diode or singly balanced mixers.

To describe qualitatively the operation of the ring mixer, we treat the diodes as switches operated by the LO. This model is adequate for a qualitative description only; it is not accurate enough for quantitative use. A quantitative understanding of the mixer's operation must be obtained, as before, via the techniques of Chapter 4; this requires full knowledge of the embedding impedances and diode parameters of the equivalent single-diode circuit.

Figure 7.26 shows a doubly balanced ring mixer. It consists of two transformers similar to the transformer hybrid discussed in Section 7.1.2 and a ring of identical diodes. The diodes are usually fabricated simultaneously on a common semiconductor substrate, and therefore have nearly identical characteristics. The RF signal is applied to the primary of one transformer, and the LO is applied to the primary of the other. The center tap of the LO transformer's secondary is grounded, and the center tap of the RF secondary serves as the IF output. (In theory, the LO center tap could be used for the IF output, but the LO-to-IF isolation, which is usually more critical than the RF-to-IF isolation, would be worse.)

The discussion of the transformer hybrid in Section 7.1.2 showed that the center taps of the transformers are virtual grounds. If two identical loads are connected in series across the entire secondary, their connection point is also a virtual ground. In Figure 7.26 the points A and A′ are therefore virtual grounds for the LO, and B and B′ are virtual grounds for the RF. Since the RF transformer's secondary is connected

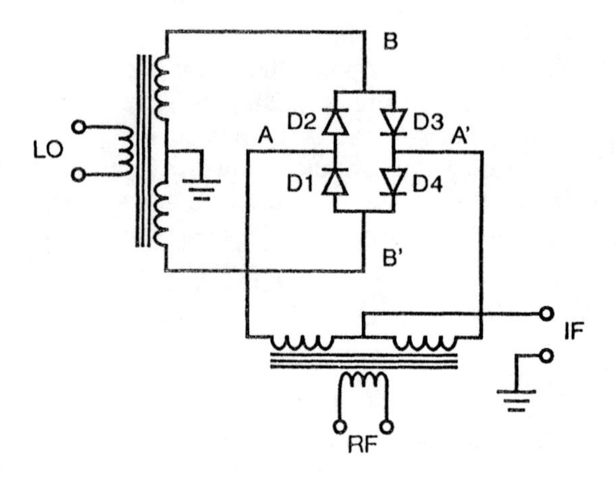

Figure 7.26 Doubly balanced diode-ring mixer.

to the LO's virtual-ground nodes, and the LO's secondary to the RF virtual grounds, the LO-to-RF isolation is (theoretically) infinite. Also, one can ignore the RF transformer while examining the LO circuit, and vice versa.

When LO power is applied, an ac LO voltage is applied to the nodes B and B′. When B is positive and B′ is negative, the diodes D3 and D4 are turned on and D1-D2 are reverse biased. They are reverse biased by a voltage equal to the forward turn-on voltage of the other pair, but this is, nevertheless, enough to make them effectively open circuits. If the junction and series resistances on the forward-conducting diodes, D3 and D4, are low enough that they can be approximated as closed switches, node A′ is connected to the RF virtual grounds at B and B′, and is therefore grounded. Node A is open-circuited, however, as is the half of the RF secondary connected to it. The result is that the right secondary of the transformer is connected to the IF output. In the next half-cycle of LO voltage, D3 and D4 are turned off and D1 and D2 are on. Node A is then grounded, node A′ is open circuited, and the other side of the RF transformer is connected to the IF.

The RF equivalent circuit is shown in Figure 7.27, with the diode pairs depicted as switches. The two switches turn on and off alternately during each half-cycle of the LO waveform. As a result, the polarity of the RF voltage, applied to the IF load, is reversed periodically at the LO frequency. This is equivalent to multiplying the RF signal by an LO-frequency square wave, and frequency conversion occurs via the product of the applied RF signal and the fundamental Fourier component of that

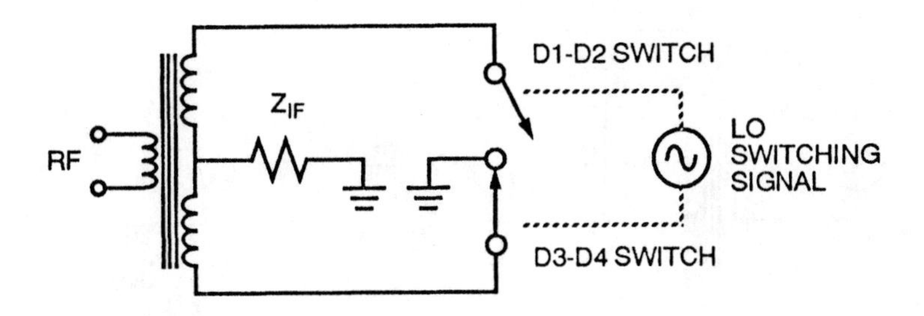

Figure 7.27 The ring mixer's equivalent circuit. This mixer is fundamentally a switching mixer.

square wave. Even if the diodes do not operate exactly as switches and their conductance waveform is not precisely a square wave, the conductance waveform has half-wave symmetry and does not contain any even-harmonic components. Therefore, the mixer has no spurious responses that involve the even harmonics, particularly the second harmonic, of the LO.

By viewing the diodes as nonlinear elements rather than switches, one can learn quite a bit more about the properties of the mixer. Figure 7.28(a) shows the mixer's LO equivalent circuit. The virtual grounds at A and A′ are shown explicitly as real ground connections, and it is clear from the figure that the structure can be divided into two separate subcircuits, each containing two antiparallel diodes. Equations (7.10) and (7.11) give the current in the diodes, as before.

From the discussion following (7.11), we conclude that the antiparallel diode pair conducts no current at even harmonics of the LO frequency. However, each individual diode has current components at all LO harmonics. Therefore, the even-harmonic currents in each diode must be equal and opposite, so each diode effectively short-circuits the other at even-harmonic frequencies, and these frequency components simply circulate in the loop created by the two diodes. This phenomenon can be modeled by making the embedding impedance zero at all even-harmonic frequencies. The odd-harmonic embedding impedances are twice the output impedance of each half of the transformer secondary. If this equivalent circuit is analyzed, it will be found that the diode voltage has the same symmetrical waveform as an antiparallel pair.

Figure 7.28(b) shows the RF equivalent circuit of the mixer. The upper pair of diodes, D1 and D2 in Figure 7.26 and Figure 7.27, has conductance waveform $g(t)$, and the lower pair, D3 and D4, has the waveform $\overline{g(t)}$, where the overline indicates that the waveform has the same shape as $g(t)$, but is shifted in time by one-half

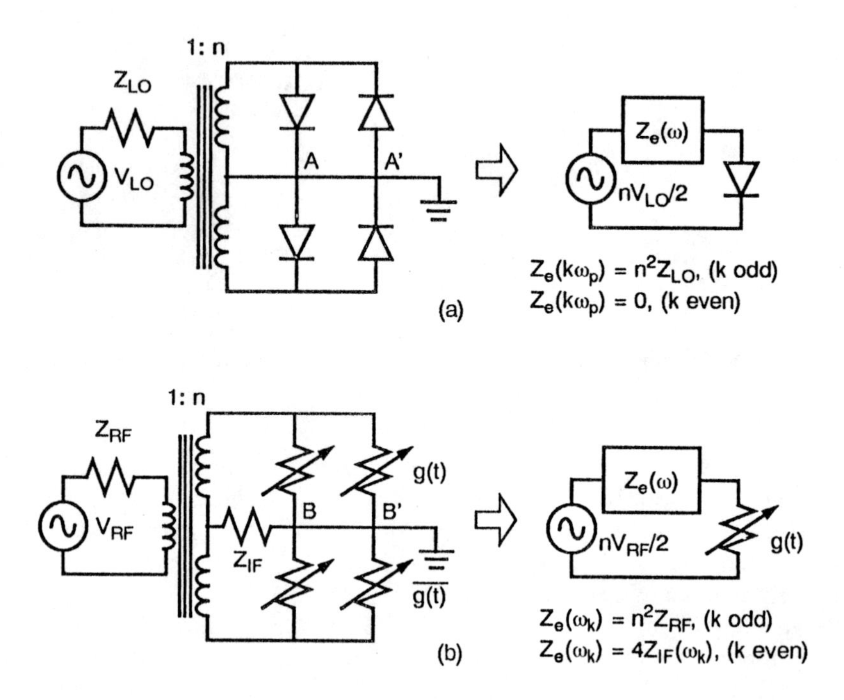

Figure 7.28 Single-device equivalent circuits of the ring mixer: (a) LO equivalent circuit; (b) small-signal (RF/IF) equivalent circuit.

period. Again we can use the approach described after Equation (7.11) to generate a single-diode equivalent circuit, and to find the equivalent embedding impedances. At mixing frequencies ω_k where k is odd, the currents in the upper and lower diode pairs are in phase, the point B-B′ is a virtual ground, and the RF excitation is simply applied to the diodes through the transformer. There is no current in Z_{IF} at these frequencies. At ω_k, where k is even (including ω_0), the currents are opposite in the diode pairs. All the current at these frequencies circulates in opposite directions in the transformer secondaries and in the same direction through Z_{IF}.

At the mixing frequencies where k is even, the difference between a transformer (used in the ideal circuit) and a microwave balun (used in real circuits) becomes evident. In the ideal circuit, the even-mode currents in the transformer secondaries generate no voltage, so the transformer is effectively a short circuit in series with

Z_{IF}. Thus, in the ideal (transformer) circuit, the embedding impedance is Z_{IF}. Virtually all microwave baluns are open circuits to such excitation, however, so in the real circuit, the embedding impedance is virtually infinite.

Obviously, an open circuit in series with Z_{IF} is unacceptable, if only at the output frequency ω_0. Therefore, ring mixers using baluns must include some sort of a structure to bypass the IF current around the balun. This is a complication of the design of such mixers, and inevitably limits the IF bandwidth. Higher-order mixing products ω_k with k greater than zero and even, however, may still have open-circuit (or at least very high) embedding impedances. It is important to note that much of the early theoretical work on such mixers, which led to the conclusions about such things as their minimum conversion loss, were based on the transformer model, and may not be applicable to real microwave mixers using baluns.

Finding the rest of the single-diode equivalent circuit is straightforward. Because all four of the diodes are in parallel at the IF, the IF load in the single-diode equivalent circuit is $4Z_{IF}$. In the ideal circuit, all ω_k with k even have this embedding impedance. For odd k, the embedding impedance is merely twice the value of Z_{RF} transformed through half the transformer (i.e., twice $n^2Z_{RF}/2$).

The RF embedding impedance of the single-diode equivalent circuit is n^2Z_{RF}, where n is the turns ratio of the transformer. We noted in Chapter 6 that the optimum RF source impedance for an untuned single-diode mixer is approximately 50Ω. If a transformer is used, best mixer performance dictates that the transformer's turns ratio should be unity; however, $n = 2$ is more easily realized, and is commonly used, so the embedding impedance is 200Ω. For the standard port impedance of 50Ω, the IF load impedance of the single-diode equivalent circuit is also 200Ω. These embedding impedances are substantially higher than optimum in most cases, and therefore the conversion loss of the ring mixer is often unspectacular, usually 6 to 9 dB. If a balun is used instead of a transformer, the designer has greater control over the RF load impedance, but not the IF.

7.3.2 Star Mixer

The star mixer is another commonly used doubly balanced mixer. The diodes of the star mixer each have one end connected to a common node, which is used as the IF output, and the other terminals connected to a set of transformers and quarter-wave parallel-strip transmission lines. Figure 7.29 shows the circuit.

Figure 7.29 makes the star mixer look very much like a ring mixer. The orientation of the transformers is similar, and, for the same reasons as in the ring mixer, the terminals of the RF transformer's secondary are virtual ground points for the LO, and vice versa (we shall see in Chapter 8 that in the most practical balun

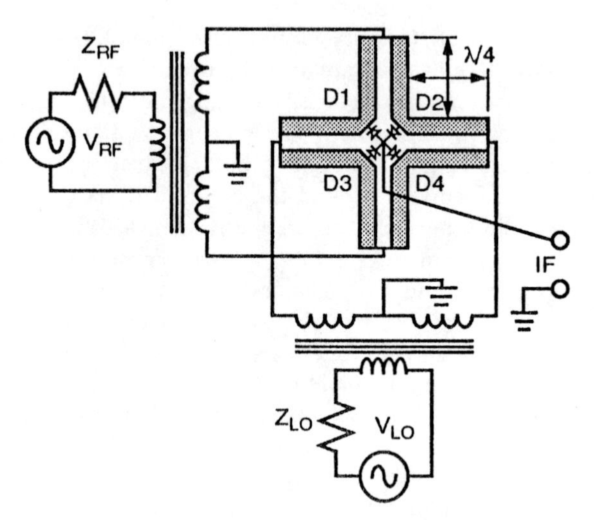

Figure 7.29 Ideal star mixer.

structure for star mixers these points are real, not simply virtual, grounds). The center of the diode "star" is a virtual ground for both the RF and LO.

The star mixer operates as a polarity-reversing switch, precisely as the ring mixer does. When the LO is applied, the diode pairs D1-D2 and D3-D4 are alternately switched on and off. When D1 and D2 are on, D3 and D4 are off, and the top terminal of the RF secondary is connected to the IF output. During the second half of the LO cycle, D3 and D4 are on, D1 and D2 are off, the bottom terminal is connected to the IF port, and the polarity is reversed. Thus, as in the ring mixer, the IF output is obtained, as shown in Figure 7.27, by multiplying the RF by a square-wave switching function at the LO frequency.

The diodes are connected to the transformers through four sets of parallel-strip transmission lines (shaded in Figure 7.29). These lines are connected to each other at the ends nearest the diodes and shorted at the point where they connect to the transformers. The strips allow the connection points of the diodes to be shorted for RF or LO excitation, but disconnected for LO or RF, respectively. They do this because each pair of strips operates as a single strip for even-mode excitation and as a shorted stub for odd-mode excitation. Thus, the upper strip, for example, is driven in an odd mode by the LO and allows LO voltage to be applied to diodes D1 and D2 while identical RF voltages are applied to the same diode terminals.

The single-diode equivalent circuit of the star mixer is the same as that of the ring mixer. These equivalents are shown in Figure 7.28.

The real value of the star configuration is in circuits realized with microwave baluns, where remarkably simple and elegant structures are possible. Because in real mixers the ends of the parallel-strip transmission lines are real grounds, this structure requires no special IF bypassing, and it consequently has the lowest parasitic IF series reactance of any balanced mixer. This low reactance results in a very broad IF bandwidth. Furthermore, the IF is dc coupled; we shall see in Chapter 8 that it is nearly impossible to make ring mixers at microwave frequencies that have this property.

7.3.3 High-Level Doubly Balanced Mixers

The point has been made more than once in the previous chapters that the only way to improve the large-signal handling capability of a diode mixer is to increase the ratio of the LO drive level to the RF signal level without overdriving the diodes. Applied LO power cannot be increased indefinitely for a given mixer without risking damage to the small diodes necessary for high-frequency operation.

One way to increase a mixer's power handling is to use two or three diodes in series in place of each single diode in the balanced mixer. The problem in this approach is that the mixer's performance—especially port-to-port isolation and spurious rejection—requires that all the series-connected diodes have identical characteristics, an improbable situation. A better way to realize a large-signal mixer is to combine two diode rings in a single mixer as shown in Figure 7.30. This "double-doubly balanced mixer" divides the RF and LO powers between twice as many diodes as a doubly balanced mixer, and it thereby achieves an improvement of 3 dB in dynamic range (compared to a doubly balanced mixer). The cost is its greater complexity and, of course, 3 dB greater LO power.

Figure 7.30 should not be taken too literally. A close inspection of the figure shows that the RF transformer is driven in an even mode by the LO. Because an ideal transformer becomes a short circuit under even-mode excitation, the RF transformer appears to short-circuit the LO. In real microwave mixers, of course, baluns would be used instead of transformers, and these are open circuits under even-mode excitation.

Another high-level mixer circuit is shown in Figure 7.31. This mixer consists of two balanced mixers combined by 90-deg hybrids in a scheme much like that of the 90-deg singly balanced mixer. Unlike the singly balanced diode mixer, however, the LO and RF VSWRs of this mixer are very good, because the reflections from the RF and LO ports of the individual mixers cancel in the hybrid. The port-to-port isolations are enhanced by the hybrids as well. The disadvantages of this structure

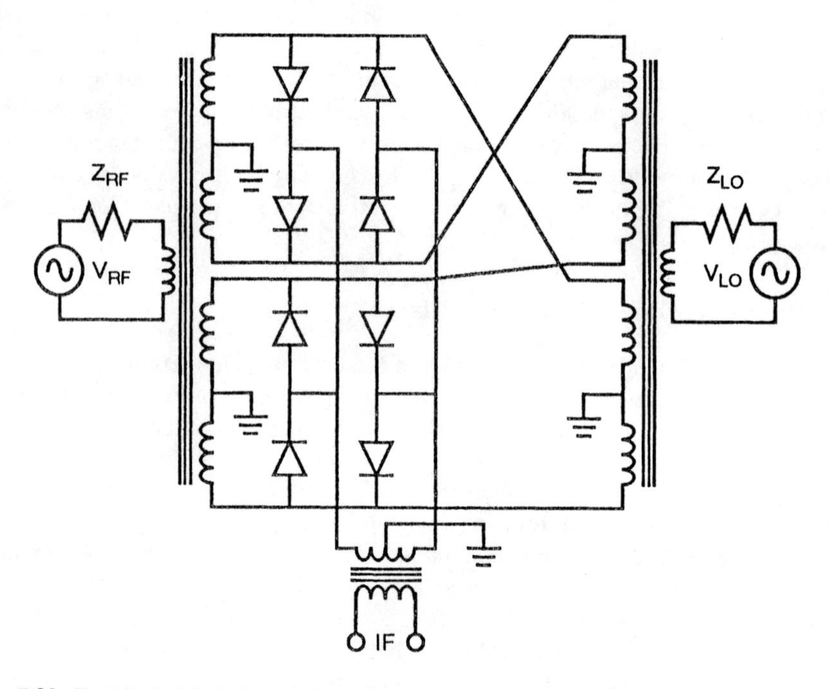

Figure 7.30 Double doubly balanced ring mixer.

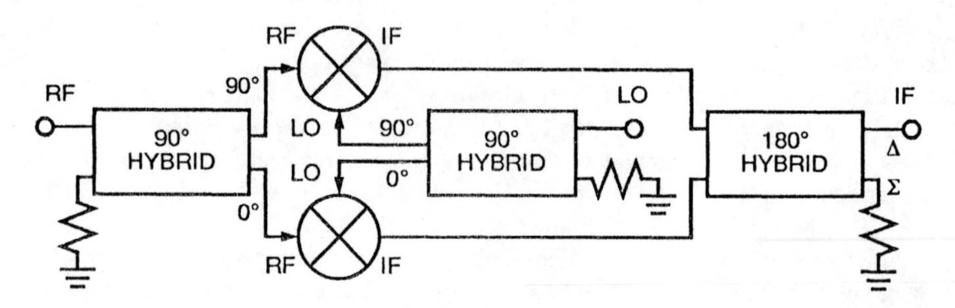

Figure 7.31 High-level mixer consisting of two "power-combined" balanced mixers. The 90-deg hybrids at the RF and LO ports provide good port VSWR and enhance the RF-to-LO isolation.

are its complexity, the signal loss in the quadrature hybrids, and the increased LO-power requirements.

7.3.4 Subharmonically Pumped Mixers

For many applications, it is expensive, inconvenient, or even impossible to generate a fundamental-frequency LO. The conversion loss and noise performance of a millimeter-wave mixer may be limited by the lack of adequate LO power or by excessive LO noise, rather than by the inherent capabilities of the mixer. In these cases, it may be wise to use a mixer that is pumped at half the LO frequency, and to mix the RF signal with the second harmonic of the junction's conductance waveform. Such *subharmonically pumped* mixers have remarkably good conversion performance, often only a decibel or two worse than comparable fundamental mixers. Even with greater conversion loss, a subharmonically pumped mixer often provides the best performance when all these other factors, especially LO noise and power, are considered.

It is possible to achieve subharmonic operation with a single-diode mixer. However, in such mixers the fundamental mixing response is usually greater than the second-harmonic response and is a source of interfering signals and downconverted LO noise. It is also an additional loss mechanism, because a large fraction of the RF input power is converted to the mixing frequencies near the LO and radiated from the LO port. Unless the IF frequency is unusually high, it is impossible to filter out this response without rejecting the LO frequency as well. Hence, single-diode subharmonic mixers are rarely used in low-noise receivers. They are sometimes used in systems where high conversion loss is tolerable and the

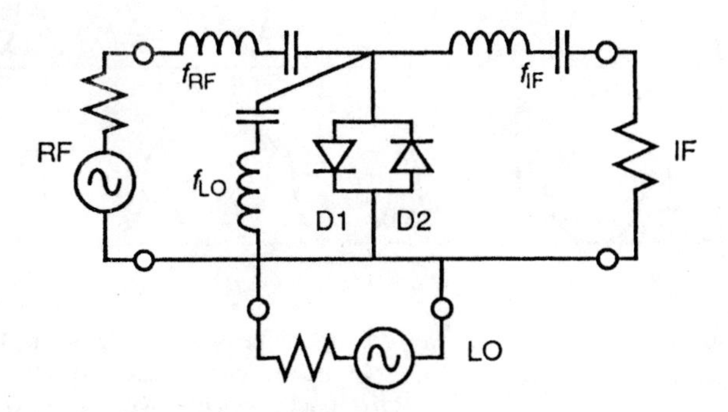

Figure 7.32 Subharmonically pumped mixer using an antiparallel diode pair.

ability to generate responses with a wide range of LO harmonics is necessary. Such mixers are called *harmonic mixers*, and their applications include frequency synthesizers and the input circuits of spectrum analyzers.

A better method for realizing a subharmonically pumped mixer is shown in Figure 7.32. If the diodes are identical, this circuit has no fundamental mixing response. The key to its operation is the use of an antiparallel diode pair. The RF and LO voltages are applied to the diode pair, and filters are used to separate the different frequencies, much as if it were a single-diode mixer. Since the RF and LO frequencies differ by approximately a factor of two, the filters are rarely difficult to realize.

The diode-junction conductance waveforms are shown in Figure 7.33, where $g_1(t)$ and $g_2(t)$ are the conductance waveforms of D1 and D2, respectively. The conductance waveform of the combination is $g_1(t) + g_2(t)$, which clearly has no fundamental-frequency component; it has only even harmonics. Like the diode pairs

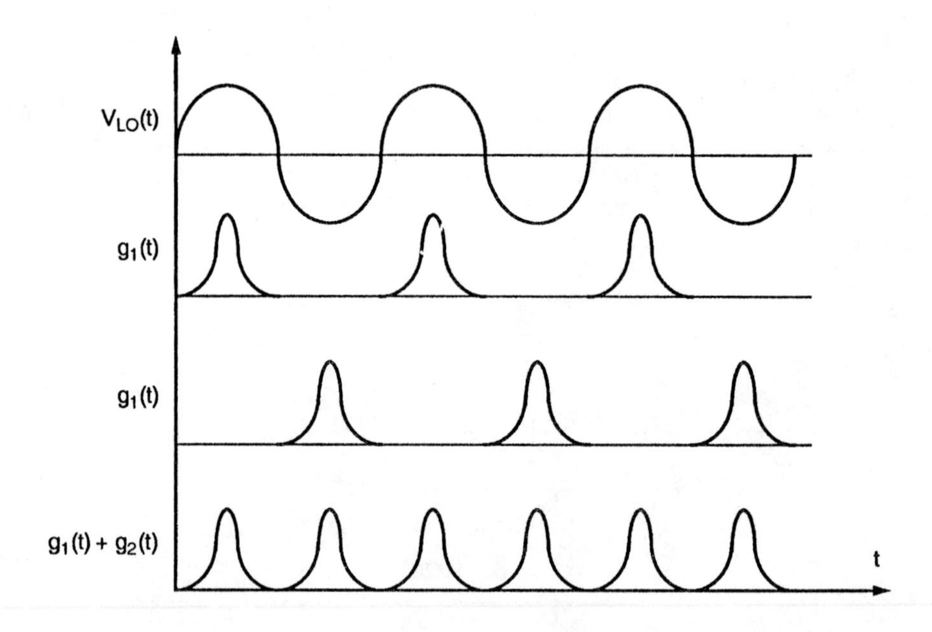

Figure 7.33 Conductance waveforms in the subharmonically pumped mixer. $V_{LO}(t)$ is the LO voltage across the diode pair, $g_1(t)$ is the conductance of D1, and $g_2(t)$ is the conductance of D2. The combined conductance waveform has no fundamental-frequency component.

in the ring mixer (Figure 7.28), each diode terminates the other in a short circuit at all even LO harmonics and at all frequencies $mf_{RF} + nf_{LO}$, where $|m| + |n|$ is even. Most important of these is, of course, the fundamental mixing response $f_{RF} \pm f_{LO}$, which is short-circuited; thus, no fundamental mixing is possible. It is important to recognize that the second-harmonic component of $g_1(t)$ or $g_2(t)$, which does the mixing, is essentially that which would be obtained in a single-diode subharmonically pumped mixer having appropriately scaled embedding impedances. The conversion loss of the antiparallel pair is therefore the same as that of the single-diode subharmonically-pumped mixer terminated in a short circuit at the appropriate frequencies.

The single-diode equivalent circuit consists of a short circuit at the odd-harmonic mixing frequencies (i.e., ω_k, k odd) and a short circuit at even LO harmonics. At the odd LO harmonics and the even-harmonic small-signal mixing frequencies, the embedding impedances is twice the impedance seen by the diode pair.

As with other balanced mixers, successful operation of the subharmonically pumped mixer requires very good circuit balance; in particular, the diodes, and all the parasitics associated with them, must have identical characteristics. At very high frequencies, this balance may be difficult to obtain; for example, a slight difference in the lengths of the contacting whiskers of the two diodes may seriously affect the performance of a millimeter-wave mixer. Furthermore, mounting and contacting two dot-matrix diodes in a single waveguide in an antiparallel configuration presents difficult mechanical design problems. For these reasons, millimeter-wave beam-lead diodes are preferred for subharmonically pumped mixers. Some manufacturers have produced and marketed antiparallel beam-lead diode pairs, consisting of two diodes having a common set of leads and nearly identical characteristics.

Subharmonically pumped mixers reject LO noise near the fundamental mixing frequency, but not at the RF frequency. LO noise is usually very low near the oscillator's second harmonic, however, and is rarely a concern. A subharmonically pumped mixer also rejects spurious responses associated with the odd harmonics of the LO and with even harmonics of the RF signal. Because it exhibits some of the properties of more conventional balanced mixers, a subharmonically pumped mixer is often considered to be a type of balanced mixer.

The design of subharmonically pumped mixers and calculation of their conversion loss is similar to that of single-diode mixers. The main differences are in the embedding impedances described earlier, and the fact that the peak value of the diode's conductance waveform is substantially less than in a fundamental mixer. As a result, the optimum RF source and IF load impedances are somewhat higher than in a fundamental mixer. Other aspects of the design procedure for unbiased single-diode mixers are generally valid for subharmonically pumped mixers. It is important to note that one deg of freedom in adjustment is lost, because diode bias cannot be applied. To compensate for this loss, greater tuning margin should be provided in the RF and IF matching circuits.

As a final note, subharmonically pumped mixers using antiparallel diode pairs are occasionally operated at the fourth harmonic of the LO. In such mixers, the mixing products near the second harmonic must be reactively terminated; this is usually a simple task, since the second LO harmonic is well separated from the fundamental LO frequency. These mixers often exhibit surprisingly good performance. A 60-GHz mixer having a 15-GHz LO and 14-dB conversion loss is typical.

7.3.5 Image-Rejection Mixers

In Section 4.6.1, we examined image-enhanced mixers, in which an image-rejection filter is used to reduce conversion loss. Because the image is a type of spurious response, it is often desirable to eliminate a mixer's image response, regardless of the effect on its conversion loss. Often the IF frequency is low and the image is too close to the RF and LO frequencies, the LO must be broadly tunable, or the RF and image bands overlap; in these instances, filtering the image is impossible, and an image-rejection mixer may be useful. The image-rejection mixers described here do not achieve any performance improvement; the energy of the image is not "recycled." It is possible to make image-rejection mixers that do provide image enhancement; such mixers are often called *image-recovery mixers* [20], [22].

Figure 7.34 shows the circuit of an image-rejection mixer. It consists of two mixers, usually doubly balanced mixers, connected to quadrature hybrids at their

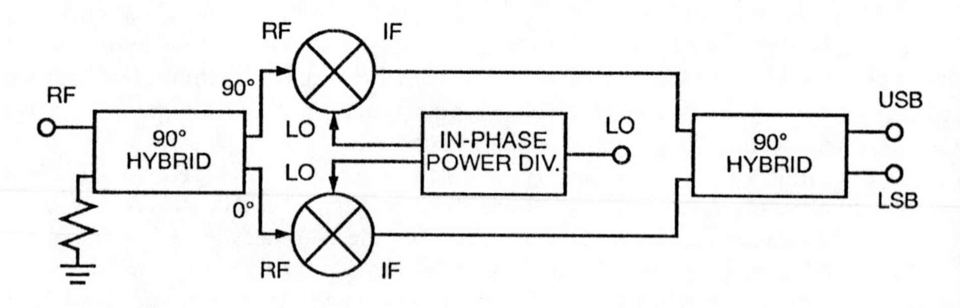

Figure 7.34 Image-rejection mixer.

inputs and outputs. The LO is split via a conventional power divider and applied in-phase to the two mixers. The RF signals above the LO frequency (*upper-sideband* (USB) frequencies) appear at one output of the IF hybrid, and lower frequencies (*lower-sideband* (LSB) frequencies) appear at the other output.

It may at first seem strange that the image-rejection mixer works at all. The configuration, at first glance, looks exactly like a quadrature-hybrid coupled amplifier, and it might be expected to have similar properties: good input and output VSWR, but no significant image rejection. In a mixer, however, the positive LSB frequency, when downconverted, becomes negative, and the negative USB becomes positive. Hence, the LSB and USB undergo different phase shifts at the RF hybrid and the IF hybrid.

Figure 7.35 illustrates the operation of an image-rejection mixer. The positive and negative frequency components of the RF input are shown separately. Note that f_L is the LSB component, and f_u is the USB component; f_{Li} and f_{ui} are the corresponding downconverted components at the IF frequency. The signals at point A in the diagram undergo no phase shift and are downconverted in the conventional manner. The signals at the other output of the RF hybrid, at point B, undergo a 90-deg delay relative to those at point A. This delay is manifest by a –90-deg phase shift in the positive frequency components, and a +90-deg shift in the negative frequency components. Phase difference is preserved in the frequency transformation, so at the IF the components have the same phase relationships as the original RF components. However, when these are subjected to the 90-deg delay in the IF hybrid, the negative IF frequencies undergo a +90-deg shift and the positive frequencies undergo a –90-deg shift. The result is shown in the lower part of the figure: the IF components from the USB terms have been shifted 180 deg, and the LSB terms again have zero phase. When these are combined with the signals at point C, the USB components cancel, and the unshifted LSB terms remain. It can easily be shown, by the same approach, that the other output of the IF hybrid contains the USB terms.

Image rejection is complete if the conversion losses and phase shifts in the mixers are identical, and the phase and amplitude balances of the hybrids are perfect. Perfection is, however, a scarce commodity, so the image rejection is generally limited. The image rejection, as a function of overall amplitude and phase imbalance, is

$$R_I = -10 \log \left(\frac{1 - 2\sqrt{G} \cos(\theta) + G}{1 + 2\sqrt{G} \cos(\theta) + G} \right) \tag{7.16}$$

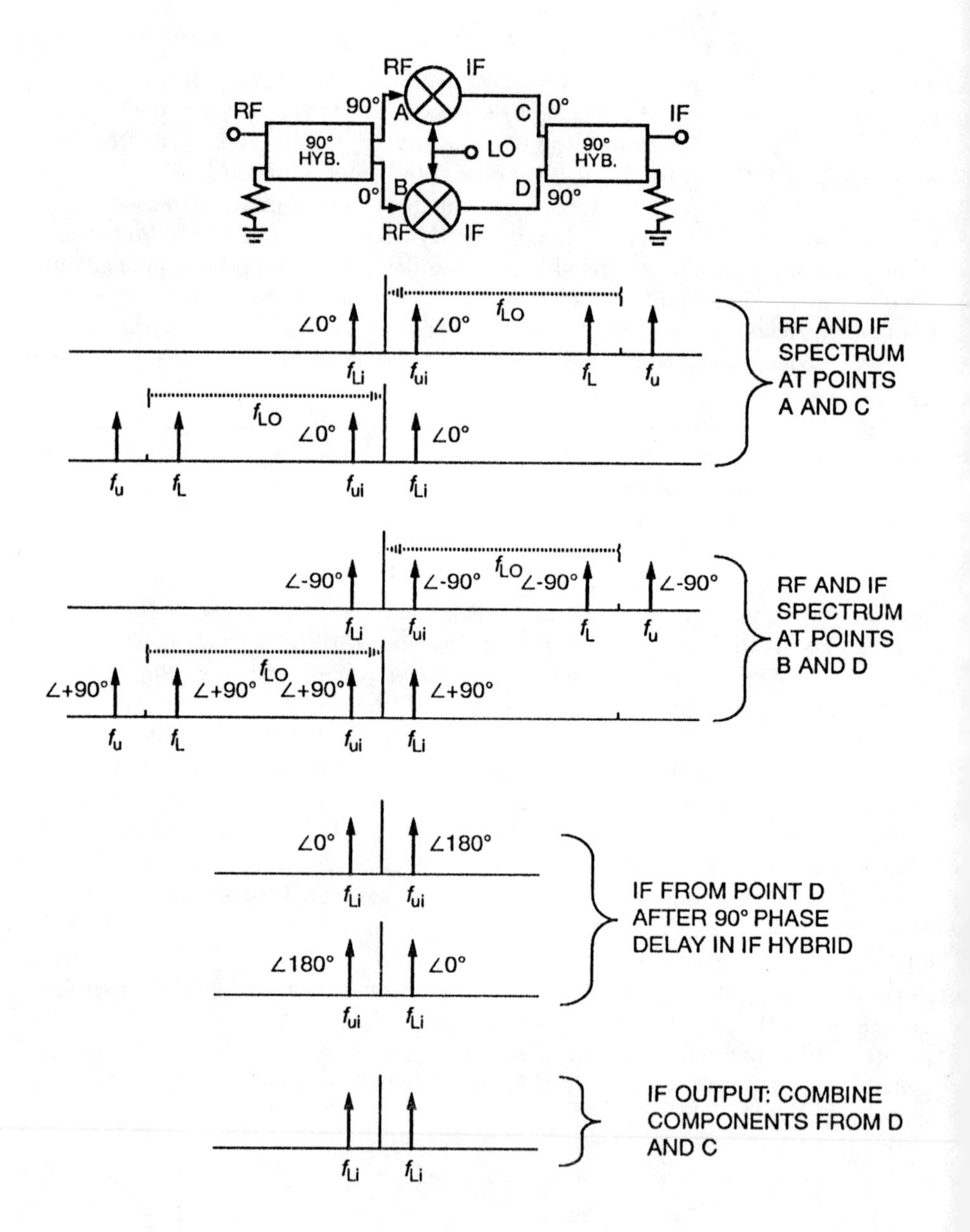

Figure 7.35 Frequency spectra in an image-rejection mixer.

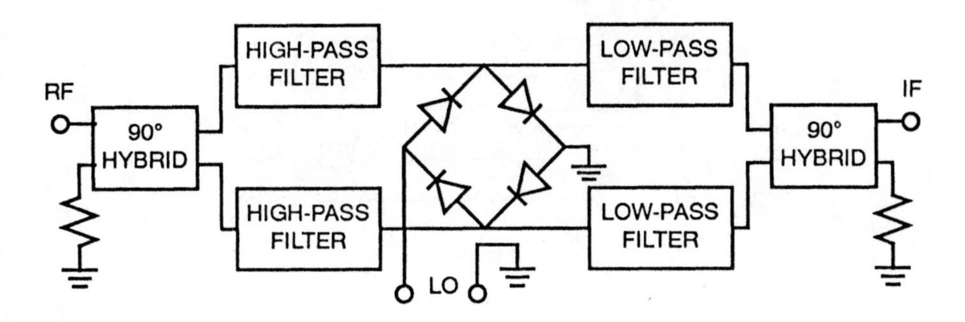

Figure 7.36 Image-rejection mixer using a diode ring.

where θ is the phase imbalance and *G* is the gain imbalance. Achieving 20 dB of image rejection requires that the phase error be kept below 10 deg and the gain imbalance below 1 dB. Practical, moderate-bandwidth mixers usually achieve image rejection of approximately 20 dB. Very broadband mixers, or those at high frequencies, where tight phase balance is difficult to achieve, have progressively less image rejection; low-frequency narrowband mixers sometimes achieve greater rejection. Careful selection and design of the hybrids in an image-rejection mixer can do much to ensure good performance. For example, coupled-line quadrature hybrids usually exhibit an accurate 90-deg phase shift over very broad bandwidths, especially if they are implemented in a TEM transmission medium such as stripline.

The image-rejection mixer can be implemented precisely as shown in Figure 7.34 with separate mixers and hybrids. A simpler circuit, however, shown in Figure 7.36, does not require an LO power divider or a pair of complete balanced mixers, but needs only a single-diode ring. It does, however, require filters to achieve good port isolation and LO rejection.

7.3.6 SSB Modulators

A doubly balanced mixer can be used as a balanced modulator. The modulating waveform is applied to the IF port, the carrier to the LO port, and the modulated waveform appears at the RF port. The modulated waveform is a *double-sideband suppressed-carrier* (DSBSC) signal. The circuit in Figure 7.37 can be used to generate a *single-sideband suppressed-carrier* (SSBSC) signal without the use of

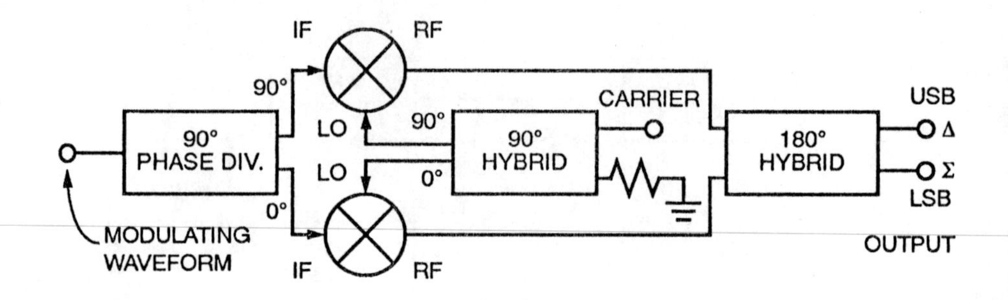

Figure 7.37 SSB modulator.

filters. In many cases, such circuits can be used to generate SSB directly at the RF frequency without requiring further upconversion. The baseband modulating signal is applied to the IF ports of the mixers through a 90-deg phase-shift network; the mixers are operated as upconverters. The modulating waveform is usually a low-frequency signal having a bandwidth of at most a few megahertz, so a conventional quadrature hybrid cannot be used to achieve the phase shift; instead, lumped-element networks are used. The LO also has a 90-deg phase difference at the mixers, and the outputs of the mixers are derived at the summing port (LSB) or subtracting port (USB) of the hybrid. The carrier suppression is determined by the LO-to-RF isolation of the mixers. The unwanted sideband suppression depends on the gain and phase balance of the two channels.

One of the problems in the implementation of this circuit is the achievement of a constant 90-deg phase shift over the entire bandwidth of the modulation signal. The network that generates this phase shift must also have flat amplitude response. Unlike the microwave hybrid, which has a flat 90-deg phase shift over all frequencies, baseband phase shift networks having a precise 90-deg phase shift over a wide frequency range are theoretically impossible to design. It is possible, however, to design lumped-element networks that approximate the desired response over a limited bandwidth with adequate gain and phase match.

7.3.7 Conclusions

The selection of a type of mixer for any application involves a trade-off between a large number of factors: port VSWR, port-to-port isolation, LO noise and spurious-signal rejection, intermodulation and spurious-response rejection, compression point, and LO requirements, as well as the all-important conversion loss and noise temperature. Table 7.1 compares these properties for the basic mixer types discussed in Chapters 6 and 7. Note that the descriptions–good, poor, moderate, low, high–are relative; a "good" IF VSWR, for example, for a mixer may be in the range of 1.5 to 2.0, which might not be considered "good" for many other microwave components. Also, be careful not to be led astray by simplistic characterizations. For example, Table 7.1 seems to indicate that it would be foolish ever to use a single-diode mixer in preference to a ring-type doubly balanced mixer, yet the single-diode mixer may be the only alternative for low-noise operation at millimeter wavelengths.

Another limitation of Table 7.1 is that it applies only to diode mixers. The story will change appreciably when FET mixers are considered in Chapter 9. A wide variety of single-FET mixers and balanced mixers are possible. Virtually any diode mixer circuit has analogs in FET mixers, often with better intermodulation characteristics and noise figure and lower LO power requirements.

Table 7.1
Mixer Comparison Chart

Mixer Type	*VSWR*			*Port-to-Port Isolation*			*LO AM*	*LO*
	RF	*LO*	*IF*	*RF/IF*	*LO/RF*	*LO/IF*	*Noise rej.*	*Spur. Sig. Rej.*
Single-diode	Depends on matching circuits	Depends on matching circuits	Depends on matching circuits	Depends on filters	Depends on filters	Depends on filters	None	None
Singly balanced (180 deg)		Same as single-diode		Depends on filters	Good; equal to hybrid isolation	Depends on filters	Good	Good
Singly balanced (90 deg)	Good	Good	Same as single-diode	Depends on filters	Poor	Depends on filters	Good	Good
Doubly balanced (ring/star)	Good	Good	Good	Good	Good	Good	Good	Good
Subhar-monically pumped		Depends on matching circuits		Depends on filters	Good	Depends on filters	Good	Good
Image rejection	Good	Good	Good	Good	Good	Good	Good	Good

Table 7.1
(Continued)

Mixer Type	*Low-Order Spurious-Response Rejection*	*LO Power Requirements*	*Third-Order IM Intercept*	*1-dB Compression*
Single-diode	None	Low	Low	Low
Singly balanced (180 deg)	(2,2): good (2,1): good if Δ port is the LO (1,2): good if Σ port is the LO	Moderate	Moderate	Moderate
Singly balanced (90 deg)	(2,2): good (2,1): none (1,2): none	Moderate	Moderate	Moderate
Doubly balanced (ring/star)	(2,2): good (2,1): good (1,2): good	High	High	High
Subharmonically pumped	Rejects all mixing with odd LO harmonics	Moderate	Low	Low
Image rejection	Same as doubly balanced mixer	High to very high (depends on type)	High	High

7.4 REFERENCES

[1] Collin, *R., Foundations1 for Microwave Engineering,* New York: McGraw-Hill, 1966

[2] C. L. Ruthroff, "Some Broadband Transformers," *Proc. IRE,* Vol. 47, 1959, p. 1337.

[3] March, S., "A Wideband Stripline Ring Hybrid," *IEEE Transactions Microwave Theory and Technique.*, MTT-16, 1968, p. 361.

[4] Ohta, I., H. Kinoshita, and K. Fujiwara, "Rectangular Disk 3-dB Hybrids," *IEEE MTT-S International Microwave Symposium Digest of Papers*, p. 235, 1989.

[5] Page, M. J., and S. R. Judah, "A Microstrip Planar Disk 3-dB Quadrature Hybrid," *IEEE MTT S International Microwave Symposium Digest of Papers*, p. 247, 1989.

[6] Gupta, K. C., and M. D. Abouzahra, "Analysis and Design of Four-Port and Five-Port Microstrip Disk Circuits," *IEEE Transactions Microwave Theory and Techniques*, MTT-33, 1985, p. 1422.

[7] Matthai, G. L., L. Young, and E. M. T. Jones, *Microwave Filters, Impedance Matching Networks, and Coupling Structures,* Dedham, MA: Artech House, 1980.

[8] Bryant T., and A. Weiss, "Parameters of Microstrip Lines and Coupled Pairs of Microstrip Lines," *IEEE Transactions Microwave Theory and Technique.*, MTT-16, 1968, p. 1021.

[9] Alexopoulos, N., and S. A. Maas, "Characteristics of Microstrip Directional Couplers on Anisotropic Substrates," *IEEE Transactions Microwave Theory and Techniques*, MTT-30, 1982, p. 1267.

[10] Tajima, Y., and S. Kamihashi, "Multiconductor Couplers," *IEEE Transactions Microwave Theory and Techniques*, MTT-26, 1978, p. 795.

[11] Djordjevic, A., et al., "Matrix Parameters for Multiconductor Transmission Lines," Artech House, Norwood, MA 1989.

[12] Kajfez, D., Z. Paunovic, and S. Pavlin, "Simplified Design of Lange Coupler," *IEEE Transactions Microwave Theory and Techniques*, MTT-26, 1978, p. 806.

[13] Presser, A., "Interdigitated Microstrip Coupler Design," *IEEE Transactions Microwave Theory and Techniques*, MTT-26, 1978, p. 801.

[14] Ou, W. P., "Design Equations for an Interdigitated Directional Coupler," *IEEE Transactions Microwave Theory and Techniques*, MTT-23, 1975, p. 253.

[15] Parisi, S. J., "A Lumped-element Rat-Race Coupler," *Applied Microwaves*, Aug./Sept. 1989. p. 84.

[16] Ho, C. Y., "Design of Lumped Quadrature Couplers," *Microwave J.*, Vol 22, No. 9, 1979, p. 67.

[17] Staudinger, J., and W. Seely, "An Octave Bandwidth 90-degree Coupler Topology Suitable for MMICs," *Microwave J.*, Vol 33, No. 9, 1990, p. 117.

[18] Marchand, N., "Transmission Line Conversion Transformers," *Electronics,* Vol 17, no 12, 1944, p. 142.

[19] Cloete, J. H., "Exact Design of the Marchand Balun," M*icrowave J.*, Vol 23, No. 5, 1980, p. 99.

[20] Hallford, B. R., "A Designer's Guide to Planar Mixer Baluns," Microwaves, Dec. 1979, p. 52.

[21] Jokanovic, B., and V. Trifunovic, "A Ku-Band SSB Mixer," *Microwave J.*, Vol 32, No. 6, 1989, p. 153.

[22] Hallford, B. R., "Low Conversion Loss X Band Mixer," M*icrowave J.*, Vol 21, No. 4, 1978, p. 53.

Chapter 8
Balanced Mixer Circuits

In this chapter we examine some of the more commonly encountered mixer designs. It is necessarily incomplete, because the number of different designs for balanced mixers is very great and always increasing. Nevertheless, virtually all balanced mixers are variations on those discussed in Chapter 7: singly balanced mixers using 90- or 180-deg hybrids, or doubly balanced mixers in a ring or star configuration. Keeping this point in mind will help considerably in the understanding of the operating principles of apparently different designs.

In the previous chapter we discussed these basic circuits and some of their properties. This chapter focuses on the practical aspects of the design of such mixers and on the hybrids and baluns that are their indispensable components. The emphasis in this chapter is on mixer designs in common use. New designs appear frequently, and some may have advantages over these in cost, ease of fabrication, planar instead of nonplanar construction, and, occasionally, performance. One new and very important phenomenon is the rapid progress of monolithic mixers; realizing these basic circuits in practical, monolithically integrable structures is a great challenge to our mixer-design skills. One possible response to this challenge, the use of mixers based on FET instead of diode technologies, will be the subject of the last chapter.

8.1 WAVEGUIDE MIXERS

8.1.1 "Magic Tee" Hybrid Mixer

In Chapter 7 we noted that the "magic tee" waveguide junction is a type of 180-deg hybrid. A singly balanced mixer can be realized by connecting two single-diode mixers to mutually isolated ports of such a hybrid and by applying the RF and LO to the other two ports. Usually, the mixers are connected to ports 1 and 2 of the hybrid

(Figure 7.10); the symmetry of the junction ensures that the mixers will see equal source impedances and good balance will be achieved. Because it normally has a better VSWR, port 3 is normally used as the RF input. It may be desirable, however, to use port 4 for the RF and port 3 for the LO input in order to reject certain spurious responses (Section 7.2.2).

Figure 8.1 shows the most common configuration of a magic-tee mixer. Instead of the hybrid shown in Figure 7.10, a folded tee is used. The folded tee is a conventional magic tee with the waveguides from ports 1 and 2 folded back so that they are parallel, and the ports are adjacent. This places the IF ports of the two mixers close together and allows them to be connected in parallel.

The individual mixers connected to ports 1 and 2 may be of any appropriate single-diode design and may be image-enhanced. The only difference between the mixers used here and those described in Chapter 6 is that the parallel connection of the IF ports causes the IF load impedance of each single-device mixer to be twice that of the balanced mixer. The IF filters and matching circuits must therefore be designed to match each diode to twice the mixer load. This situation is generally beneficial: the mixer's IF load is usually 50Ω, but the optimum IF load impedance for each diode is usually closer to 100Ω. If this situation is not desirable (e.g., if a balanced mixer is produced from two existing single-diode mixers that were not designed for use in a balanced circuit), an IF hybrid can be used to combine the outputs, and each diode will see the hybrid's input impedance, usually 50Ω.

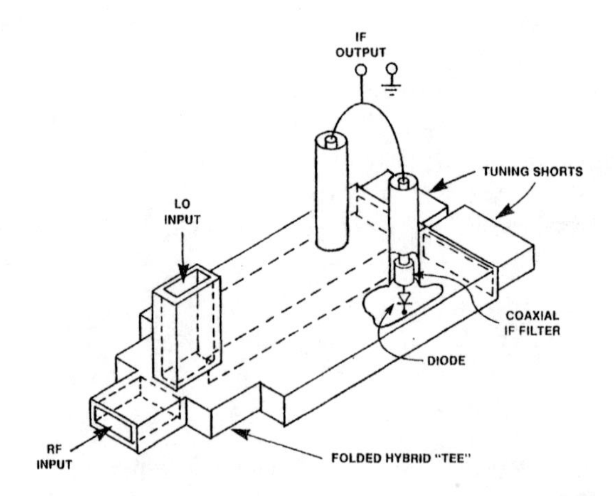

Figure 8.1 Waveguide balanced mixer using a 180-deg "folded tee" hybrid. The cutaway shows the diode mounting.

The conversion loss of the magic-tee mixer is essentially the same as that of the individual mixers, which may be very low. The loss of the hybrid is usually very low, much lower than a filter diplexer used in a single-device mixer. Image enhancement of a balanced mixer is often difficult, because the required image tuning must be performed identically on both single-diode mixers. It is generally not possible to tune both single-diode mixers individually, since in operation they share a common IF terminal. An alternative to tuning the two mixers individually is to locate a single image filter at the RF input port of the hybrid. In this case, however, the filter may be physically too far from the diodes to allow wide image-enhancement bandwidth.

One characteristic of the magic-tee mixer is that the 180-deg phase shift in the hybrid is caused by the geometry of the junction, not by frequency-sensitive structures. Thus, the hybrid's balance is usually very good over a wide frequency range. However, the junction may be difficult to match over a wide bandwidth, so the mixer's bandwidth is usually much less than the full waveguide band. Use of an unmatched tee is rarely an acceptable way to increase bandwidth; the combination of the tee's poor VSWR and the inevitably high input VSWR of the mixer causes ripples in the mixer's passband. A better choice for large bandwidths is the crossbar mixer.

8.1.2 Crossbar Mixer

A crossbar mixer is shown in Figure 8.2(a). The figure is cut away to show the diode mounting more clearly. The diodes are in series across the RF waveguide, each connected between the waveguide top or bottom wall and a thin, flat metal strip that runs perpendicular to the E-field across the center of the guide. The diodes must have the polarity shown in the figure. Because of its orientation, the strip does not couple to the fields in the RF guide, so RF-to-LO isolation is very good; no filter or hybrid is needed. Energy is coupled directly into the diodes, just as in a single-diode waveguide mixer. The strip crosses the LO waveguide in the E-field direction, so it couples to the LO waveguide E-field. Backshorts are used for tuning in both waveguides.

The RF and LO voltages are applied to the diodes as shown in Figure 8.2(b); comparing this to Figure 7.21 shows that the phases are those of a 180-deg hybrid mixer. The LO is not coupled into the RF waveguide in the dominant TE_{10} mode; it will not couple to other modes as long as the guide's height is below one-half of the free-space wavelength at the LO frequency. LO harmonics may couple into the RF waveguide, however, causing passband "glitches" or poor balance over very narrow frequency ranges.

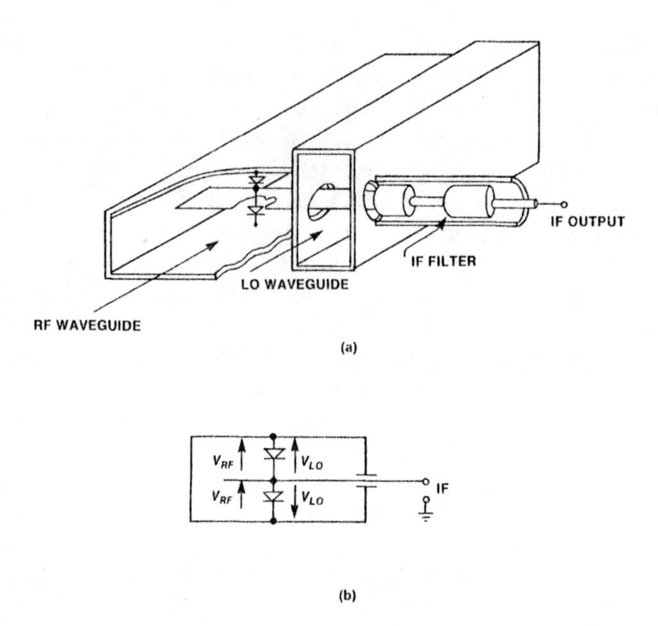

Figure 8.2 The crossbar mixer: (a) structure; (b) phases of the RF and LO at the two diodes.

An important property of the crossbar mixer is that, like the magic-tee mixer, the RF and LO phase relationships are achieved entirely by geometrical means and do not require frequency-sensitive circuits. Unlike the magic-tee mixer, the structure can be matched easily over a full waveguide bandwidth. It is also much more easily image-enhanced. For these reasons, the crossbar mixer has become a very popular design for microwave and millimeter-wave applications.

The major disadvantage of the crossbar mixer is that it requires packaged diodes. Because of the diodes' orientations, it is very difficult to use whiskered dot-matrix diodes in a crossbar mixer. The most serious limitation of packaged diodes (Section 2.5.3) is that the chips used in such diodes must have relatively large junction diameters in order to allow bonding a wire or ribbon to them. One partial solution to this problem has been to use beam-lead diodes supported by a substrate that includes the LO and IF circuits. This is a low-cost approach, which often results in high performance. Another is to realize the diodes and the LO-IF circuit monolithically on GaAs. This approach may make crossbar mixers more practical at very high frequencies, where the parasitics of packaged devices or even beam-lead diodes would limit performance. The monolithic crossbar mixer is not a low-cost approach at present, however, because a large amount of GaAs substrate area is needed. Another disadvantage of the crossbar mixer is that it is difficult to bias the diodes; most crossbar mixers are operated without dc bias.

The crossbar mixer is not particularly difficult to design. Because the diodes are in series across the RF waveguides, each sees half the waveguide impedance; otherwise, the diode mount is identical to the single-diode mixer mounting structure. Hence, the design techniques for the single-diode mixer (Section 6.2.2) can be applied directly to the crossbar mixer; the only change is to use a waveguide height (*a* dimension) in the single-diode design that is half the height of the actual RF waveguide. As with other singly balanced mixers, the diodes are in parallel at the IF frequency, so each sees an IF load impedance that is twice that of the entire mixer. The strip to which both diodes are connected can be modeled at the IF as a piece of stripline. Because the RF coupling to the strip is very low (typically –20 dB or less), the IF filter needs to be designed to reject only the LO frequency band. As with other singly balanced mixers, the IF circuits must be designed to match the actual mixer load impedance to the parallel combination of the diode IF impedances (e.g., if the mixer load is 50Ω and the optimum diode load is 120Ω, the IF is designed to match the 50Ω load to the parallel combination of the diodes, or 60Ω). This is a different situation from that encountered in the magic-tee mixer. In all other respects the design of the IF filter is the same as that of the single-diode waveguide mixer.

If a variable-frequency LO used, the LO circuit must be designed to provide uniform pumping of the diodes across the entire LO band. The LO port's input VSWR is critical in broadband mixers; the combination of a poor LO input and source VSWR may cause poor diode pumping at certain frequencies. Tuning of the LO port is usually empirical; an adjustable backshort, tuning screws, or other tuning elements may be used in the LO waveguide to optimize the match.

8.1.3 Fin-Line Mixers

The availability of low-cost, beam-lead diodes usable into the millimeter-wave region has stimulated the development of fin-line (often called *E-plane*) circuits for mixers. These are low in cost and easy to assemble. The low cost comes from realizing virtually all the mixer's circuitry in a microwave strip transmission medium, which can be reproduced photolithographically, and the use of beam-lead diodes, which do not entail the labor-intensive assembly required of dot-matrix diodes. Fin-line consists of two strip conductors mounted on a dielectric substrate in the center of a rectangular waveguide; the wave is guided along the gap between the conductors. Fin-line is similar in concept to slotline, but because it is surrounded by waveguide, it is not subject to radiation loss. Fin-lines are invariably realized on soft substrates, such as polytetrafluoroethylene-glass composites, clamped between the halves of the split waveguide. Fin-line mixers probably have the lowest cost-to performance ratio of all types of millimeter-wave mixers.

A popular fin-line mixer design is shown in Figure 8.3(a). Because the RF voltages are applied to the diodes out of phase, and the LO voltage is applied in phase, this is a type of 180-deg hybrid mixer (Figure 7.21). All the mixer circuitry is in fin-line or in microstrip; the waveguide serves only as a conduit for the RF and LO signals and as support for the dielectric substrate. The RF is coupled to the fin-line through a tapered transition. The LO is applied from the opposite end of the waveguide, through a transition, into the microstrip LO circuit. This part of the circuit has a ground plane indicated by the crosshatched area. There is no ground plane behind the fin-line part of the RF circuit.

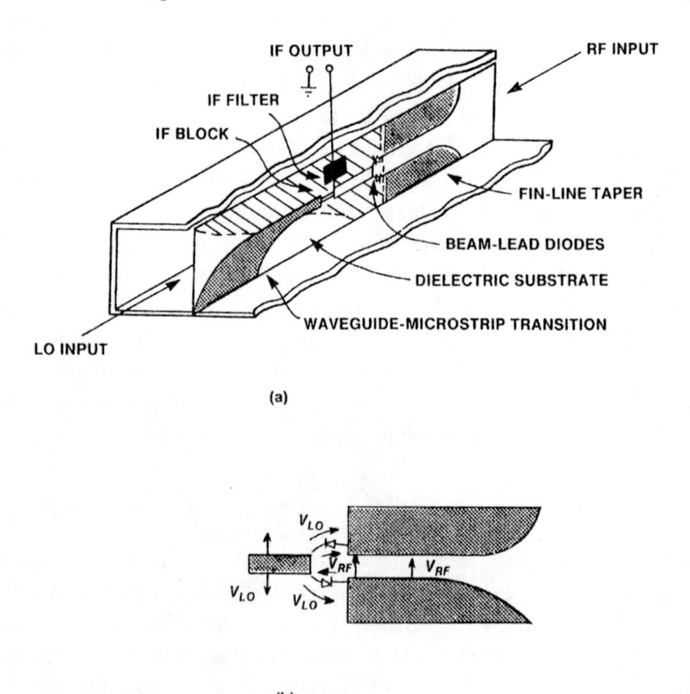

Figure 8.3 Fin-line singly-balanced mixer (a). The shaded area is the metallization on the side viewed; the cross-hatched area is the metalized area on the opposite side of the substrate; (b) detail of the diodes, showing RF and LO phases.

A detail of the diode mounting is shown in Figure 8.3(b). The fin-line is a balanced transmission line, and the connecting point between the diodes is a virtual ground. Because there is no RF voltage between the microstrip line and ground at this point, no RF signal is impressed on the microstrip; as long as good symmetry is maintained in the structure, the RF-to-LO isolation is inherently very good. For the same reason, no LO voltage is excited on the RF fin-line. The IF can be filtered most conveniently from the LO microstrip line. The IF block is designed as a simple interdigital dc block and prevents the IF from being shorted to ground through the waveguide-to-microstrip transition. This type of block works well if the IF is less than approximately 10% of the LO frequency. If it is higher, a more complex filter can be used, or an E-plane probe transition can be used in the LO waveguide.

As with the crossbar mixer, the RF and LO phases and the port-to-port isolations are related to the geometry of the circuit. No frequency-sensitive structures are used to provide phase shifts, and the LO and RF waveguide-to-microstrip transitions are inherently broadband. Hence, the fin-line mixer is capable of broad bandwidth, often a full waveguide band. Also, as with the crossbar mixer, the diodes are in series at the RF frequency and in parallel at the IF and LO. Hence, each diode sees half the fin-line characteristic impedance at the RF and twice the IF load impedance. Figure 8.3(a) shows no tuning or matching circuitry. This situation is not usual; fin-line mixers are usually designed to achieve adequate performance at minimum cost, so the mixer is designed to use diodes having parasitics that are small enough to allow reasonable performance without tuning. Fin-line mixers designed according to this rationale are limited to about 100 GHz by the uncompensated diode parasitics, although as better beam-lead diodes become available, similar untuned mixers may be designed at higher frequencies. Most fin-line mixers are nearly impossible to bias. Image-enhancement of a fin-line mixer is possible; however, image-enhancement usually offers little improvement unless the nonenhanced mixer is capable of less than 5 dB conversion loss. Most fin-line mixers do not meet this criterion.

The design of the tapered transitions is easier than it might seem. The designs are based on the fundamental microwave principle that almost anything works if it is done gradually enough. The transitions are designed to taper, over a few wavelengths, from the waveguide to the strip transmission medium. The microstrip taper is the more difficult of the two, but it works well if the region from the crossover point of the top and ground plane metallizations to the point where the microstrip line begins is a relatively long and smooth curve. A cosinusoidal taper usually works well. The fin-line taper is even less critical; as long as the curves are smooth and it is few wavelengths long, it will work well. If it is absolutely necessary have an analytical technique, the method presented by Beyer and Wolff [1] is simple and practical. A more important and difficult problem concerns the calculation of the characteristic impedance and effective dielectric constant of the fin-line. Few existing techniques are accurate, available to the practical designer, and are not numerically involved. References [2] – [4] may be helpful in this regard.

8.2 STRIPLINE AND MICROSTRIP MIXERS

8.2.1 Hybrid Ring (Rat-Race) Mixer

The ring or "rat-race" hybrid was described in Section 7.1.2 as a type of 180-deg hybrid, and the use of this and other 180-deg hybrids in diode mixers was the subject of Section 7.2.2. These implied that the design of a rat-race mixer involved little more than coupling two diodes to isolated ports of the hybrid. Nevertheless, there are some practical considerations that are important in achieving a successful mixer design.

Not all applications are appropriate for this simple type of mixer. Doubly balanced mixers are available at frequencies to at least 24 GHz, and their cost and performance are generally equal to or better than those of a microstrip ring-hybrid mixer. The doubly balanced mixers have better spurious-response rejection and often have inherently wider bandwidth. At higher frequencies, the ring-hybrid mixer has some advantages: it is easier to apply dc bias, easier to include tuning and matching (thus reducing conversion loss) and is in many cases smaller. Finally, since it uses half as many diodes, LO power requirements are usually lower, especially if the diodes are biased.

An example of a ring-hybrid mixer is shown in Figure 8.4, which shows the circuit of a 30-GHz microstrip mixer having approximately 1-GHz RF bandwidth, a fixed 26.5-GHz LO, and a 3.5-GHz IF frequency. This mixer was used in an integrated low-noise receiver. The RF is applied to the sigma port and the LO to the delta port of the hybrid. GaAs beam-lead diodes are used, and are dc biased. The IF output is filtered directly from the ring, and low-frequency blocks are used on the RF and LO lines to prevent short-circuiting the IF or dc bias through these ports. The substrate is 0.01-inch thick fused silica, with 75-microinch chrome-gold metallization. In order to minimize radiation loss, the substrate is mounted in a channel 0.2 inches wide by 0.6 inches long; the channel has a metal cover.

A few subtleties in the design of the mixer are responsible for its success. The diodes are mounted so that the ends not connected to the substrate are grounded or are connected to bypass capacitors within a few mils of the substrate. The capacitor to which one diode is connected has a series-resonant frequency very near the RF frequency and has a high enough value, 20 pF, to be effectively a short circuit at the

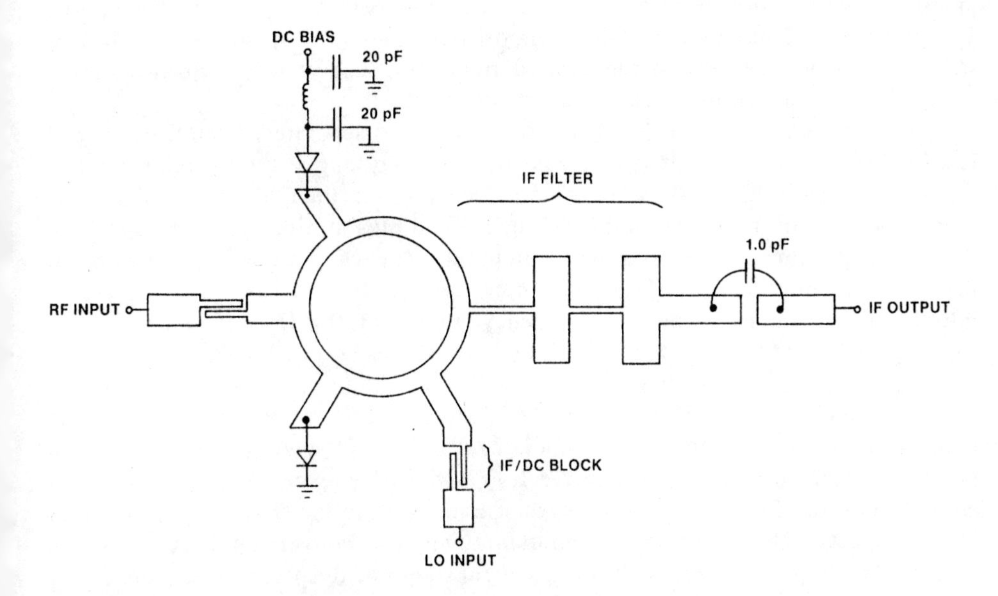

Figure 8.4 30-GHz microstrip hybrid ring mixer with dc bias.

IF frequency. Its thickness is the same as that of the substrate, 0.01 inch, and its width and length are 0.02 inch and 0.04 inch, respectively. The grounded end of the other diode is connected to a metal block having the same dimensions. The diodes are selected to provide good conversion performance with no tuning, although some improvement in both conversion loss and LO power requirements can be achieved with minimal empirical tuning. The center frequency of the ring hybrid is the LO frequency, rather than a frequency between the RF and LO. This way, the diodes are pumped precisely 180 deg out of phase, so spurious rejection is maximized and LO leakage from the RF port is minimized. This is particularly important because of a (2, 2) spurious response near the RF passband, and to prevent dynamic range reduction in the RF amplifier that precedes the mixer.

Because of some uncertainty in the actual electrical length of the ring, it was expected that the center frequency of the hybrid initially would not be exactly as it was designed. The center frequency of the hybrid was checked by measuring its LO-to-RF isolation without diodes installed; the transmission between these isolated ports has a sharp null at the hybrid's center frequency; this null is a much more precise indication of center frequency than insertion loss between nonisolated ports. The first hybrid fabricated was 450 MHz off frequency (1.7%), so a second hybrid was scaled from the first to the desired frequency. The center frequency of the second hybrid was within a few megahertz of 26.5 GHz.

The IF filter was designed, constructed, and tested separately from the mixer. It is a Chebyshev low-pass filter; its stubs are a few deg long at the IF frequency and are therefore capacitive. At the LO frequency, however, the stubs are 90 deg long. These short-circuit the IF at the LO frequency and thus minimize LO leakage from the IF. The design was based on the assumption that each diode was a 100Ω load at the IF frequency. The IF filter also compensates for the parasitic reactances introduced by the ring and low-frequency blocks at the IF frequency. The IF capacitor is part of the matching circuit and also serves as a dc block.

The conversion loss of the mixer is 7.0 dB, flat within 0.2 dB, and IF VSWR is below 2.0 over the entire IF band. The LO-to-IF and RF-to-IF isolations are greater than 34 dB, and the LO-to-RF isolation is 29 dB. The LO power required is 6 dBm, and the dc bias voltage and current are 1.3V and 4 mA, respectively. Although the conversion loss of this mixer is not spectacular, its performance is very good in all other respects. These are more important than low conversion loss, since the receiver in which the mixer is used has several stages of RF preamplification.

8.2.2 Slotline Rat-Race Mixer

Figure 8.5 shows an unusual form of the rat-race mixer. The ring in this mixer is realized in slotline, and the LO and RF are coupled to the slot via appropriate transitions [5], [6]. The ring is a circular slotline on the ground-plane side of a microstrip substrate and is 1.5 wavelengths in circumference. The diodes are mounted one-quarter wavelength from the RF transition and one-half wavelength from the LO slot. The IF port is connected through a low-pass filter to the metallization in the center of the structure.

The LO slot is connected to the ring in such a way that the LO voltage splits into two waves having opposite phase, one propagating around the ring in a clockwise direction and the other counterclockwise. These waves meet at the RF transition, and since they have 180-deg phase difference, that point is a virtual ground for the LO. The diodes are mounted one-quarter wavelength away from this point, where the shunt impedance of the quarter-wave section of line is infinite. The waves coupled to the ring from the RF transition are in phase, and the LO junction is therefore an open-circuit point for the RF. The diodes are mounted one-half wavelength from this point, at other open-circuit points.

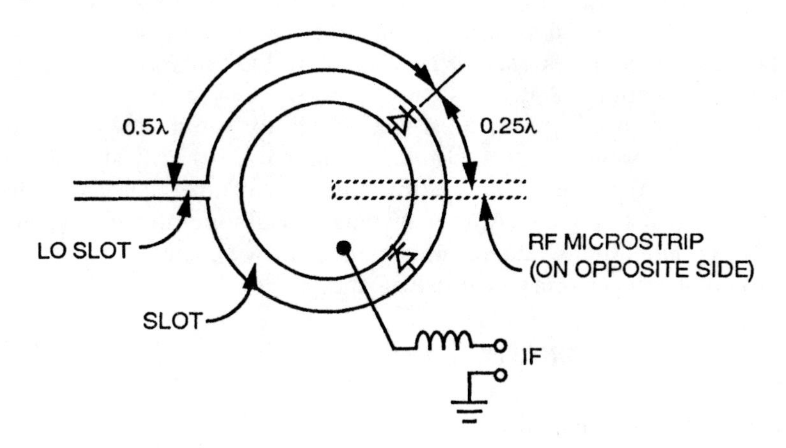

Figure 8.5 Slotline rat-race mixer.

Because the phase reversal results from the geometry of the circuit, not from an extra half-wavelength of transmission line, this mixer has inherently greater bandwidth than the classical rat-race microstrip mixer. Unfortunately, it is difficult to design broadband microstrip-to-slotline transitions, so these transitions may limit the bandwidth more than the inherent properties of the mixer.

8.2.3 Quadrature Hybrid Mixers

Quadrature hybrid mixers are used in microwave systems more often than they deserve. Because of their inherently poor LO-to-RF isolation, poor spurious-response rejection, and the fact that a poor RF-source VSWR at the LO frequency can unbalance the mixer, they are distinctly inferior to 180-deg mixers. They do, however, have the practical advantage that the microstrip-line impedances used to realize a branch-line hybrid are easy to achieve, and the branch line is slightly smaller than a ring hybrid at the same frequency. The RF and LO ports of the branch-line hybrid mixer are on the same side of the hybrid, which is often an advantage in creating a compact circuit. When their bandwidth is adequate, their simpler fabrication makes branch-line hybrids preferred over Lange couplers for these mixers.

The design considerations for quadrature hybrid mixers are much the same as those for 180-deg hybrid mixers in terms of selection of diodes; RF, IF, and LO embedding impedances; and matching-circuit design. The RF and LO embedding circuits should be designed with care, since the input return loss of the matched diodes at these frequencies establishes the RF-to-LO isolation of the complete balanced mixer (Section 7.2.3).

Quadrature-hybrid mixers are most appropriate for use in low-cost, moderate-performance applications, where LO-to-RF isolation is not critical. These include low-cost receiver front ends for terrestrial microwave links, consumer electronic equipment, amateur radio, and AM modulators. Quadrature mixers have also been used in GaAs monolithic integrated circuits (MMICs), where size and layout advantages outweigh performance disadvantages.

8.3 DOUBLY BALANCED MIXER CIRCUITS

8.3.1 Low-Frequency Ring Mixers

A common approach to the realization of a transformer-hybrid ring mixer is shown in Figure 8.6. This mixer is a realization of the circuit of Figure 7.26 and consists of

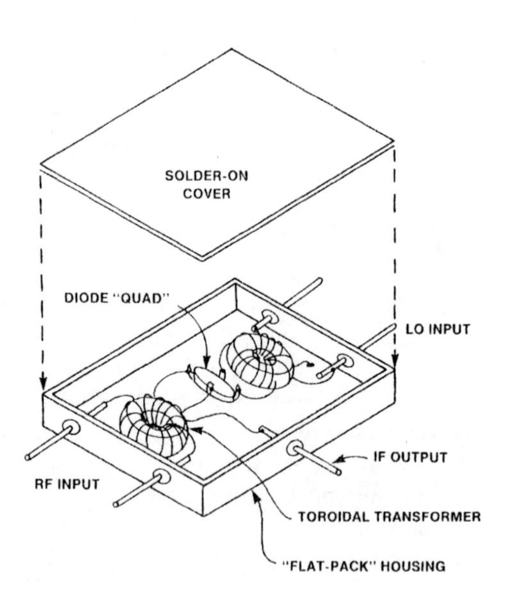

Figure 8.6 Doubly balanced ring mixer using toroidal transformers, mounted in a "flat pack." Because of its lumped-element construction, this circuit is useful to at most 1 GHz.

two toroidal transformers and one epoxy-encapsulated diode ring, or "quad." These components are epoxied to the bottom of a "flat-pack," a flat hybrid metal package having a soldered-on cover and glass hermetic feedthroughs for the electrical connections. All internal interconnections are short lengths of wire. This type of mixer is very inexpensive, but its stray capacitances and inductances limit it to frequencies below approximately 1 GHz. Its performance at frequencies of a few hundred MHz or below may be very good, however, where the diode's parasitics are negligible and the transformers are nearly ideal.

If a simple trifilar toroidal transformer with a 2:1 turns ratio is used (Figure 7.4), and the RF and LO source and load impedances are the standard 50Ω, a 200Ω source impedance is presented to each diode. This value is close enough to the optimum to allow reasonably good conversion loss and VSWR. If the mixer's load impedance is 50Ω, each diode sees 200Ω at the IF. This is, again, close to the 100Ω to 150Ω optimum. Therefore, the design of such mixers requires little more than designing a transformer and selecting an appropriate diode. Typically, such mixers have a conversion loss of 6 to 7 dB and port VSWRs of 2.0.

At higher frequencies microstrip interconnects and more complex transformers must be used. Unencapsulated beam-lead diode quads may be needed to minimize parasitic capacitances. Even with these improvements, however, the transformer hybrid mixer is limited to frequencies below approximately 2 GHz.

8.3.2 Microwave Doubly Balanced Ring Mixers

Broadband microwave doubly balanced mixers use a pair of baluns instead of transformers. Because good planar baluns are very difficult to design successfully, nonplanar baluns are often used. The use of nonplanar circuitry (circuits with metal patterns instead of a ground plane on the underside of the substrate) may present some difficulty in integration with such planar circuits as microstrip, but the performance advantages are usually worth the trouble.

The mixer in Figure 8.7 uses a parallel-line balun to realize a transition from microstrip to parallel-strip balanced transmission line. Often the ground-plane conductor is tapered to minimize the discontinuity at its input. Identical baluns are used for the LO and RF, and their outputs are connected directly to the diode ring. Because the diodes must be connected to the top and bottom of the substrates, some type of gap or hole in the substrate must be provided. Beam-lead quads are generally too fragile to survive the contortion necessary to connect their leads to both baluns, so epoxy-encapsulated quads are usually preferred. The RF connection points to the quad are virtual grounds for the LO, and vice versa. Therefore, good LO-to-RF isolation is ensured.

Section 7.1.3 and Figure 7.18 describe the parallel-line balun. The baluns are best treated as broadside-coupled lines mounted in a housing; the housing is the ground plane. Equation (7.8) gives the odd-mode impedance of the coupled lines; their even-mode impedance should be as high as possible, at least ten times the odd-mode impedance, and the even- and odd-mode phase velocities should be as well matched as possible (in practice, a substrate having a low dielectric constant should be used). Each balun is terminated by four diodes in a series-parallel arrangement, so Z_L in (7.8) is the impedance presented to each diode by the RF or LO balun.

As illustrated in Figure 7.17, the even-mode output impedance of a parallel-line balun is an open circuit; this contrasts with a transformer, which has a short-circuit even-mode output impedance and a center tap that can be used conveniently for the IF. Thus, in a microstrip ring mixer, we must use some type of special structure for the IF port. One technique, illustrated in Figure 8.7, is to use a long, fine wire or strip connected to both sides of the RF input line. The wires are made one-quarter wavelength long at the RF frequency, so the pair of wires realizes a shorted, high-impedance quarter-wave stub. Because the input impedance of this structure is very high at the RF frequency, it has minimal effect on the RF circuit. To prevent shorting the IF to ground at the lower end of its frequency range, dc blocks—often series capacitors—are used. The inductance of this interconnection clearly limits the IF bandwidth. Still, this is a simple, low-cost, broadband approach to the design of a

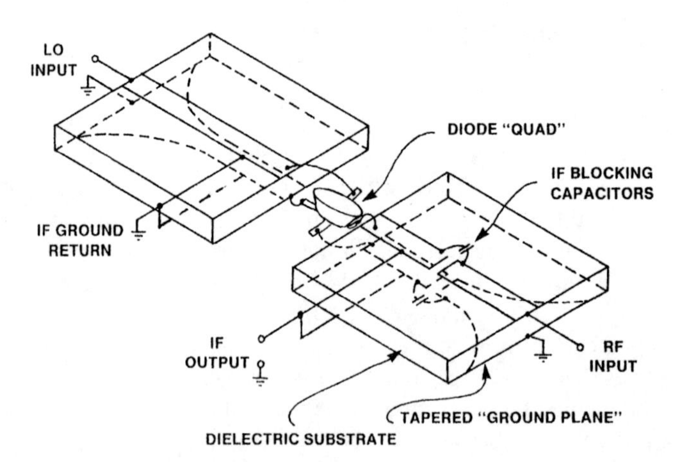

Figure 8.7 Doubly balanced diode-ring mixer using parallel-line baluns. The capacitors act as IF blocks, and an IF ground return is needed at the LO balun. The ground-plane side of the LO balun can be used as the other IF return; here a separate ground return is used to maintain balance.

doubly balanced mixer, and mixers having 2 to 26 GHz RF and LO bandwidths have been produced in this manner.

Figure 8.8(a) shows another design for a ring mixer [7]. This mixer is completely planar, allowing it to be realized as an integrated circuit. The RF and LO connections are *coplanar waveguide* (CPW), and the mixer uses transitions from CPW to slotline as LO and RF baluns. Like the other ring-diode mixer circuits, it requires a special IF-coupling circuit.

The baluns used in this mixer are not unlike Marchand baluns. The low-impedance, open-circuit section of CPW serves the same function as the open-circuit stub in a Marchand balun, and the output terminals are connected to points that are ultimately grounded. A quarter-wave shorted slot is used to prevent the ground plane on the far side of the CPW line from short-circuiting the output. The open area around the diodes is, in effect, a high-impedance shorted stub bridging the diode quad.

Figure 8.8(b) shows a detail of the diode mount. The diode quad is connected directly to the end of the slot at the LO side, but is connected to IF-blocking capacitors at the RF side. These capacitors prevent the IF from being short-circuited by the slot metallization. Fine wires, approximately one-quarter wavelength long at the center of the RF/LO band, are used for both the IF connection and the IF ground return. This mixer exhibited 5 to 8 dB conversion loss over an RF and LO bandwidth of 5 to 18 GHz; the IF was 1 GHz.

Conventional Marchand baluns could conceivably be used for ring mixers. However, the logical structure for such a mixer is almost identical to that of a star mixer and still entails the difficulty of making the IF connection. In the next section, we shall see the elegant way that the star mixer solves this problem.

8.3.3 Microwave Star Mixer

The star mixer circuit, shown in ideal form in Figure 7.29, is widely used. The star mixer has the lowest IF-circuit parasitic reactances of any type of doubly balanced mixer; thus, its IF bandwidth can be very great. Additionally, like the parallel-line balun used in doubly balanced ring mixers, the Marchand balun, which is invariably used in star mixers, can present a wide range of embedding impedances to the diodes, optimizing conversion efficiency. Finally, because the Marchand balun is very broadband, star mixers usually have extremely broad RF and LO passbands.

The Marchand balun, described in Section 7.1.3, must be modified for use in a star mixer. Figure 8.9 illustrates the evolution of the balun. Figure 8.9(a) shows a realization of the balun of Figure 7.19 in strip transmission media. The shaded area represents the metal on the top side of the substrate, and the cross-hatched area

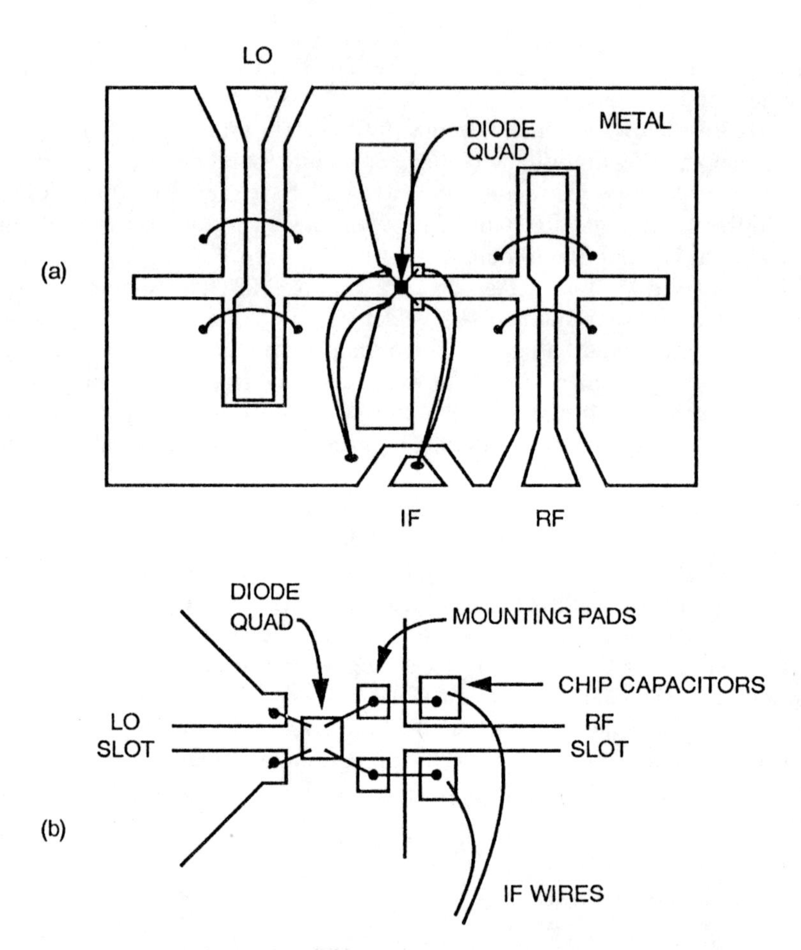

Figure 8.8 Planar diode-ring mixer: (a) mixer top view; (b) detail of the diode mount.

represents the ground plane. Part of the ground plane is removed to create the bottom-side strips. This entire structure must be mounted in a housing; as in Figure 7.19, the housing is the ground surface.

Splitting the metallization on the bottom side of the balun provides two separate outputs. This structure is shown in Figure 8.9(b). The load impedance for each of these two outputs is, of course, twice that of the single-output balun. Finally, a second split-output balun is connected to the first, and diodes are mounted as shown

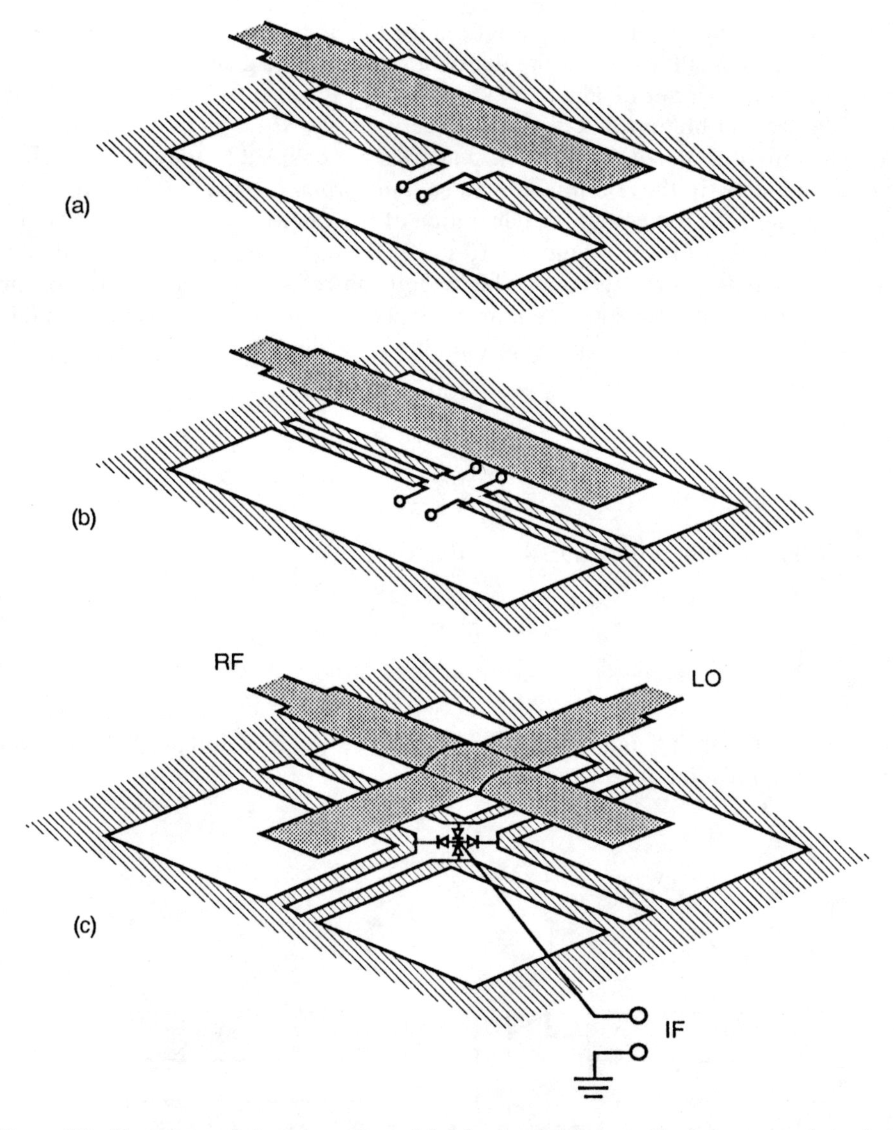

Figure 8.9 Evolution of the star mixer: (a) Marchand balun realized in suspended-substrate microstrip; the shaded area is the top conductor, and the cross-hatched is the ground plane. (b) The output conductors are split to provide two outputs. (c) Another balun identical to that in part (b) is connected at right angles to the first. The LO and RF microstrips must cross over each other without contacting.

in Figure 8.9(c). This requires a crossover in the top metallizations; the crossover is usually relatively small and therefore does not create a large discontinuity.

Comparing the circuit of Figure 8.9(c) to the ideal mixer in Figure 7.29, shows that the diodes and bottom metallization pattern are the same as those of the ideal mixer. The only difference is that the LO and RF are coupled to the cross structure electromagnetically in the real mixer, while a transformer is used in the ideal mixer.

The design of the mixer and the generation of the single-diode equivalent circuit are relatively straightforward. Figure 8.10 shows the equivalent circuit of the diodes and one of the baluns driving them. This circuit can represent either the LO or the RF balun; the second, orthogonal, balun is not shown because it is driven in an odd mode and therefore presents an open circuit to the diodes. From (7.8), the load impedance for the unsplit balun is

$$Z_L = \frac{4Z_{0o}^2}{Z_s} \tag{8.1}$$

and for each port of the split balun the load is

$$Z_L = \frac{8Z_{0o}^2}{Z_s} \tag{8.2}$$

Each output of the split balun drives two diodes in series; therefore, each diode sees an RF or LO embedding impedance of

$$Z_L = \frac{4Z_{0o}^2}{Z_s} \tag{8.3}$$

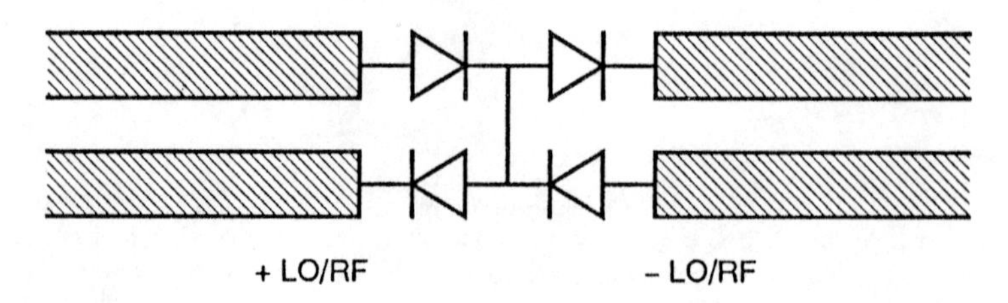

Figure 8.10 Equivalent circuit of the diodes, driven by either the LO or RF balun.

Note that Z_{0o} is the odd-mode impedance of the unsplit balun, and Z_{0e}, its even-mode impedance, must be at least three or four times Z_{0o}. This is the same as the load impedance of the simple, unsplit balun of Figure 8.9(a).

Unfortunately, the split Marchand balun used in a star mixer is difficult to model in a circuit simulator. To calculate the admittance matrix of such a balun it is necessary to find the modes on a three-conductor coupled structure having a nonhomogeneous dielectric. This requires the solution of an eigenvalue problem, a process that is probably too time-consuming to be practical. Instead, the balun can be modeled approximately as four pairs of coupled lines. This model is shown in Figure 8.11. The odd-mode impedance of each pair of these coupled lines is twice the value given by (8.3).

This model approximates the coupling between the top and bottom conductors very well, but does not account for the coupling between the two bottom-side conductors. The LO and RF excitations generate even modes on the split conductors, and the voltage between them is zero; thus, for RF or LO excitation, the elimination of the coupling has little effect. However, the RF balun is driven in an odd mode by the LO, and the LO in an odd mode by the RF. The elimination of the coupling between these lines causes them to appear to be perfect open circuits to odd-mode excitation, an ideal condition that does not exist in practice. The practical result of this idealization is that the RF-to-LO isolation is likely to be poorer than calculations indicate.

Because the four diodes of the star mixer are in parallel at the IF frequency, the

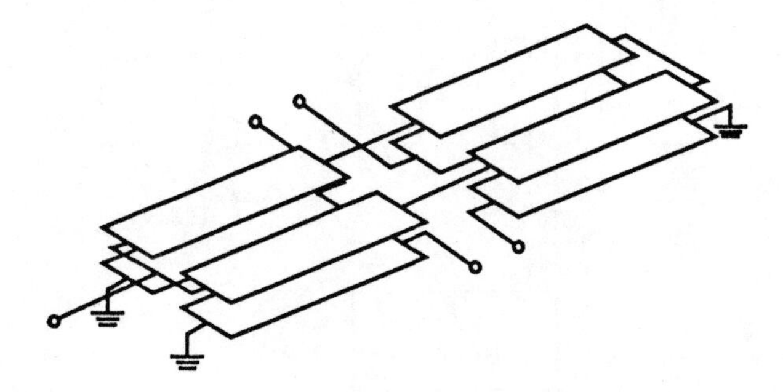

Figure 8.11 Split marchand balun modeled as a set of four broadside-coupled pairs of striplines. This model can be used in most general-purpose CAD programs. It is approximate, but is usually adequate for the design of star mixers. The odd-mode impedances of the individual coupled-line sections should be twice that given by (8.3).

load impedance seen by each diode is four times the load impedance of the complete mixer. Thus, if the load is 50Ω, each diode is effectively terminated in 200Ω. In some respects, the diodes behave like an antiparallel pair: they effectively short-circuit each other at even LO harmonics. As indicated in Section 7.3.2, the single-diode equivalent circuit is the same as that of the diode-ring mixer; the embedding impedances are summarized in Figure 7.28.

Unlike the IF connection in the ring-diode mixer, the IF connection to the star mixer does not require any special circuitry; the split strips to which the diodes are connected provide the IF ground return. Furthermore, the IF connection can be very short; it need not be longer than half the length of the balun. As a result, the IF bandwidth of a star mixer can extend from dc to the lower end of the balun's operating range. This broadband response at all ports makes the star mixer a very useful device.

8.3.4 Biasable Balanced Mixer

The circuit of Figure 8.12 includes some features of singly balanced and doubly balanced mixers. It is often called a doubly balanced mixer, but its performance characteristics are more like that of the singly balanced circuit in Figure 7.23. The

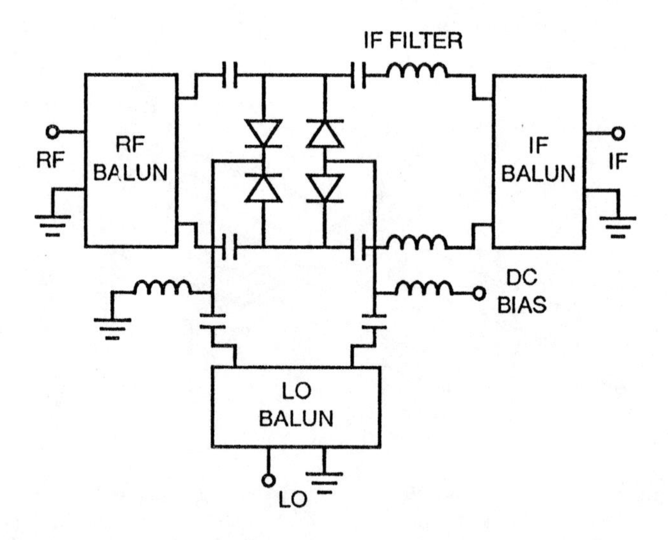

Figure 8.12 Biasable balanced mixer.

diodes are not arranged in a ring as is the ring mixer; the polarities of two of them are reversed. Nevertheless, the balanced diode arrangement provides LO-to-RF isolation, because, like the doubly balanced mixer, the RF and LO are applied at each other's virtual ground points. This mixer has no inherent RF-to-IF isolation, however, so an IF filter is necessary.

The LO pumps all four diodes in phase. The diodes operate as a switch that periodically short-circuits the RF; the RF is short-circuited when the diodes are on, and open-circuited when the diodes are off. The mixer's spurious response and LO noise rejection properties are most like those of the 180-deg hybrid mixer with the LO applied the delta port.

A potential problem of this circuit—as with any balanced mixer using dc bias—is that small differences in the *I*/*V* characteristics of the diodes can unbalance the mixer. The ring of diodes is, in effect, two series-connected pairs of diodes, and the voltage applied to these pairs is exactly the same. Because of the exponential dependence of the diode's current on voltage, slight differences in the diodes may cause one pair to conduct much more heavily than the other. The resulting imbalance degrades spurious-response rejection and port-to-port isolation.

8.4 SUBHARMONICALLY PUMPED MIXERS

Subharmonically pumped (SHP) mixers are most valuable at frequencies where high-power, low-noise LO sources are difficult to obtain, usually the millimeter-wave region. Millimeter-wave SHP mixers using dot-matrix diodes have been produced, but two serious difficulties have made them relatively unpopular. The first is the near impossibility of mounting two dot-matrix diodes in such a way that they can be whiskered easily. The second is that small differences in the lengths and shapes of the whiskers, and the resulting differences in their inductances, degrades the mixer's balance and, consequently, fundamental mixing rejection. Furthermore, the development of good millimeter-wave beam-lead diodes has made such mixers largely unnecessary; the performance obtained from beam-lead diode SHP mixers is often as good as that obtained from whiskered diodes [8].

8.4.1 High-Frequency SHP Mixers Using Beam-Lead Diodes

A practical SHP mixer design is shown in Figure 8.13. This mixer is similar in concept to the single-diode mixer in Figure 6.10; the major difference is that it has an antiparallel pair of diodes instead of a single diode. RF-to-LO isolation is provided by the low-pass filter located between the LO and RF waveguides. Because the LO and RF frequencies differ by a factor of approximately two, this filter is not

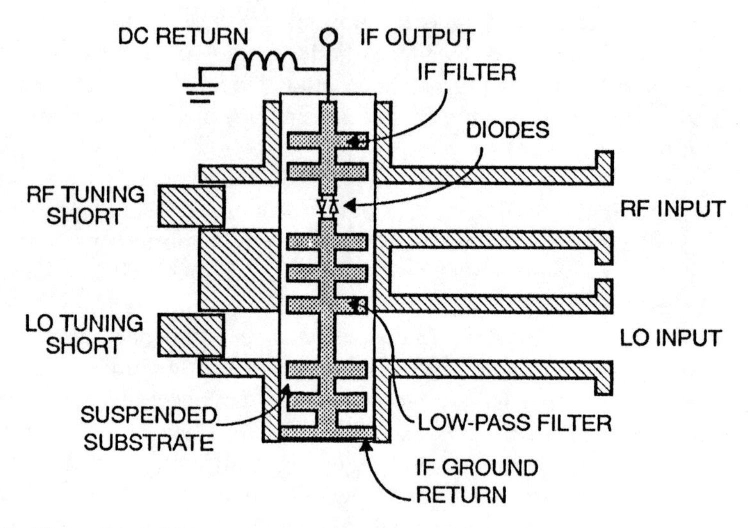

Figure 8.13 Waveguide subharmonically pumped mixer using beam-lead diodes. The antiparallel diode pair is often fabricated as a single unit.

difficult to realize.

The circuit uses millimeter-wave beam-lead diodes. These have lower series inductances than whiskered dot-matrix diodes and have parasitic capacitances that are very low, although generally greater than those of dot-matrix diodes. They may also have additional parasitic capacitances related to the individual diode's structure. Such diodes are available commercially as antiparallel pairs, consisting of two diodes fabricated simultaneously and connected by common beam leads. These devices have excellent balance when used in SHP circuits.

The diodes are mounted in series in the suspended-substrate stripline circuit. They can also be mounted in parallel, although this configuration is more difficult to implement without introducing parasitic reactances, because the diodes are usually much smaller than half the width of the stripline channel. Some RF-frequency fine tuning is available through the backshort in the RF waveguide, but best performance is achieved when the stripline circuit provides most of the matching. Because the diodes are not mounted directly in the waveguide, it is usually not necessary to use reduced-height RF waveguide.

A low-pass filter is used to connect the stripline electrically to the lower surface of the LO waveguide. It is rarely practical to ground the stripline to the waveguide bottom-wall directly via a ribbon bond; such mechanical connections are often unreliable at high frequencies. However, the lower end of this structure is physically connected to the body of the mixer to provide an IF return; this is usually acceptable at the relatively low IF frequency.

The IF port includes a dc return. This is theoretically unnecessary; if the diodes are identical and are pumped identically, their rectified current will circulate in the diode pair. However, any uneven pumping will cause the dc voltage across the pair to deviate from zero. The dc return guarantees that this will not happen.

8.4.2 Microstrip SHP Mixer

The SHP mixer shown in Figure 8.14 consists of little more than two stubs connected by an antiparallel pair of diodes. Both stubs are one-quarter wavelength long at the LO frequency, or approximately one-half wavelength at the RF; one is shorted, and the other is open. The RF and LO are applied to the connection points of the diode pair. Additional matching circuitry, if needed, can be included in the lines from the diodes to the respective ports.

Because the leftmost stub has an open-circuit termination, it presents a short circuit to the left terminals of the diodes at the LO frequency, effectively grounding the terminals. The right stub is shorted; it presents an open circuit to the diodes' right terminals at the LO frequency, allowing LO voltage to be applied at this point. At the RF frequency, the situation is precisely the opposite: the left stub is one-half wavelength long and presents an open circuit to the diodes; the right stub grounds them at the RF frequency. The IF is coupled from the RF strip, and the shorted stub provides the IF return. It may be necessary to include an IF block, not shown in the figure, and a dc return.

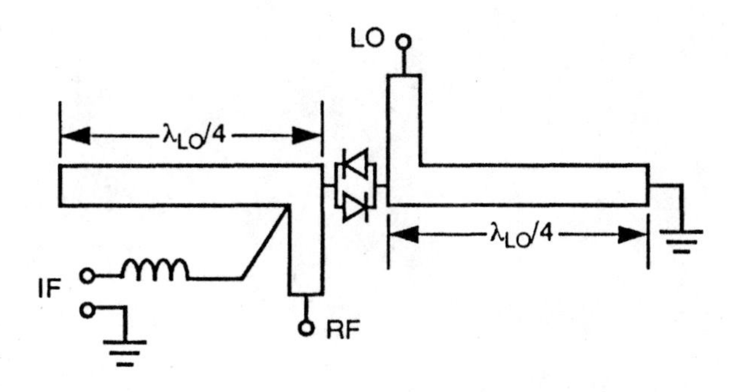

Figure 8.14 Microstrip SHP mixer.

This circuit is relatively narrowband, and its isolation is not exceptional. Nevertheless, its simplicity is attractive and other aspects of its conversion performance should be no worse than other SHP structures.

8.5 REFERENCES

[1] Beyer, A., and I. Wolff, "Finline Taper Design Made Easy," *IEEE MTT-S Int. Microwave Symp. Dig.*, 1985, p.493.

[2] Solbach, K., "The Status of Printed Millimeter-Wave E-Plane Circuits," *IEEE Trans. Microwave Theory Tech.*, Vol. MTT-31, 1983, p. 107.

[3] Sharma, A. K., and W. J. R. Hoefer, "Empirical Analytical Expressions for Fin-Line Design," *IEEE MTT-S Int. Microwave Symp. Dig.*, 1981, p. 102.

[4] Pramanick, P., and P. Bhartia, "CAD Models for Millimeter-Wave Suspended-Substrate Microstrip Lines and Fin-Lines," *IEEE MTT-S Int. Microwave Symp. Dig.*, 1985, p. 453.

[5] Schüppert, B., "Microstrip/Slotline Transitions: Modeling and Experimental Investigation," *IEEE Trans. Microwave Theory Tech.*, Vol. MTT-36, 1988, p. 1272.

[6] Schuppert, B., "Analysis and Design of Microwave Balanced Mixers," *IEEE Trans. Microwave Theory Tech.*, Vol. MTT-34, 1986, p. 120.

[7] Cahana, D., "A New, Single-Plane Double-Balanced Mixer," *Applied Microwaves*, Aug./Sept. 1989, p. 78.

[8] Carlson, E., M. V. Schneider, and T. F. McMaster, "Subharmonically Pumped Millimeter-Wave Mixers," *IEEE Trans. Microwave Theory Tech.*, Vol. MTT-26, 1978, p. 706.

Chapter 9
FET Mixers

Although silicon FET mixers have been common in VHF and UHF receivers since the mid-1960s, it is only recently that microwave GaAs FET mixers have been taken seriously. Early FET mixers achieved conversion gain, but had disappointing noise figures, no better–often even worse–than those of diode mixers*. Because diodes were more amenable to use in balanced-mixer circuits, and were less expensive, most receiver designers were slow to accept FET mixers.

Several recent developments have given receiver designers good reasons to reconsider FET mixers. First, after a bit of effort, many have learned how to make FET mixers work: the noise figures of well-designed FET mixers are now consistently lower than those of diode mixers at frequencies well into the millimeter range, and the design techniques necessary for achieving low noise are understood. Other aspects of a FET mixer's performance–distortion levels, bandwidth, and stability–are comparable to those of diode mixers, and gain is, of course, much greater. Because monolithic diodes fabricated in FET-compatible technologies are often relatively poor, MMICs favor the use of FETs; the widespread use of these circuits has generated further interest in FET mixers. Finally, the development of high-performance FET variants, such as HEMTs, promises improved noise figures and gains of FET mixers, while Schottky-diode mixers have reached the limit of their performance. FET resistive mixers offer noise and conversion loss comparable to diodes, but much lower intermodulation.

---

* This point is best illustrated by an anecdote from the author's experience. Before virtually anything had been published on the design of FET mixers, some talented engineers at a major aerospace company tried to build a microwave FET mixer. They succeeded in creating an apparently successful mixer: it had good conversion gain over the desired passband, reasonable LO power, and appeared in all other respects to be satisfactory. They then decided to measure the noise figure. Unfortunately, they were unable to complete the measurement because they could not get the noise-figure meter on scale; in spite of heroic efforts to reduce the mixer's noise, the meter on the instrument stayed pinned above 20 dB. The project was eventually abandoned.

Diode mixers do have distinct advantages in many applications, and it is unlikely that FET mixers will ever completely supplant them. Nevertheless, FETs are progressively becoming the devices of choice for many types of receivers, especially monolithic ones.

9.1 DESIGN OF SINGLE-GATE FET MIXERS

Because most low-frequency silicon FET mixers use dual-gate MOSFETs, it was natural that dual-gate devices would be the choice for early microwave FET mixers. Unfortunately, dual-gate mixers never achieved outstanding noise figures or conversion gain, so attention moved to single-gate designs. Single-gate FETs have consistently provided better noise temperature and conversion gain than dual-gate FETs, although dual-gate mixers often have slightly lower distortion. Single-gate mixers have one major disadvantage in comparison to dual-gate mixers: it is much more difficult in single-gate devices to achieve good LO-to-RF isolation.

9.1.1 Design Philosophy

For most types of communication receivers, a mixer must have a low noise figure and adequate, but not excessive, conversion gain; in a FET mixer these two properties are more or less independent. This is somewhat different from the diode mixer, where minimum noise and conversion loss usually occur together. Thus, a FET is designed primarily to achieve low noise, and secondarily to obtain high gain. Fortunately, when low noise is achieved, good gain is usually a by-product; however, the converse is not necessarily true.

Like other receiver components, a mixer must be stable. Because a FET mixer is an active component and has conversion gain, instability is at least a theoretical possibility. Fortunately, a mixer that is well designed in most other respects is usually stable as well. A mixer also must be, in general, well behaved: its performance parameters must vary gracefully with LO level, dc bias, and environmental temperature; it should be approximately unilateral (i.e., have low reverse gain) so that its performance is not unduly sensitive to source or load impedance; and it should not require excessive tuning or "tweaking" to make it work properly.

These properties are best realized by a *transconductance mixer*: one in which the time-varying transconductance is the dominant contributor to frequency conversion, and the effect of other nonlinearities is minimal. In designing such a mixer, we maximize the fundamental LO-frequency component of the transconductance waveform $g_m(t)$, and minimize the time-variation of other circuit elements, especially the drain-to-source conductance $g_{ds}(t)$. There is both empirical evidence and theoretical justification for the assertion that this mode of operation minimizes the mixer's noise figure and maximizes its gain [1], [2].

The transconductance of a FET is maximum when its drain-to-source voltage V_{ds} is maximum. Thus, as the FET is pumped by the LO, it is imperative that V_{ds} not be allowed to decrease substantially from its dc value during any part of the LO cycle. It is especially important that V_{ds} not drop below the knee of the $I_d(V_{ds})$ characteristic. Clearly, this requirement implies that the LO should not be applied to the FET's source or drain terminal; the LO should be applied to the gate. Furthermore, to remove any ac components from $V_{ds}(t)$, the drain should be short-circuited at all harmonics (including the first) of the LO frequency.

If these requirements are met, V_{ds} will remain constant at its dc value over the entire LO cycle, and the magnitude of the transconductance waveform $|g_m(t)|$ will be as great as possible. However, our goal in designing a FET mixer is not simply to maximize $|g_m(t)|$, but to maximize the fundamental-frequency component of $g_m(t)$, because this is the major contributor to frequency mixing. A few minutes experimenting with Fourier-series relations will lead one to the conclusion that, under the obvious constraint

$$|g_m(t)| < G_{m,\max} \tag{9.1}$$

where $G_{m,\max}$ is the FET's peak transconductance, $|g_m(t)|$ is maximized when the gate bias $V_{gg} = V_t$, the turn-on voltage, and $g_m(t)$ is a train of approximately half-sinusoidal pulses having peak value $G_{m,\max}$.

Failing to short-circuit the drain effectively does not only limit the transconductance. If the drain voltage drops below the knee of $I_d(V_{ds})$, $g_{ds}(t)$ increases dramatically. The mixer's conversion gain is affected strongly by the time-average value of $g_{ds}(t)$, and any increase in $g_{ds}(t)$, even over a short part of the LO cycle, has a strong and deleterious effect on gain. Because most of a FET mixer's noise is generated in the channel (hence the output circuit), reduction in gain from increased $g_{ds}(t)$ effectively increases the mixer's noise temperature. Furthermore, when V_{ds} drops below the I/V knee, the gate-to-channel capacitance C_{gd} also increases rapidly, further reducing the mixer's gain and compromising its stability. Fortunately, it is usually not difficult to short-circuit the drain effectively and prevent this from happening.

Short-circuiting the drain has another important advantage: a short circuit is a stable termination for most FETs over a very broad range of frequencies, so this termination provides good stability as well as optimum gain and noise figure. In fact, for this reason it is worthwhile to short-circuit the drain over a bandwidth as broad as possible, excluding only the IF frequency.

Figure 9.1 shows the waveforms in a FET mixer. In Figure 9.1(a), the FET is terminated in the ideal short circuit, and trajectory of the *I/V* locus in the plane of the drain *I/V* curves is a vertical line. The channel conducts in pulses, and the $g_m(t)$ waveform is an approximately sinusoidal pulse train. In Figure 9.1(b), the termination is a finite resistance, the *I/V* trajectory is a sloped line (not unlike the load line in a power amplifier), and the $I_d(t)$, $g_m(t)$, and $g_{ds}(t)$ waveforms are as shown. Most striking is the collapse of $g_m(t)$ and the rise in $g_{ds}(t)$ as the FET leaves its current-saturated mode of operation. The reduction in the fundamental-frequency component of the transconductance and the increase in the average value of $g_{ds}(t)$ are apparent.

The FET's optimum terminations at unwanted small-signal mixing frequencies $\omega_n = \omega_0 + n\omega_p$, like those at the LO harmonics, should be short circuits at both the gate and drain. This prevents the downconversion of noise at any of these frequencies to the IF. Although the gate termination at the unwanted LO harmonics is not critical (as long as the drain is properly terminated), the gate termination at the IF is somewhat more important. The pumped FET's time-average transconductance is nonzero; therefore, the pumped FET can operate as an amplifier, as well as a mixer, and will amplify any IF-frequency noise that is applied to the gate. Such noise can come from preamplifiers and even occasionally from the mixer's bias circuit. Short-circuiting the gate at the IF frequency reduces the mixer's "amplifier-mode" gain and prevents it from amplifying this noise.

For much the same reasons as with the LO, it is best to short-circuit the drain at unwanted mixing frequencies. This termination provides optimum gain and is necessary (and usually, but not always, sufficient) for stability. The termination at the IF frequency is a bit more interesting. In single-gate active FET mixers, the IF output impedance is usually real and quite high, and it is often impossible to conjugate-match in practice (this phenomenon has been recognized almost since the beginning of the use of FET mixers [1], [2]). Therefore, an appropriate, unmatched termination must be chosen for the mixer. The best load impedance is a real impedance that provides adequate, but not excessive, gain; usually a value in the range 50Ω to 100Ω is adequate. In most mixers the input is conjugate matched at the RF frequency. It is possible that mismatching the input of a FET mixer may improve the noise figure in some cases (as is the usual practice in amplifiers), but there is very little experimental evidence. It is not likely that significant improvement in noise figure can be obtained by mismatching the input, especially in mixers having low conversion gain: even in FET amplifiers the optimum source impedance approaches a conjugate match when the maximum available gain is low.

A final consideration in the design of single-gate (and to some degree dual-gate)

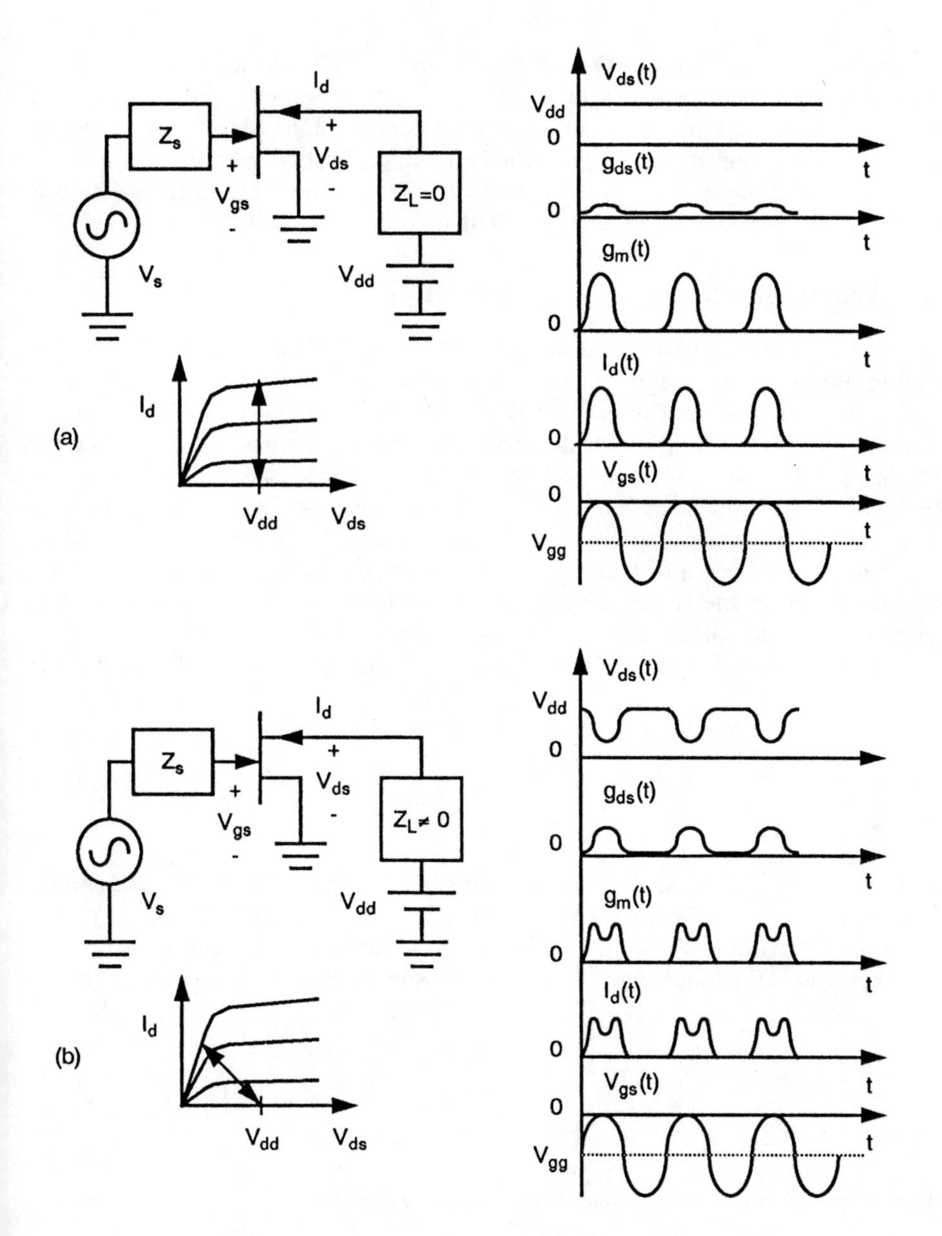

Figure 9.1 Voltage and current waveforms when (a) the FET's drain is short-circuited at the LO fundamental frequency and harmonics; and (b) the drain has a nonzero load impedance.

FET mixers is the problem of LO amplification by the FET. The amplification of the LO causes the LO-to-IF isolation of a FET mixer to be inherently worse than that of a diode mixer, which performs no such amplification. It is usually necessary to provide extra filtering in the output of a FET mixer, beyond what is necessary to provide a short-circuit termination, to reject the LO leakage. LO-to-IF isolation is one parameter that is clearly poorer in FET than in diode mixers.

9.1.2 Approximate Design of Single-Gate Mixers

It is possible to use approximate techniques to design FET mixers successfully. The resulting approximate design can be optimized by means of a harmonic-balance calculation, or can be used directly, with a little judicious "tweaking," to realize a mixer. An important condition for the validity of this approximate design is that the FET have the optimum gate and drain terminations—short circuits—at all significant LO harmonics and unwanted mixing frequencies. Clearly, it is a difficult task to realize these terminations exactly; fortunately, the design of a FET mixer leaves some room for error in this regard. As might be expected, the most important terminations are at the lower LO harmonics and lower-order mixing frequencies. Greater error can be tolerated in terminating the higher-order products.

A good first-order estimate of the input impedance at either the RF or the LO frequency is

$$Z_{\mathrm{in}}(\omega) = R_s + R_i + R_g + \frac{1}{j\omega C_{gs0}} \tag{9.2}$$

where C_{gs0} is the FET's zero-voltage gate-to-source capacitance, and R_i, R_s, and R_g are the resistances in the FET's input loop (Section 3.2.1 and Figure 3.7). If a lumped model of the FET is not available, the input impedance can be estimated from the device's S parameters. The input reflection coefficient is approximately

$$\Gamma_{\mathrm{in}} = S_{11} - \frac{S_{12}S_{21}}{1 + S_{22}} \tag{9.3}$$

and the impedance can be calculated from the usual relation,

$$Z_{in}(\omega) = Z_0\left(\frac{1+\Gamma_{in}}{1-\Gamma_{in}}\right) \tag{9.4}$$

where Z_0 is the normalizing impedance for the S parameters, usually 50Ω. These relations are based on the conventional expressions for the input impedance or reflection coefficient of a linear two-port having a short-circuited output. They are surprisingly accurate for pumped FETs as well.

When the drain of the FET mixer is shorted, and the gate is biased near V_t, the FET mixer is represented approximately by the circuit in Figure 9.2. The mixer's conversion gain can then be found in a straightforward manner. If $Z_s(\omega_{RF}) = Z_{in}^*(\omega_{RF}) = R_s + R_i + R_g + j/\omega C_{gs}$, the conversion gain is

$$G_c = \frac{G_{m,\max}^2 R_L}{16\omega_{RF}^2 C_{gs}^2 (R_s + R_i + R_g)} \tag{9.5}$$

This relation can also be used to select a value of R_L. Finally, the minimum required LO power (which is achieved when $Z_s(\omega_p) = Z_{in}^*(\omega_p)$) is

$$P_{LO} = \frac{1}{2}(V_{g,\max} - V_{gg})^2 \omega_p^2 C_{gs}^2 (R_s + R_i + R_g) \tag{9.6}$$

where $V_{g,\max}$ is the gate voltage at which $G_{m,\max}$ occurs, and V_{gg} is the gate bias

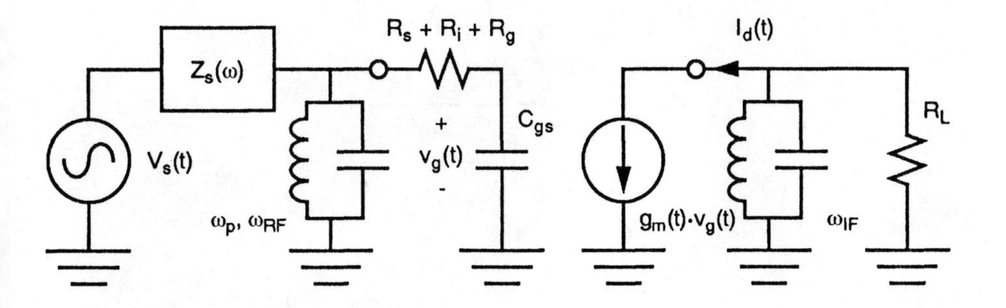

Figure 9.2 Simplified small-signal equivalent circuit of a FET mixer.

voltage. Usually $V_{gg} \sim V_t$. The above expressions do not include circuit losses; these are often significant and must be included in any estimate of conversion loss. Also, it is often impossible to match the gate at both the RF and LO frequencies; in this case, it is best to optimize the RF match at the expense of the LO match. The resulting LO mismatch increases the required LO power, but has no other disadvantages.

Equation (9.5) implies that a FET mixer's conversion gain can be made arbitrarily high if R_L is made great enough. This is not far from the truth. Indeed, the gain of a FET mixer can be made quite high in this manner; however, at some point the gain is limited by considerations of stability, bandwidth, intermodulation, practical realizability, and the fact that the pumped FET's output impedance, although high, is nevertheless finite.

In any case, high gain in a mixer is not always desirable; in most communication receivers the optimum mixer gain is at most five or six dB. Higher gain results in greater intermodulation distortion and spurious responses; lower gain increases the receiver's noise figure. In general, a receiver front end should have no more gain than necessary to provide an adequate noise figure.

9.1.3 Design Example

As an example, we shall examine the design of a 7.9- to 8.4-GHz single-gate FET mixer. The mixer's IF passband is 0.9 to 1.4 GHz, and the LO frequency is fixed at 7.0 GHz. The desired gain is in the range of 5 to 6 dB. The first order of business is to select an appropriate FET; the ubiquitous 0.5 × 300-μm Ku-band small-signal device is a good place to begin. The parameters of such a device are given in Table 9.1.

Table 9.1
FET Parameters

Parameter	*Value*
R_s	3Ω
R_i	3Ω
R_g	1.4Ω
C_{gs0}	0.3 pF
$G_{m,\max}$	0.04 mS

Using (9.5), we estimate that a load resistance of approximately 60Ω should provide 5.5 dB of gain; this load should be relatively easy to realize over the required 500-MHz bandwidth. We note that for minimum noise it may be necessary to use an LO level slightly less than the value that provides optimum gain. Consequently, 4.5 to 5.0 dB is a more realistic estimate of the conversion gain. Because the RF and LO frequencies are so far apart, it will not be possible to match the gate at the LO frequency. Thus, in this case, Equation (9.6) does not provide a useful estimate of the required LO power.

We first design the IF circuit. A conventional low-pass filter provides a short circuit over the required bandwidth, from the lower end of the image band at 5.6 GHz to the upper end of the RF band, 8.4 GHz. The termination at the second LO harmonic is not quite as good, but probably adequate. Because the filter is based on an asymmetrical prototype, its input impedance at the IF frequency is not 50Ω; instead, it is closer to 65Ω, the desired value. The shunt capacitances of the filter are realized by sections of low-impedance microstrip line, and the inductive sections by high-impedance lines. These are approximated by the relations

$$\begin{aligned} \omega C &= Y_0 \tan(\theta) \\ \omega L &= Z_0 \tan(\theta) \end{aligned} \tag{9.7}$$

which work best when the characteristic impedance and admittance Z_0 and Y_0 are high, and electrical length θ is small. The final step in the design of the IF circuit is to optimize it on the computer. Because of the approximations inherent in (9.7), it is usually necessary to modify the initial design to achieve the best possible performance.

Many FET mixers reported in the literature use simple quarter-wave stubs to short-circuit the drain. These are rarely adequate: they do not short-circuit the LO harmonics, and do not have great enough bandwidth or low enough loss to provide a good short circuit. A filter structure usually gives much better results and is not much more difficult to design or implement.

We use (9.4) to estimate the RF input impedance. The RF input-matching circuit consists of a simple shunt-C, series-L structure; the capacitance is realized by a pair of stubs, and the inductance by a series line. A series L-C structure consisting of a high-impedance microstrip and a chip capacitor provides the IF short circuit at the gate; the high-impedance line is approximately one-quarter wavelength long at the center of the combined RF and LO bands, so it does not affect the RF or LO matching significantly. This stub serves as an input for the dc gate bias as well.

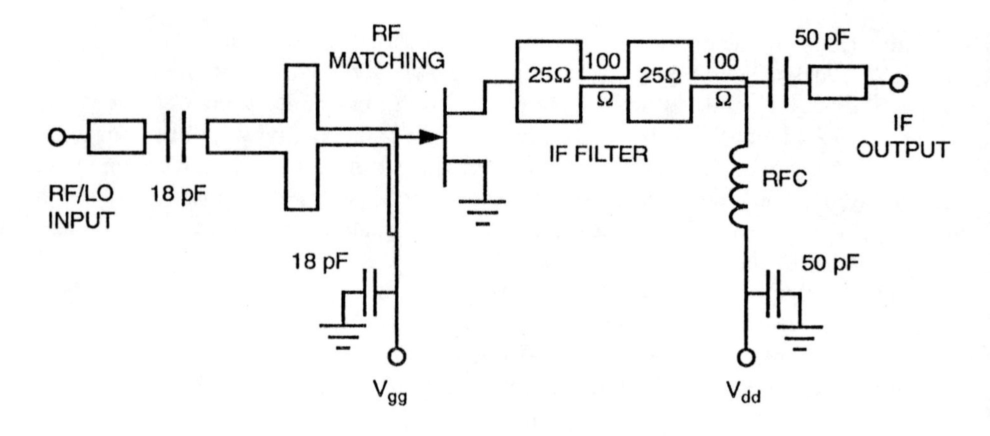

Figure 9.3 Single-gate microstrip FET mixer of the design example.

Figure 9.3 shows the circuit of the complete mixer. Figure 9.4 shows the mixer's conversion gain and Figure 9.5 shows the noise figure. In both cases, the LO is fixed at 7.0 GHz and the IF frequency is allowed to vary with the RF. The mixer's minimum noise figure occurs at an LO level of approximately 0 dBm, and its conversion loss is maximum at 6 dBm. The mixer's input VSWR is below 2.0 across the RF band, and at the 6-dBm LO level, its third-order intercept point, in terms of output power, is 12 dBm. The dc bias is the same for both LO levels.

For test purposes the RF and LO are combined by an external diplexer. In practice, two of these single-gate mixers would be used in a balanced structure. A balanced mixer would have nearly the same conversion gain and noise figure (these would be degraded approximately 0.3 dB by losses in the hybrid) and a 3-dB greater intercept point. We shall examine balanced FET mixers further in Section 9.3.

9.1.4 Numerical Analysis of FET Mixers

Large-signal harmonic-balance and small-signal, time-varying methods, similar to those described in Chapter 4, are often used to analyze FET mixers. Although the theory presented in Chapter 4 is oriented toward the analysis of diode mixers, the extension to FETs is straightforward. The general approach is also analogous: the mixer first undergoes a large-signal analysis under LO excitation only, and subsequently a small-signal, time-varying analysis is performed.

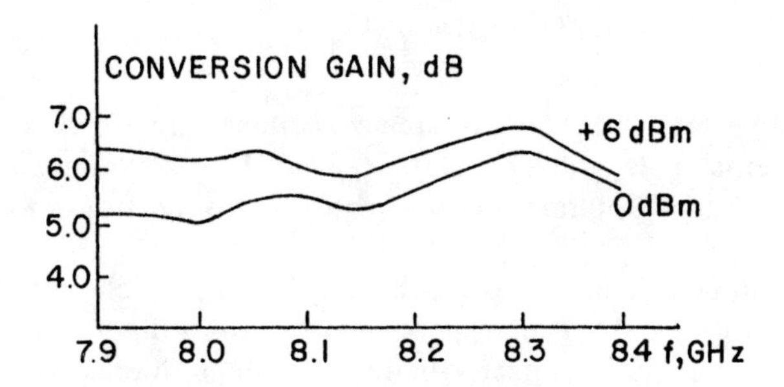

Figure 9.4 Conversion gain of the FET mixer at two LO levels.

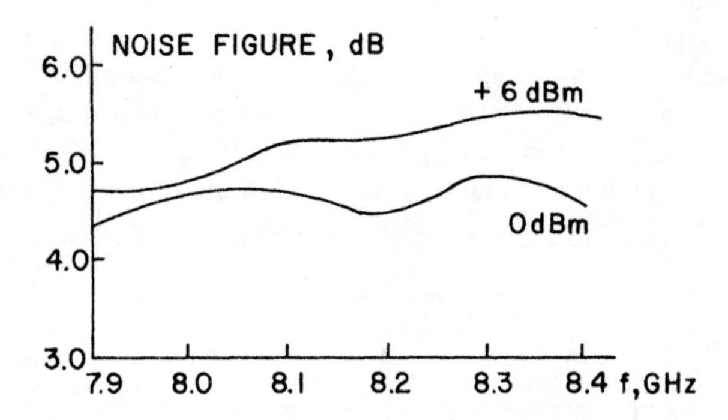

Figure 9.5 Noise figure of the FET mixer.

Figure 3.7 shows the nonlinear equivalent circuit of a FET. It includes five nonlinear elements: a controlled current source I_d, two nonlinear capacitors C_{gs} and C_{gd}, and two diodes representing the resistive part of the gate-to-channel junction. In a well-designed FET mixer, the drain voltage does not drop low enough for the FET to enter its linear region (Section 3.1.2); current saturation is maintained in the channel throughout the entire LO cycle. In this case, the gate-to-drain diode is never forward biased, the nonlinearity of C_{gd} is negligible, and $V_d(t) \cong V_{dd}$ (as part of this approximation we are ignoring the small dc voltage drops across R_s and R_d). Then

$$
\begin{aligned}
I_d(V_g, V_d) &\equiv I_d(V_g, V_{dd}) \\
C_{gs}(V_g, V_d) &\equiv C_{gs}(V_g, V_{dd})
\end{aligned}
\tag{9.8}
$$

that is, the drain current and gate-to-source capacitance are functions of only one ac voltage, the internal gate voltage $V_g(t)$. V_{dd} can be substituted for $V_d(t)$ in the expressions for I_d and C_{gs}, turning these equations into functions of a single control voltage $V_g(t)$.

The equivalent circuit is then given by Figure 9.6, which includes source and load impedances $Z_s(\omega)$ and $Z_L(\omega)$, respectively. As with a diode mixer, this circuit is partitioned into linear and nonlinear subcircuits. The partitioned circuit is shown in Figure 9.7; the source and load impedances are absorbed into the linear subcircuit. The large-signal analysis is almost identical to that described in Section 4.2.7. Virtually all of the solution algorithms described in Section 4.2.3 have been applied at one time or another to the task of analyzing FET mixers.

The small-signal equivalent circuit is shown in Figure 9.8. The small-signal analysis, like the large-signal, can be handled in a manner similar to that of a diode mixer. It is necessary first to generate the conversion matrices of the individual circuit elements and combine them with the admittance matrix of the linear subcircuit. The conversion loss can be calculated in the following manner:

1. Determine $Z_s(\omega_n)$ and $Z_L(\omega_n)$ at all small-signal mixing frequencies ω_n (4.1).

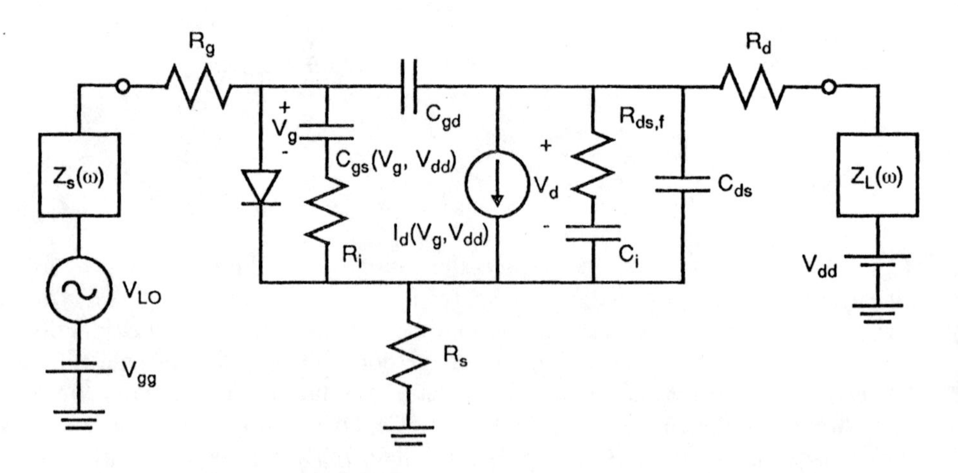

Figure 9.6 Large-signal (LO) equivalent circuit of an active FET mixer.

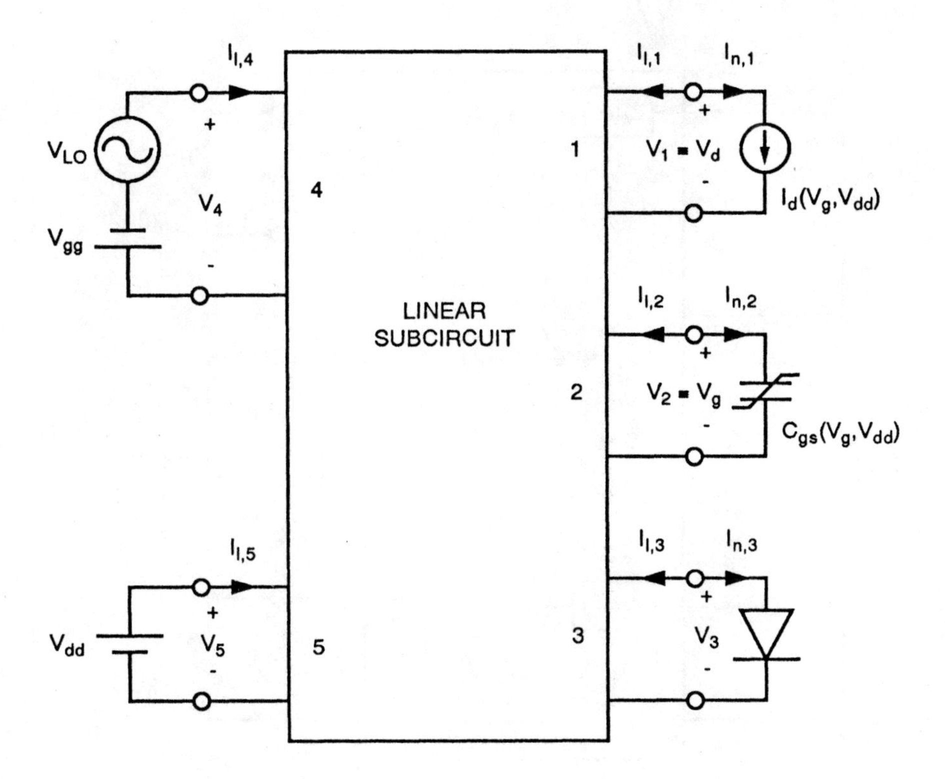

Figure 9.7 Partitioned large-signal equivalent circuit of the FET mixer in Figure 9.6.

2. Form the admittance matrix **Y** of the linear subcircuit at all ω_n. This is arranged as a matrix of diagonal submatrices:

$$\mathbf{Y} = \begin{bmatrix} \mathbf{Y}_{11} & \mathbf{Y}_{12} & \mathbf{Y}_{13} & \mathbf{Y}_{14} & \mathbf{Y}_{15} \\ \mathbf{Y}_{21} & \mathbf{Y}_{22} & \mathbf{Y}_{23} & \mathbf{Y}_{24} & \mathbf{Y}_{25} \\ \mathbf{Y}_{31} & \mathbf{Y}_{32} & \mathbf{Y}_{33} & \mathbf{Y}_{34} & \mathbf{Y}_{35} \\ \mathbf{Y}_{41} & \mathbf{Y}_{42} & \mathbf{Y}_{43} & \mathbf{Y}_{44} & \mathbf{Y}_{45} \\ \mathbf{Y}_{51} & \mathbf{Y}_{52} & \mathbf{Y}_{53} & \mathbf{Y}_{54} & \mathbf{Y}_{55} \end{bmatrix} \tag{9.9}$$

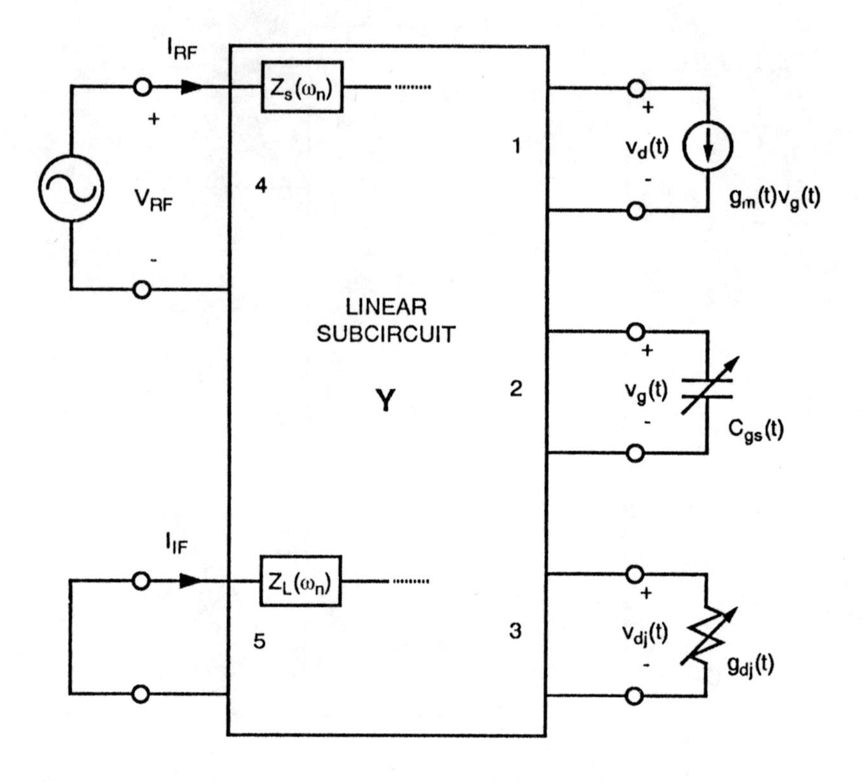

Figure 9.8 Small-signal equivalent circuit of the FET mixer.

where

$$\mathbf{Y}_{i,j} = \mathrm{diag}(Y_{i,j}(\omega_n)) \tag{9.10}$$

arranged, like Equation (4.68), with ω_{-N} at the top left position and ω_N at the bottom right.

3. Form the conversion matrices of the transconductance $\mathbf{G}_m$, the gate-to-source capacitance $\mathbf{C}_{gs}$, and the resistive diode junction $\mathbf{G}_d$ (Section 4.3.2).
4. Form the conversion matrix $\mathbf{Y}_c$ of the entire circuit of Figure 9.8 by augmenting the linear matrix $\mathbf{Y}$ with the conversion matrices; that is,

$$\mathbf{Y}_{c,12} = \mathbf{Y}_{12} + \mathbf{G}_m$$
$$\mathbf{Y}_{c,22} = \mathbf{Y}_{22} + j\Omega\mathbf{C}_{gs} \tag{9.11}$$
$$\mathbf{Y}_{c,33} = \mathbf{Y}_{33} + \mathbf{G}_d$$

where Ω is given by (4.67). This process absorbs the time-varying elements into $\mathbf{Y}_C$.

5. In the manner described in Section 4.3.3, eliminate ports 1 through 3. First, form $\mathbf{Z}_C = \mathbf{Y}_C^{-1}$. Because the elements terminating the ports in Figure 9.8 have been absorbed into the network, ports 1 to 3 are now open-circuited and the currents in $\mathbf{V} = \mathbf{Z}_C\mathbf{I}$ at these ports are zero. Therefore, set the port-current subvectors $\mathbf{I}_n = \mathbf{0}$, $n = 1,...,3$. This reduces $\mathbf{Z}_C$ to an impedance-form conversion matrix that relates the voltage and current vectors at only ports 4 and 5 in Figure 9.8. We call this matrix $\mathbf{Z}_{c,2}$.
6. Finally, invert $\mathbf{Z}_{c,2}$ to obtain $\mathbf{Y}_{c,2}$. The RF and IF ports are short-circuited at all mixing frequencies except ω_{RF} at port 4. Thus, a single term of $\mathbf{Y}_{c,2}$, $Y_{c,IR} = I_{\text{IF}}(\omega_{\text{IF}})/V_{\text{RF}}(\omega_{\text{RF}})$, relates the short-circuit IF output current $I_{\text{IF}}(\omega_{\text{IF}})$ in Figure 9.8 to the RF excitation voltage $V_{\text{RF}}(\omega_{\text{RF}})$; the conversion loss L_c is

$$L_c = \frac{1}{4|Y_{c,IR}|^2 \,\text{Re}\{Z_s(\omega_{\text{RF}})\}\ \text{Re}\{Z_L(\omega_{\text{IF}})\}} \tag{9.12}$$

7. Port impedances can be found in essentially the same manner as in Chapter 4. The input impedance is simply

$$Z_{\text{in}} = \frac{1}{Y_{c,RR}} - Z_s(\omega_{\text{RF}}) \tag{9.13}$$

and

$$Z_{\text{out}} = \frac{1}{Y_{c,II}} - Z_s(\omega_{\text{IF}}) \tag{9.14}$$

where $Y_{c,RR}$ and $Y_{c,II}$ are the components of $\mathbf{Y}_{c,2}$ relating input and output voltages and currents, respectively.

General-purpose harmonic-balance simulators are used frequently to analyze mixers. These often can be set to operate in one of two modes. The first is one in which all LO harmonics and mixing frequencies are used in the harmonic-balance process, and a nonperiodic time-to-frequency transform is used in each Newton iteration to relate the time- and frequency-domain quantities. The second is to perform the combined large-signal and small-signal analyses in a manner similar to that described above. The latter mode is much more efficient and accurate than the former, and should be viewed as the preferred method for calculating a mixer's port impedances and conversion loss or gain.

9.2 DUAL-GATE MIXERS

Dual-gate FET mixers have one major advantage over single-gate: the LO and RF signals can be applied to separate gates and, because the capacitance between the gates is low, the mixer has good RF-to-LO isolation. Thus, it is often practical to use a single-device dual-gate FET mixer in applications where a balanced mixer would otherwise be needed (e.g., in integrated circuits, where the elimination of a hybrid or a filter saves a significant amount of expensive substrate area). Balanced dual-gate FET mixers are also possible and are used in applications where the spurious-response and LO-noise rejection of a balanced mixer are valuable. Balanced dual-gate mixers usually require hybrids for all three ports, and thus may be relatively complicated circuits.

9.2.1 Operation of Dual-Gate Mixers

Dual-gate FETs were examined briefly in Section 3.3. Such devices are best modeled as two single-gate FETs in series; the parameters of the dual-gate device are obtained from measurements of equivalent single-gate devices. In particular, the *I*/*V* characteristics of a dual-gate device can be derived from the *I*/*V* characteristics of two single-gate devices; an example is shown in Figure 3.10. The *I*/*V* characteristics are found by applying two constraints: (1) the sum of the drain-to-source voltages of the individual devices must equal the drain voltage of the dual-gate device, and (2) the drain current must be the same in both devices. Because of the usefulness of this representation, we shall view the dual-gate FET, in the following description, as a pair of series-connected, single-gate devices. We shall also assume that the LO is applied to the gate of the upper device (gate no. 2), and the RF to the gate of the lower device (no. 1). This is the usual mode of operation, and is illustrated in Figure 9.9.

An important property of dual-gate FETs is that both devices in the series-

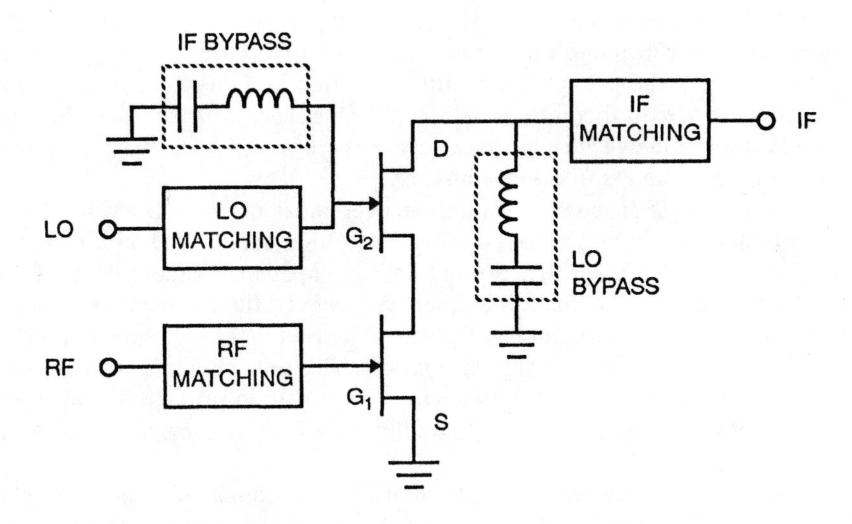

Figure 9.9 Dual-gate FET mixer. The dual-gate device is represented by two single-gate FETs in series.

connected pair can remain in current-saturated operation over only a narrow range of gate-bias voltages. This point is evident from Figure 3.10(a). Both devices are simultaneously in current saturation when both V_{ds1} and V_{ds2} are greater than approximately 1V; this condition occurs only when $V_{gs1} \sim V_{gs2}$. However, applying an LO voltage to gate no. 2 (the gate of the uppermost FET in Figure 3.9) varies V_{gs2} over a very wide range of voltages, so it is impossible for both FETs to remain current-saturated throughout the LO cycle. Furthermore, with the LO applied to the second gate, the transconductance of the lower FET cannot be varied in the same manner as with the single-gate mixer. Thus, the type of transconductance mixer that was optimum for single-gate FET mixers is not realizable with a dual-gate device.

It has been well established [3] that the best mode of operation of a dual-gate FET mixer is one in which the LO drives the lower FET (no. 1) into and out of current saturation over the LO cycle. This occurs as the lower FET's drain voltage is forced alternately low and high by the LO. When the drain voltage of the lower FET is low, its transconductance is low and its drain-to-source conductance is relatively great. When this voltage rises, the lower FET enters its current-saturated region; the transconductance is then relatively great and drain-to-source conductance low. The region of the *I/V* curve in which the device operates is shown by the shaded region in Figure 3.10(b). The pumping of these two parameters—g_{ds} and g_m—provides frequency mixing in the lower FET. A little reflection will convince the reader that this is the only mode of operation possible in such a device: if the lower FET were to remain in saturation over the entire LO cycle, there would be no way to pump any of this device's parameters to achieve mixing.

The upper FET is in current saturation over most of the LO cycle. Thus, it operates simultaneously as a source-follower amplifier for the LO and a common-gate amplifier for the IF. Ideally, the LO source impedance should be conjugate-matched to gate number 2 at the LO frequency; however, the LO input impedance at this gate is often too great to match in practice, and the only recourse is to drive it from a real impedance that is as high as possible. This gate should also be grounded at the IF frequency. Often, a simple inductor is adequate to provide this bypassing; however, if the IF is not very low compared to the LO frequency, a series-resonant structure may be needed.

As with the single-gate mixer, the drain of the dual-gate device should be short-circuited at the LO frequency. As with a single-gate mixer, this short circuit keeps the drain voltage of the pair of FETs constant and guarantees that the upper FET remains in saturation over most of the LO cycle.

The operation of a lower FET as both a conductance and transconductance mixer has a cost. First, in Section 9.1.1 we saw that allowing the FET to enter current saturation caused the average value of $g_{ds}(t)$ to be relatively great. This average conductance is in parallel with the lower FET's channel and causes power loss. Second, because $V_{ds1}(t)$ never reaches zero, the peak value of $g_m(t)$ is not as great as in a single-gate mixer, and its waveform is also very different; as a result, the fundamental-frequency component of $g_m(t)$ is much smaller than in a single-gate mixer. Although the operation of the upper FET, as a common-gate IF amplifier, may make up for these deficiencies to some degree, the loss and thermal noise introduced by the conductance and the degradation of the transconductance are invariably deleterious.

Dual-gate FET mixers are beset by another fundamental weakness. The upper FET operates as a common-gate amplifier, and such amplifiers invariably have poor stability. The unavoidable use of the upper FET as a common-gate amplifier introduces the possibility of unstable operation. It is not unusual for the RF input impedance of the FET at the lower gate to have a negative real part, regardless of the mixer's IF load impedance. Often, the only way to stabilize the mixer in such circumstances is to add a resistance in series with the FET's source and to suffer a reduction in the mixer's conversion gain.

9.2.2 Approximate Small-Signal Analysis of Dual-Gate FET Mixers

Figure 9.10 shows an approximate small-signal equivalent circuit of a dual-gate FET mixer. The waveforms $g_{m1}(t)$ and $g_{ds1}(t)$ have the Fourier series

$$\begin{aligned} g_{m1}(t) &= G_{m0} + G_{m1}\cos(\omega_p t + \phi_{m1}) + G_{m2}\cos(2\omega_p t + \phi_{m2}) + \ldots \\ g_{ds1}(t) &= G_{d0} + G_{d1}\cos(\omega_p t + \phi_{d1}) + G_{d2}\cos(2\omega_p t + \phi_{d2}) + \ldots \end{aligned} \tag{9.15}$$

The fundamental-frequency components of $g_m(t)$ and $g_{ds}(t)$ are dominant in providing conventional downconversion mixing. The IF current in the drain of FET number 1 is

$$I_{\mathrm{IF}}(t) = g_{m1}(t)v_{g1}(t) + g_{ds1}(t)v_{d1}(t) \tag{9.16}$$

where $v_{d1}(t)$ is the small-signal drain-to-source voltage of the lower FET. Substituting the equations of (9.15) into (9.16) and separating the IF component gives

$$I_{\mathrm{IF}}(\omega_0) = \frac{1}{2}\left(G_{m1}V_{g1}(\omega_1) + G_{d1}V_{d1}(\omega_1)\right) + G_{d0}V_{d1}(\omega_0) \tag{9.17}$$

where the frequency notation is given by Equation (4.1). We note that $V_{d1} \sim G_{m1}V_{g1}/G_{d0}$, so the second term is smaller than the first by a factor of at least G_{d1}/G_{d0}. In fact, it is smaller than this because of the loading effect of C_{gs2} and R_{in2}

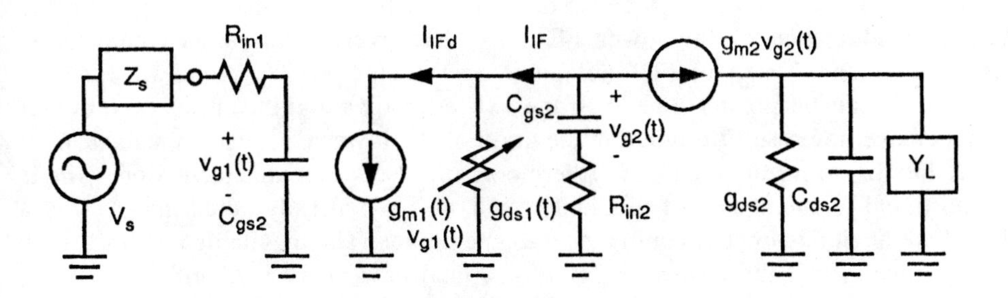

Figure 9.10 Simplified small-signal equivalent circuit of a dual-gate FET mixer. Both $g_{m1}(t)$ and $g_{ds1}(t)$ are pumped by the LO; $v_{g1}(t)$ and $v_{g2}(t)$ are small-signal voltages. Y_L is the IF load admittance. R_{in} is the sum of R_g, R_s, and R_i of each device.

(R_{in} is the sum of the source, intrinsic, and gate resistances of each device). Thus, the second term in (9.17) is usually negligible, and only $g_{m1}(t)$ provides significant mixing. If the input is conjugate matched, we can obtain

$$\left|\frac{I_{\mathrm{IF}d}(\omega_0)}{V_s(\omega_1)}\right| = \frac{G_{m1}}{4\omega_1 C_{gs1} R_{in1}} \tag{9.18}$$

Similarly, we assume that g_{ds2} is negligible ($g_{ds2} \ll \mathrm{Re}\{Y_L(\omega_0)\}$) and the load susceptance resonates C_{ds2}. A straightforward analysis then gives the output current,

$$\left|\frac{I_L(\omega_0)}{V_s(\omega_1)}\right| \approx \frac{\dfrac{G_{m1} g_{m2}}{G_{d0} + g_{m2}}}{4\omega_1 C_{gs1} R_{in1} \left(1 + \left(\dfrac{C_{gs2}\omega_0}{g_{m2}}\right)^2\right)^{0.5}} \tag{9.19}$$

Note that g_{m2} is time-invariant. Finally, the conversion gain is

$$G_c = \frac{4\,\mathrm{Re}\{Z_s(\omega_1)\}}{\mathrm{Re}\{Y_L(\omega_0)\}} \left|\frac{I_L(\omega_0)}{V_s(\omega_1)}\right|^2 \tag{9.20}$$

Of the above equations, (9.19) is most telling. It implies that the high average output conductance of the lower FET G_{d0} effectively reduces the "conversion transconductance" G_{m1} of FET number 1 by a factor $g_{m2} / (g_{m2} + G_{d0})$. Because G_{d0} easily can be greater than g_{m2}, the existence of a significant average output conductance, the result of pumping the drain of the lower FET, causes a substantial reduction in conversion gain. Furthermore, the reduction in conversion gain is proportional to the square of this factor, and we have already noted that G_{m1} in a dual-gate device is lower than in a single-gate device. The inevitable result of these inequalities is that the conversion gain of a dual-gate mixer is significantly lower than that of a single-gate mixer. Reported experimental results support this observation: while conversion gains of 6 to 10 dB at X band are not unusual in single-gate mixers, it is unusual to see more than a decibel or two of conversion gain, under comparable conditions, in dual-gate mixers. The noise figures of dual-gate mixers are also generally inferior to those of single-gate mixers.

Although the noise and conversion efficiency of dual-gate FET mixers is generally worse than single-gate mixers, the intermodulation characteristics are often moderately, but not dramatically, better. The drain current as a function of gate voltage V_{gs1} is usually more linear in a dual-gate FET than in a single-gate device. Figure 9.11 shows the gate-to-drain *I*/*V* characteristic of one such device; the linearity of this characteristic is obviously better than that of a single-gate FET, which has a distinctly quadratic appearance.

9.3 BALANCED FET MIXERS

Balanced FET mixers can be realized with either single-gate or dual-gate devices. Although either type of device is useful for singly balanced mixers, dual-gate mixers are especially well suited for doubly balanced circuits. Unlike diodes, however, FETs cannot be "reversed"; the consequence of this rather obvious characteristic is that balanced FET mixers often require IF hybrids, while diode mixers do not. This is a disappointing but unavoidable situation: multiple hybrids significantly increase the size of a mixer and limit its practicality for monolithic applications.

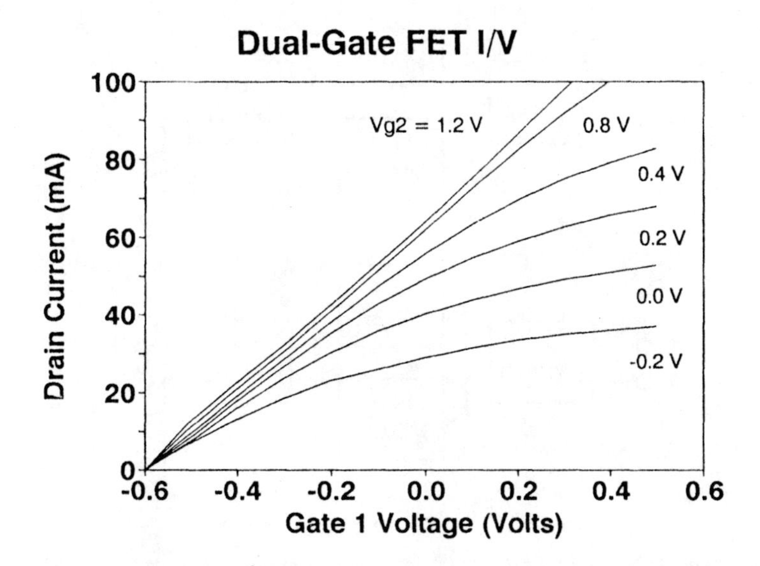

Figure 9.11 Gate-to-drain *I*/*V* characteristic of a dual-gate device. V_d = 3V.

9.3.1 Single-Gate Balanced FET Mixers

Figure 9.12 shows the two fundamental types of singly balanced single-gate FET mixers. Except for the need of a 180-deg IF hybrid, they are analogous to diode mixers, and their properties are similar to those of a singly balanced diode mixer having identically oriented diodes and an output hybrid (Section 7.2.2, page 265). Specifically, these mixers exhibit the same inherent LO-to-RF isolation, spurious-response rejection, and LO noise rejection characteristic of singly balanced diode mixers. Like singly balanced diode mixers, the intermodulation-rejection properties of the 180-deg hybrid mixer are significantly better than the quadrature mixer, and the balance of the 180-deg mixer is much less sensitive to source VSWR at its RF and LO ports. In both mixers, the IF is always the delta port of the output hybrid.

Like balanced diode mixers, the conversion gain and noise figure of an ideal balanced FET mixer are identical to that of a single-device mixer, and the output power and intermodulation intercept points are increased 3 dB by the "power combining" effect of the two devices. In real mixers, the loss and imbalance of the hybrids slightly degrade the conversion loss and noise figure and fundamentally limit the rejection of even-order spurious responses and intermodulation products.

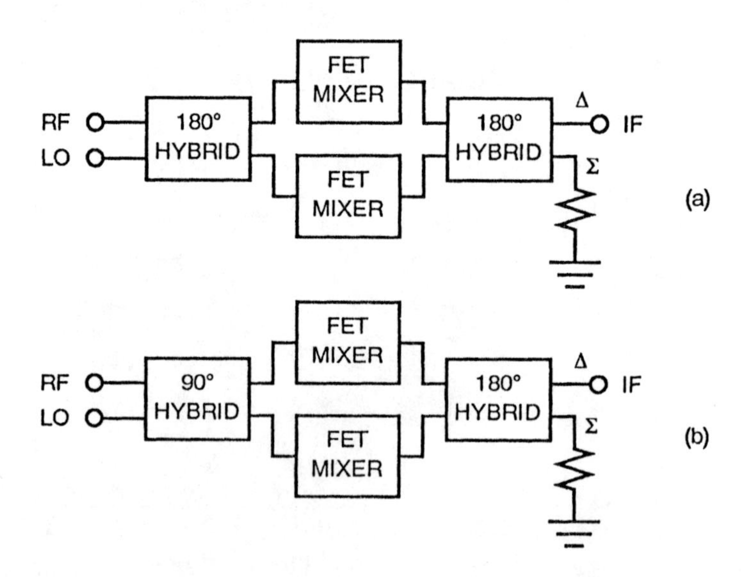

Figure 9.12 Singly balanced FET mixers: (a) 180-deg hybrid mixer; (b) quadrature-hybrid mixer.

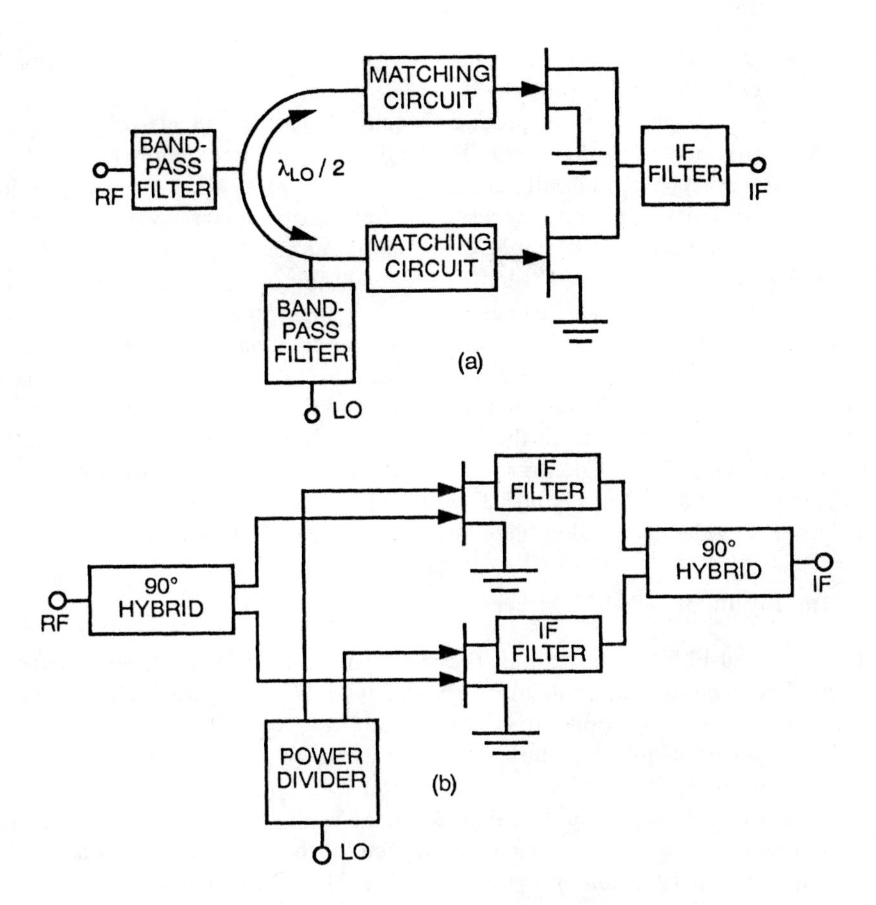

Figure 9.13 (a) Subharmonically pumped FET mixer and (b) image-rejection mixer.

Like the diode mixer in Figure 7.23, the spurious-response rejection of this mixer is the opposite of that given in Table 7.1.

In the 180-deg diode mixer, the IF-connection node is a virtual ground for the signal applied to the delta port of the input hybrid. Because of this, the diode mixer exhibited a limited degree of inherent LO- or RF-to-IF isolation. Unfortunately, this situation usually exists only theoretically in a FET mixer; if the IF hybrid's passband includes the LO, as it might in a modulator or upconverter, such rejection is possible; however, in conventional downconverters, the IF hybrid rarely has adequate bandwidth to provide such isolation. Because the LO-to-IF isolation of a

FET mixer is inherently poor, the inability to improve LO-to-IF isolation by applying the LO to a selected port is a disappointment.

Although it is possible to produce doubly-balanced, image-rejection, and subharmonically pumped single-gate FET mixers, the large number of hybrids involved often makes such circuits impractical. Sometimes, however, it is possible to create such circuits that are not excessively complex or that are even simpler than their diode counterparts. An example is the subharmonically pumped mixer shown in Figure 9.13(a). In this mixer, the half-wavelength transmission line causes the LO voltages at the gates of the two devices to be out of phase; the RF voltages, however, are in phase. The resulting fundamental conversion components are out of phase at the IF, so the devices effectively short-circuit each other at the fundamental mixing frequency. The mixing products between the RF and the second LO harmonic are in phase, however, and combine in the IF. Figure 9.13(b) shows an image-rejection mixer realized via a pair of dual-gate FETs. Because such mixers often have very low IF frequencies and the LO and RF frequencies are nearly the same, the LO-to-RF isolation provided by the dual-gate devices is especially valuable in this circuit.

9.3.2 Doubly Balanced FET Mixers

Doubly balanced FET mixers exhibit the same performance advantages, compared to singly balanced or single-device mixers, as doubly balanced diode mixers: inherent port-to-port isolation, broad bandwidth, and rejection of all even-order spurious responses. However, like many other types of FET mixers, they usually require hybrids at all ports and their circuits can be fairly complex.

Figure 9.14(a) shows a doubly balanced mixer using dual-gate FETs. It consists of four devices connected in a manner reminiscent of the classical Gilbert multiplier used in bipolar-transistor analog multipliers. The RF and LO signals are applied to the gates and the IF is extracted from the drain via baluns or microwave hybrids; virtually every type of balun or hybrid described in Chapter 7 has been used at one time or another for these baluns. Like other dual-gate mixers, this mixer requires dc bias and gate and drain matching circuits, which are not shown in the figure. Fortunately, the drains of the FETs are virtual ground points for the RF and LO; therefore, in contrast to the single-device dual-gate mixer, no special circuit is required to provide an RF/LO short at the drains. Such mixers normally include matching circuits at least for the RF and, of course, dc bias circuits; these are not shown in the figure.

Figure 9.14(b) illustrates the operation of the mixer. The LO voltage $V_{LO}(t)$ is applied to the upper gate of all four devices with the phases shown. Two devices are driven by each output of the LO balun; the transconductance of each device varies in phase with the LO signal. The RF voltage $V_{RF}(t)$ is applied similarly to the lower gates, but with a different set of phases. The IF output is proportional to the product of the transconductance and the RF voltage; thus, the IF currents in each pair of FETs are opposite and must be subtracted by an IF output balun. The same type of analysis used for diode mixers in Section 7.2.2 shows that IF currents resulting from even-order IM and spurious responses are in phase at all the drains and cancel in the IF

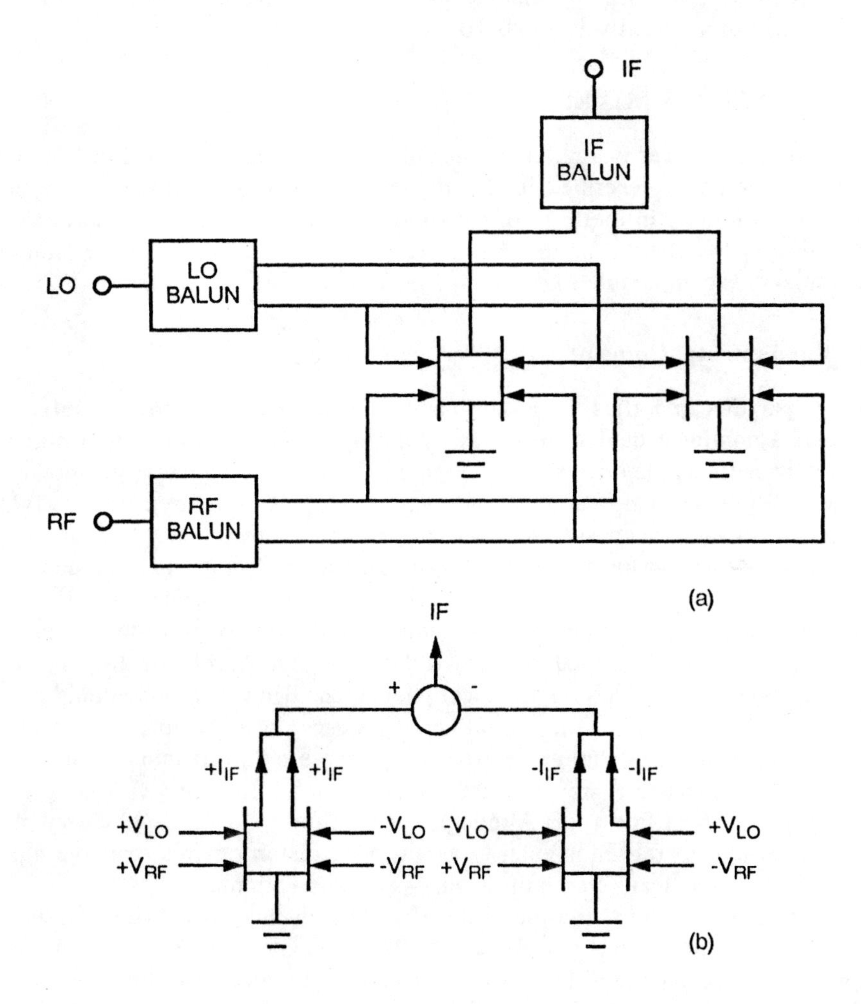

Figure 9.14 (a) Doubly-balanced dual-gate FET mixer and (b) equivalent circuit.

balun.

The baluns, as well as the use of separate gates, provide isolation between the RF and LO. Because the circuit is symmetrical, coupling from the LO to the RF (and RF to LO) must be the same in all devices. The LO leakage through the FETs is coupled equally to both the +RF and –RF terminals of the balun; the resulting LO output at the RF port is (ideally) zero. Even if the balance of the RF balun is poor at the LO frequency, the use of separate gates for the LO and RF alone provides significant isolation, usually 15 to 20 dB.

9.4 FET RESISTIVE MIXERS

The FET resistive mixer is a relatively new idea. It was first described in [4], and a balanced version was described in [5]. Since then, many such mixers have been reported, occasionally in the form of commercial products. The advantages of such mixers are very low distortion, low 1/f noise, and no shot noise; the conversion loss of such mixers is comparable to diode mixers, around 6 dB.

9.4.1 Fundamental Concept

A strange paradox underlies the entire business of designing microwave mixers: a mixer uses a nonlinear device to realize a fundamentally linear circuit. A mixer is theoretically a linear device; shifting a signal from one frequency to another is unquestionably a linear operation. This operation is performed by a time-varying, linear circuit element such as a time-varying resistor. Unfortunately, we normally create this time-varying linear element by applying a large-signal LO to a nonlinear element.

As long as we are forced to use a nonlinear device to perform the mixing operation, mixers are doomed to have relatively high levels of IM, spurious responses, and other undesirable nonlinear phenomena. However, if we could obtain this time-varying element without nonlinearity, we could use it to realize a mixer having no distortion. A "thought experiment" serves as an example: purchase a rotary potentiometer and remove its limit stops. Attach it to a very fast motor, and spin it at about 60 billion rpm. Although such a mixer would have a very short lifetime, it would operate momentarily as an ideal, distortion-free resistive mixer. All we need is a practical version of the high-speed potentiometer.

Now we examine the ubiquitous MESFET. The channel of a MESFET, at low drain-to-source voltages, is a very linear resistor. It becomes significantly nonlinear only when the drain voltage becomes great enough to accelerate the electrons to their saturated drift velocity. In most FETs, this occurs at a few tenths of a volt to one volt, depending on the gate voltage. Thus, at normal small-signal voltages of a few millivolts, the FET's resistive channel is very linear.

The resistance of this linear channel can be varied by applying an LO signal to the gate. The LO voltage changes the depth of the depletion region under the gate and thus the resistance of the entire channel. When the gate voltage drops below V_t, the FET's turn-on voltage, the resistance is virtually infinite; when the gate voltage reaches its maximum value (just below the value that causes gate-to-channel conduction, about 0.5V), the channel resistance is very low, usually a few ohms (this value depends on the geometry of the device). This range of resistances is entirely adequate to achieve good conversion performance in a resistive mixer; it is, in fact, not very different from the junction conductance of a diode mixer.

A mixer that uses the resistive channel of a MESFET to provide frequency conversion is called a *FET resistive mixer.* To realize such a mixer we must do the following:

- Apply the LO to the FET's gate, along with dc gate bias.
- Apply the RF to the drain.
- Filter the IF from the drain.
- Above all, *apply no dc bias to the drain*!

Of course, appropriate filtering is required to separate the RF from the IF and to prevent LO leakage from pumping the drain conductance. This latter point is more important than it might at first appear: because the drain is unbiased, the FET's gate-to-drain capacitance C_{gd} is much greater than it would be in a more conventional application, such as an amplifier. Figure 3.8 shows that C_{gd} of a resistive FET can be several times its current-saturated value. Therefore, to prevent significant LO voltage from being coupled to the drain, the RF or IF matching circuit must be designed to short-circuit the drain at the LO frequency. Similarly, the gate should be short-circuited at the RF frequency to prevent the RF voltage from introducing nonlinearity by varying the channel conductance.

9.4.2 Single-Device FET Resistive Mixer

Figure 9.15 illustrates the basic, single-device circuit. The LO, RF, and IF are applied as specified above, and provision is included for dc bias at the gate. The gate bias voltage is usually approximately V_t; this gives a pulsed conductance waveform that is little different from that of a diode. The drain voltage must be maintained at 0V; the use of an RF choke may be necessary to guarantee this.

Although it is possible to analyze FET resistive mixers with the use of a diode-

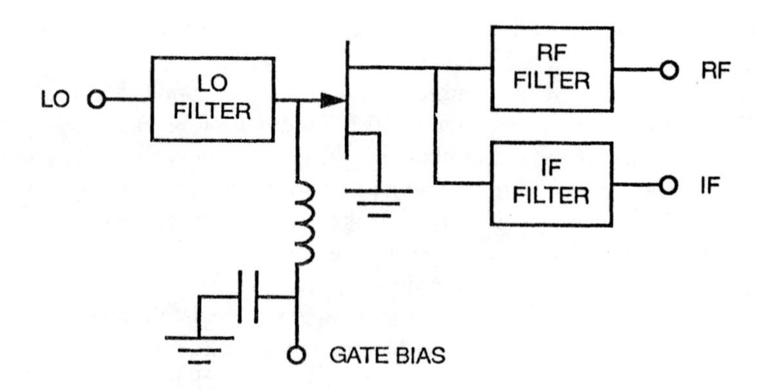

Figure 9.15 FET resistive mixer. No dc bias is used at the drain.

mixer analysis program [4], it is best to use a general-purpose harmonic-balance simulator. Section 3.2.7 describes methods for modeling the FET for such applications. The channel's RF and IF input impedances are usually surprisingly practical; the conjugate-match RF input and IF output impedances of the X-band mixer described in [4], using an ordinary 300 × 0.5-μm Ku-band FET, were $60 - j50\Omega$ and $170 - j19\Omega$, respectively. The LO input impedance is given approximately by (9.2).

Because the FET's channel is purely resistive, its noise is almost exclusively thermal. Therefore, the noise temperature of a FET resistive mixer should be almost exactly given by the relations in Section 5.1.4, with T_d equal to the mixer's physical temperature. This is somewhat lower than a diode mixer, which includes both shot and thermal noise.

9.4.3 Balanced FET Resistive Mixers

FET resistive mixers can be realized in balanced form [5]–[7]. The principles–and the advantages and disadvantages–are similar to those of most other balanced mixers. There is, however, one important consideration in the design of such mixers. We noted earlier that the mixer's embedding circuits must be designed to short-circuit the LO at the drain and the RF at the gate. Unfortunately, most microwave baluns are driven in an even mode by the waveform they are required to short-circuit, and the baluns present an open circuit to such excitation (this point was discussed in Section 7.1.3). Thus, one must be very careful in the design of balanced

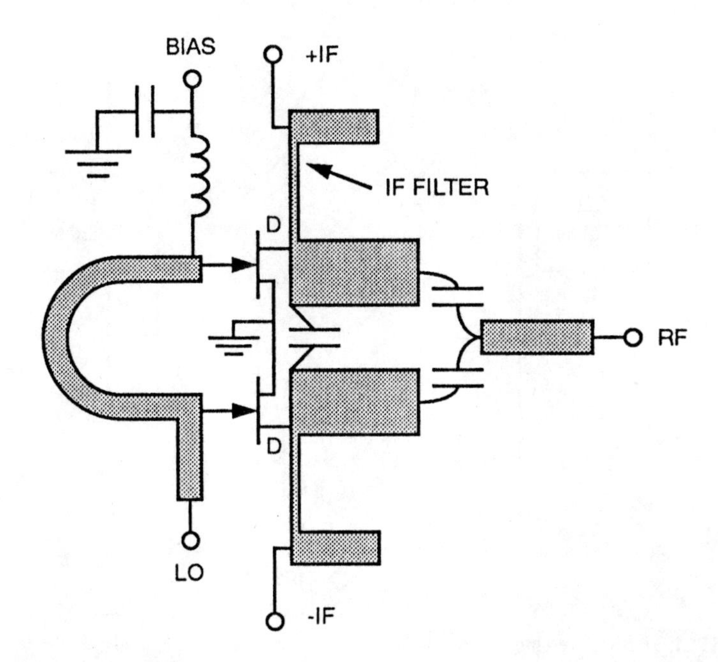

Figure 9.16 Singly balanced FET resistive mixer. The +IF and –IF terminals are connected to the IF hybrid. The "hairpin" in the LO transmission line is one-half wavelength long at the center of the LO band. The capacitors must be short circuits at the RF and LO frequencies and open circuits at the IF; thus, the transmission line from the drains to the single RF line are short-circuited together and comprise, effectively, a single transmission line.

mixers that the embedding network, consisting of a combination of baluns and filters, present the proper termination to the FETs of the mixer.

Figure 9.16 shows the microstrip circuit of a singly balanced FET resistive mixer. The LO is applied to the gates through a balun, and the RF is applied to the drains in phase. The drains are connected together at all frequencies except the IF by capacitors. The LO balun, consisting of a half-wavelength "hairpin" transmission line, does not have a very wide bandwidth. Unlike most simple microwave baluns, however, it presents a short circuit to RF leakage at the FET's gates; this leakage excites the balun in an even mode. Conversely, the connection point of the two drains is a virtual ground for LO leakage. As with active singly balanced mixers, the IF currents in the drains are out of phase; thus, an output balun or hybrid must be

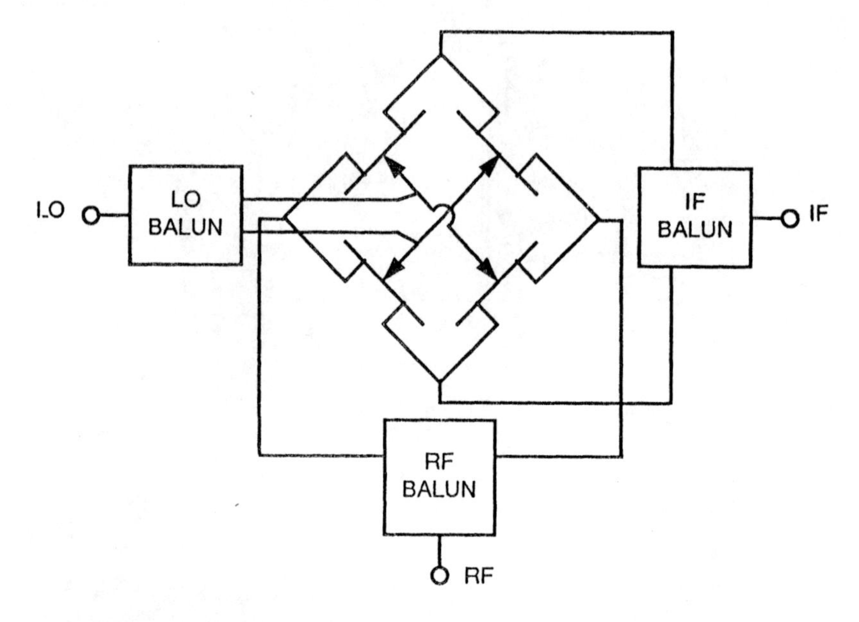

Figure 9.17 FET resistive ring mixer.

used to combine them. RF tuning can be used on the line from the capacitors to the RF terminal. LO tuning elements should not be located on the hairpin; they should be between the hairpin and the LO terminal.

Figure 9.17 shows a doubly balanced resistive FET mixer in a ring or "commutating" structure. This is a polarity-switching mixer, analogous to the doubly balanced diode ring mixer of Section 7.3.1; unlike the diode mixer, however, the FET mixer requires three hybrids instead of the diode mixer's two. The RF, LO, and IF are connected to the ring via these hybrids. All four corners of the ring are virtual grounds for the LO; the IF connection points are virtual grounds for the RF, the RF connection points are virtual grounds for the IF, and the gates are virtual grounds for both. The existence of these virtual grounds implies that the RF, LO, and IF are inherently isolated.

The mixer has the same IM rejection properties as a diode-ring mixer: all even-order IM products are rejected. Good IM rejection requires careful balance, which can be upset easily by the large number of parasitics introduced by the complex layout. Low IM levels also require hard pumping of the devices: the FETs must be driven as hard as possible, but the gate-to-channel junction must not be allowed to conduct.

Because of the inherently low IM of resistive mixers, the even-order IM rejection, and the 6-dB improvement in intercept points due to the power-combining effect of the four devices, mixers such as this should have spectacular IM performance. Unfortunately, reported mixers have not exhibited performance as good as would be expected from a simple scaling of the results in [4]; in fact, the performance of the mixer reported in [5], which used the same type of device, are

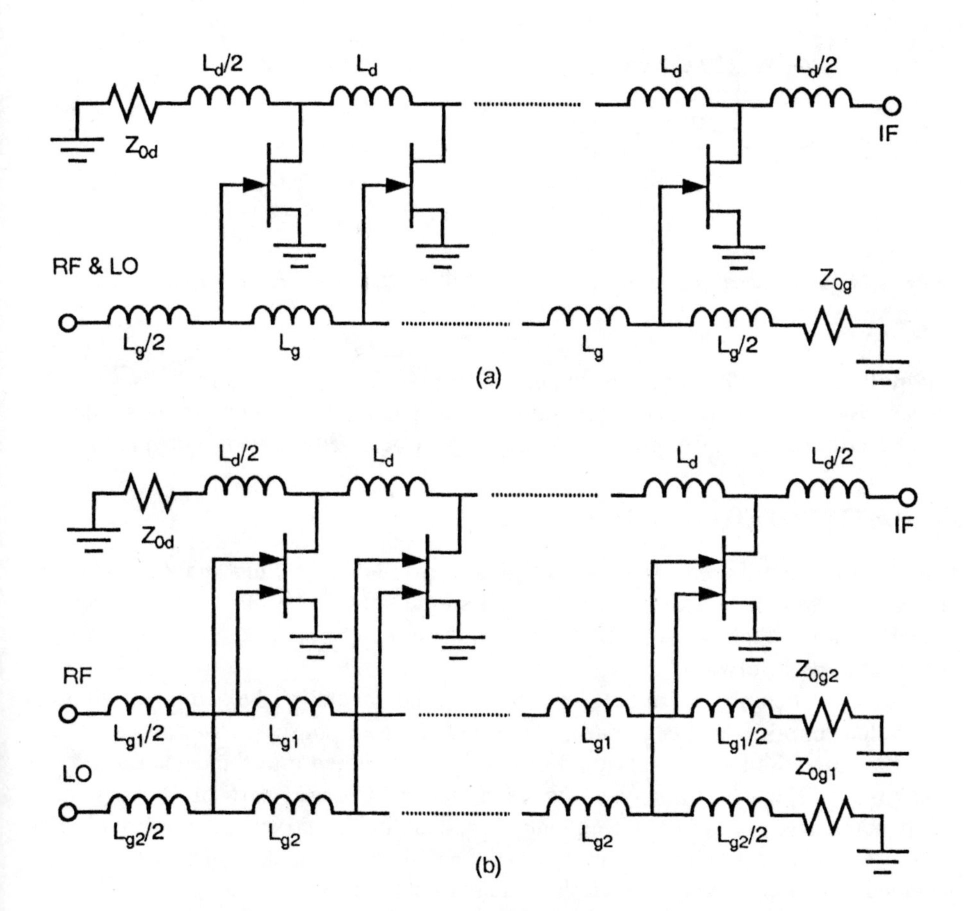

Figure 9.18 Distributed mixers using (a) single-gate and (b) dual-gate FETs. In the single-gate version, the RF and LO must be diplexed and applied to the same input. In both mixers, an LO-rejection filter is necessary at the output.

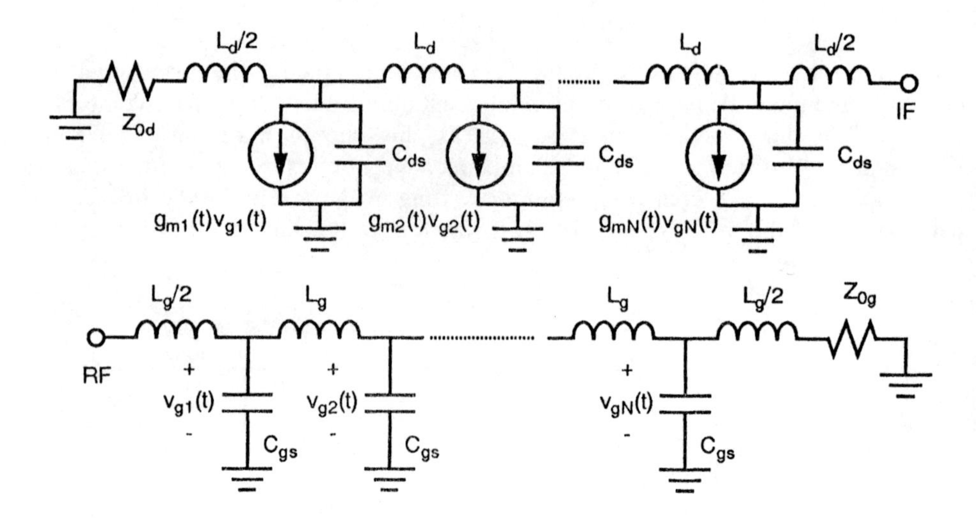

Figure 9.19 Small-signal equivalent circuit of the FET distributed mixer.

not quite as good as would be expected from [4]. The reason for this sporadic success is not clear; it may be related to an incomplete understanding of design considerations. As the design of such mixers matures, better performance is likely.

9.5 DISTRIBUTED MIXERS

The idea of a distributed mixer was originally presented by Tang and Aitchison [8] in 1985. Although their original idea used single-gate FETs, dual-gate devices are probably more practical in such circuits, and the extension of the concept to such mixers is straightforward.

The idea behind distributed mixers is fundamentally the same as that of distributed amplifiers: the gate-to-source and drain-to source capacitances, which limit the bandwidth of conventional circuits, are absorbed into the capacitance of a transmission line. In this way, the capacitances become part of a frequency-independent structure, so the frequency response of the mixer is very broad. The trade-off is that a number of devices are needed to obtain the gain that could be provided over a narrower bandwidth by a single device.

Figure 9.18(a) shows the circuit of a single-gate distributed mixer, and Figure 9.18(b) shows a dual-gate circuit. The operating principles of the mixer are essentially the same; therefore, we shall examine the single-gate circuit. Figure 9.19

shows a simplified small-signal equivalent circuit of the mixer. The gate capacitances C_{gs} of the devices and the gate inductors L_g realize an L-C ladder circuit that approximates a transmission line. Its characteristic impedance is

$$Z_{0g} = \sqrt{\frac{L_g}{C_{gs}}} \tag{9.21}$$

Its phase shift per section is

$$\gamma_g = \omega\sqrt{L_g C_{gs}} \tag{9.22}$$

and its cutoff frequency is

$$f_{cg} = \frac{1}{\pi\sqrt{L_g C_{gs}}} \tag{9.23}$$

Similarly, for the drain line, the respective quantities are

$$Z_{0d} = \sqrt{\frac{L_d}{C_{ds}}} \tag{9.24}$$

$$\gamma_d = \omega\sqrt{L_d C_{ds}} \tag{9.25}$$

and

$$f_{cd} = \frac{1}{\pi\sqrt{L_d C_{ds}}} \tag{9.26}$$

The waves propagating along the gate line undergo phase shifts of γ_d between each gate terminal; thus, at the gate of the nth device, we have

$$V_{Ln} = V_{L1} e^{-j(n-1)\gamma_{gL}} \tag{9.27}$$

$$V_{Rn} = V_{R1} e^{-j(n-1)\gamma_{gR}} \tag{9.28}$$

where V_{L1} and V_{R1} are the input LO and RF voltages at the gate of the first device and γ_{gL} and γ_{gR} are the phase shifts per section at the LO and RF frequencies, respectively. The fundamental-frequency component of the transconductance is in phase with the LO voltage and is primarily responsible for downconverting the RF. The IF current in each device, at the drain, is

$$I_{\mathrm{IF}n} = I_{\mathrm{IF}1} e^{-j(n-1)(\gamma_{gR} - \gamma_{gL})} \tag{9.29}$$

where I_{IF1} is the IF drain current in the first device.

The IF currents generated in each device combine in the drain line. Thus, at the IF output we have

$$I_{\mathrm{IF}} = \sum_{n=1}^{N} I_{\mathrm{IF}n} e^{-j(n-1)\gamma_{dI}} \tag{9.30}$$

Where N is the total number of FETs. Each term in this series represents the contribution from each FET. Ideally, these should be in phase. This will be the case when

$$\gamma_{dI} = \gamma_{gR} - \gamma_{gL} \tag{9.31}$$

This is satisfied when

$$L_g C_{gs} = L_d C_{ds} \tag{9.32}$$

Usually $Z_{0g} = Z_{0d}$. In order to satisfy this requirement and (9.32) simultaneously, we must have $L_g = L_d$ and $C_{gs} = C_{ds}$. The latter is not satisfied in general; it can be satisfied by adding capacitance to the drain line in parallel with C_{ds}, which is usually lower than C_{gs}. Often, however, some mismatch can be tolerated in the gate- and drain-line phase shifts, and capacitive loading is not necessary.

The conversion gain of the mixer is

$$G_c = \frac{P_{L,\,\mathrm{IF}}}{P_{a,\,\mathrm{RF}}} \tag{9.33}$$

where $P_{L,\mathrm{IF}}$ is the power delivered to a load at the IF port and $P_{a,\mathrm{RF}}$ is the power available from the RF source. If the input and output are matched, and (9.31) is satisfied,

$$G_c = \frac{N^2 G_{m1}^2 Z_{0g} Z_{0d}}{16} \tag{9.34}$$

where G_{m1} is the fundamental-frequency component of the transconductance waveform. If (9.31) is not satisfied, [8] gives

$$G_c = \frac{N^2 G_{m1}^2 Z_{0g} Z_{0d}}{16} \left| \frac{\sin(\frac{n\theta}{2})}{\sin(\frac{\theta}{2})} \right|^2 \tag{9.35}$$

where

$$\theta = \gamma_{gR} - \gamma_{gL} - \gamma_{dI} \tag{9.36}$$

(typographical errors in the above two equations in [8] have been corrected here).

It is important to recognize an important approximation in the above equations: the input and output resistances of the FET have not been included. The effect of these is to add loss to the gate and drain transmission lines. This loss has several important effects: first, the loss on the gate line causes the RF signal driving the last devices along the line to be very weak; in a mixer having many devices, the latter devices do not contribute significantly to conversion gain. Similarly, because of the drain-line loss, the first devices also contribute little. Thus, in contrast to the implication of (9.34) and (9.35), the gain does not increase indefinitely with the number of devices N. Because of this loss, there is an optimum number of devices for any distributed mixer; increasing N beyond this optimum value actually decreases the mixer's conversion gain and increases its noise figure. Second, the loss changes the phase shift per section to a value different from the quantities given by (9.22) and (9.25), so (9.36) may be much more difficult to satisfy, especially over a broad bandwidth, than the earlier discussion implies. Even so, the ideal of loading the drain line with extra capacitance to satisfy (9.32) is still helpful.

Because the gate and drain lines are cascades of image-matched sections, image-parameter theory is well suited to the task of analyzing distributed mixers. This theory has been applied to distributed amplifiers, and the principles involved are essentially the same for mixers. For a very readable treatment of this process, see [9]; for a treatment of image parameters, see [10] or any other very old book on network theory.

Even from this simple analysis, it should be possible to see several clear advantages in using dual-gate devices in distributed mixers. The first is the obvious improvement in RF-to-LO isolation. Less obvious, perhaps, are the extra degree of freedom in satisfying (9.31) and, because the input impedance of the second gate of a dual-gate mixer is very high, very low losses along the LO line. Because of these low LO losses, the individual FETs in a dual-gate distributed mixer are pumped more uniformly than in a single-gate mixer.

9.6 REFERENCES

[1] Pucel, R. A., D. Masse, R. Bera, "Performance of GaAs MESFET Mixers at X-Band," *IEEE Trans. Microwave Theory Tech.*, Vol. MTT-24, 1976, p. 351.

[2] Maas, S. A., *Theory and Analysis of GaAs MESFET Mixers*, Ph. D. Diss., University of California, Los Angeles, 1984.

[3] Tsironis, C., R. Meirer, and R. Stahlmann, "Dual-Gate MESFET Mixers," *IEEE Trans. Microwave Theory Tech.*, Vol. MTT-32, 1984, p. 248.

[4] Maas, S. A., "A GaAs MESFET Mixer with Very Low Intermodulation," *IEEE Trans. Microwave Theory Tech.*, Vol. MTT-35, 1987, p. 425.

[5] Maas, S. A., "A GaAs MESFET Balanced Mixer with Very Low Intermodulation," *IEEE MTT-S International Microwave Symposium Digest*, 1987, p. 895.

[6] Chang, K. W., B. R. Epstein, E. J. Denlinger, and P. D. Gardner, "Zero-Bias GaInAs MISFET Mixers," *IEEE MTT-S International Microwave Symposium Digest*, 1989, p. 1027.

[7] Weiner, S., D. Neuf, and S. Spohrer, "2 to 8 GHz Double Balanced MESFET Mixer with +30 dBm Input 3rd Order Intercept," *IEEE MTT-S International Microwave Symposium Digest*, 1988, p. 1097.

[8] Tang, O. S. A., and C. S. Aitchison, "A Very Wideband Microwave MESFET Mixer Using the Distributed Mixing Principle," *IEEE Trans. Microwave Theory Tech.*, Vol. MTT-33, 1985, p. 1470.

[9] Beyer, J. B., et al., "MESFET Distributed Amplifier Design Guidelines," *IEEE Trans. Microwave Theory Tech.*, Vol. MTT-32, 1984, p. 268.

[10] Ruston, H., and J. Bordogna, *Electric Networks: Functions, Filters, Analysis*, New York: McGraw-Hill, 1966.

Chapter 10
Monolithic Mixers

We have occasionally mentioned MMICs in earlier chapters: we discussed them in Chapter 9 with regard to FET mixers and we examined diodes for monolithic circuits in Chapter 2. However, the importance of monolithic circuits in the field of microwave engineering is so great that they deserve their own chapter.

Monolithic mixers can be used in a relatively wide range of applications. They are most useful in applications where small size and light weight are critical; we examine a few of these below. Monolithic circuits can be produced in large quantity at low cost, allowing their use in prosaic consumer applications. The primary limitation to such use at present is not in the fabrication technology but in the high cost of testing and packaging these circuits.

Focused on the technology of 1992, this chapter has more of the flavor of a survey paper than the earlier chapters of this book. This is, at first, a little embarrassing; one of the "ground rules" for the preparation of this book was the avoidance of a survey-paper outlook. Here, however, we are dealing with a subject that is so new that fundamentals do not exist; the technology is in such a state of ferment that it is difficult to identify anything that could be called a body of basic theory. We shall leave those fundamentals, perhaps, for the next edition of this book.

10.1 APPLICATIONS

Monolithic circuits are used in much the same types of applications as *hybrid* circuits (circuits fabricated from discrete components, usually on ceramic or composite substrates). Monolithic circuits are especially well suited to applications where small size or large quantities of identical circuits are needed. Some of these applications are examined below.

10.1.1 Communication Receivers

Probably the strongest motivation for the use of microwave monolithic circuits is in receivers; indeed, most applications for MMICs involve reception of UHF or microwave signals. Such circuits must usually be made as small as possible; thus, they are ideal candidates for monolithic realization. There is a great deal of interest in MMICs among manufacturers of cellular telephones, pagers, hand-held communication equipment, and mobile radio. In Europe and Japan (less so in the United States) there is also interest in receiving equipment for broadcasting from satellites [1]–[3]. Such systems will probably not be practical without MMIC circuits in the ground station receivers.

10.1.2 Television

A surprisingly large effort has gone into the development of MMICs for television tuners; some of the earliest work reported on GaAs monolithic circuits was devoted to this application [4]–[6]. Conventional TV tuners usually use a single dual-gate MOSFET in a downconverting configuration. MMIC TV tuners can use multiple-device balanced mixers having much better rejection of IM and spurious responses, perennial problems in TV receivers. Furthermore, it is possible with MMICs to use upconverting mixers and high IF frequencies; this removes "blank spots" in the receiver's frequency coverage and eliminates certain perpetually troublesome low-order spurious responses. Upconverting architectures are commonly used in commercial systems; MMIC technology offers the possibility of extending the use of these architectures to consumer electronics [7].

10.1.3 Radar

Any system that uses a large number of identical elements is a likely application for monolithic circuits. One such application is in phased-array radars. Phased arrays use a large number of individual *transmit-receive* (TR) modules, each consisting of an antenna element, a medium-power transmitter, a receiver, and phase shifters for steering the antenna's beam. TR modules must be low cost, easily produced in large quantities, and, of course, must have reasonable performance.

A similar situation exists in the design of seekers for missiles and so-called "smart weapons." The radars in such devices must be very small, light, and rugged. Additionally, because of the limited size of their antennas, missile radars must operate at very high frequencies. There are plans to used phased-array radar technologies in these systems.

10.2 CHARACTERISTICS OF THE MONOLITHIC MEDIUM

10.2.1 General Characteristics

The most attractive feature of monolithic circuits is the potential (not always realized) for making large numbers of circuits at low cost. Thus, economics is an important factor—arguably the most important factor—in the production of monolithic circuits. Because the cost of processing a substrate is largely independent of the number of chips on the wafer, the cost of a MMIC chip is inversely proportional to the number of working chips obtained from a single substrate. There is a strong incentive for a designer to fit as many chips as possible onto a single wafer and to obtain satisfactory performance after a single design iteration. Not all chips produced on a single wafer are expected to be operable; unavoidable imperfections in processing cause some chips to be defective. The cost of a chip is inversely proportion to *yield*, the fraction of chips that are operable, and thus a designer will do well to avoid structures that cannot be produced reliably. The latter include FETs having very short gates, other structures having small dimensions, or structures requiring nonstandard processes.

Perhaps the most fundamental property of the monolithic medium is also the most obvious; nevertheless, it must be stated: the fabrication of a monolithic circuit is the end of its design cycle. There is no opportunity to adjust, modify, or otherwise "tweak" a circuit to optimize it. Therefore, it is inevitable that the performance of a monolithic circuit is lower than that of a hybrid circuit.

Aside from the technological and economic differences between monolithic and hybrid circuits, there is one additional factor, related most closely to human psychology. The mindset of the monolithic circuit designer must be precisely the opposite of that of the hybrid circuit designer: while the hybrid circuit designer attempts to optimize a circuit's performance (within some obvious constraints, some of them economic) to meet the needs of a system, performance and cost constraints force the MMIC designer to create a system that works adequately with lower-performance components. Thus, the needs of the system define the characteristics of hybrid components; the limitations of MMIC components define the system in which they are used.

10.2.2 Materials

Although silicon is sometimes used, MMIC circuits are almost invariably fabricated on GaAs or heterojunction substrates. GaAs substrates are significantly more expensive than silicon, and the small dimensions of microwave circuits necessitate special—and expensive—processing. GaAs (and related heterojunction materials) are preferred over silicon because of superior performance, caused by its higher electron mobilities and saturation velocities. GaAs is also preferred because its bulk conductivity, when it is undoped, is much lower than silicon's. The conductivity of silicon is so high as to make it virtually unusable for circuits having strip transmission lines. However, silicon is used occasionally for low-frequency circuits that do not require such transmission lines [8]. Also, heroic efforts in fabricating low-conductivity silicon substrates have occasionally made silicon MMICs practical for certain applications [9]. When it can be used, the low cost of silicon substrates and the high degree of maturity of silicon technology is attractive and may in some cases outweigh the performance penalty.

The low thermal conductivity of GaAs and related materials (much lower than silicon) presents difficulties in designing monolithic circuits. This property makes it difficult to obtain high power from GaAs ICs. Although most mixers are inherently low-power circuits, they are often fabricated on the same chip as a large number of small-signal devices or even power-amplifier circuits, and the power dissipation of the chip may be substantial. The power dissipation of even a simple chip may cause its temperature to rise significantly, increasing the noise figures of the devices and reducing their transconductances.

10.2.3 Circuit Elements

Because of the high cost of GaAs "real estate," it is necessary to make each chip as small as possible. There is a strong incentive to avoid structures, especially transmission lines and other distributed elements, that occupy a large amount of substrate area. The need for small circuits, and the inevitable distaste for distributed matching, has brought about a reexamination of the use of lumped elements in microwave circuits. Indeed, the small sizes of such elements as capacitors and spiral inductors tend to move their resonances—the problem that obviates their use in hybrid circuits—to higher frequencies. The lumped-element hybrids and baluns described in Chapter 7 are frequently used in MMIC mixer circuits, as are lumped-element matching circuits. Capacitors are used almost invariably for dc blocking, bypassing, and other applications where their values and parasitic inductances are not critical.

When transmission lines or other distributed elements are used in MMICs, they are necessarily much smaller than in hybrid circuits and the metallizations are much thinner. As a result, the dissipation losses are quite high, and high-Q components simply cannot be realized. Thus, low-loss filtering and matching circuits and high-Q resonators normally cannot be fabricated. When such elements are necessary, they are best fabricated off-chip.

Monolithic circuits often have circuit elements that are not commonly used in hybrid circuits. One such element is the *via hole*, a metallized hole through the substrate that provides a ground connection for an element at the top of the substrate. These are used often to connect a FET's source lead to ground. Air bridges (unsupported metal jumpers) are used as crossovers. Resistors are fabricated as thin-film elements using a thin sheet of metal as a resistor, or bulk resistors, in which a region of semiconductor serves as a resistor. Capacitors may be parallel-plate structures separated by a dielectric or may be realized by multiple coupled lines. Low-value inductors are realized by short lengths of transmission lines; higher values can be obtained by the use of round or square spirals.

As with silicon integrated circuits, the values of capacitors and resistors depend on fabrication processes that are difficult to control precisely. As a result, these elements often have very loose tolerances: ±30% or greater is not unusual. MMIC circuits must be designed to tolerate circuit elements that do not have precise values. This need for "low-sensitivity" circuit design is an important aspect of the design and fabrication of monolithic circuits.

10.2.4 Models

The various lumped and distributed elements used in a monolithic circuit are often difficult to model. Especially troublesome are discontinuities in strip transmission lines: steps, crosses, gaps, T-junctions, etc. Modeling relatively complicated elements such as capacitors or spiral inductors is another concern; rarely is a simple lumped capacitor or inductor adequate to describe the microwave properties of these components. Because the models for such elements included in microwave circuit-analysis programs are often poor, circuit designers often use "standard elements" that have been fabricated, measured, and modeled.

There are two fundamental approaches to the modeling of passive circuit elements. The first is to measure a large number of standard elements, and to store their S parameters in a large database. If the measurements are performed properly, the resulting models can be very accurate. Unfortunately, however, not all designs can be planned around the standard set of element models, and using a nonstandard element becomes a significant problem. The second approach is to develop analytical models for a class of elements. Such models are less accurate, but allow a virtually infinite variety of elements to be used in designs. Practical monolithic circuit design usually involves both types of models.

A fundamental quandary in the use of models for transmission-line discontinuities is that the most accurate models—numerical calculations—are not efficient enough for use in computer circuit analysis, and "fast" models (developed in a number of different ways) are often not accurate enough. There is currently a wide gulf between these extremes: the need for models that are accurate and can be calculated rapidly is very great.

The modeling of active devices in the monolithic medium is not significantly different from that of hybrid circuits. One additional factor is that monolithic circuits offer great freedom in adjusting the geometries of FET devices to optimize them for a particular application. The most useful practice is to adjust the total gate width and number of gate segments. Designers need a way to model all such devices; the most common approach is to model a basic device and use a set of scaling rules to obtain models for FETs having other geometries.

10.2.5 Testing

Testing monolithic circuits is an important concern. It is rarely practical to "dice" a MMIC wafer into individual components, mount them in packages, and test the packaged circuits. Instead, monolithic circuits are best tested with RF "wafer probes" designed to make temporary connections directly to the substrate. Probes having adequate performance to 50 GHz or greater are available. Calibration standards can be fabricated on the MMIC wafer, so the circuit can be measured accurately using a vector network analyzer. Similarly, individual FETs, resistors, and capacitors are often placed on the wafer, as well as complete circuits, as test and process-control elements.

These techniques can be used to perform accurate measurements of monolithic circuits. Unfortunately, such measurements, although accurate, are not cheap. The cost of testing large numbers of monolithic circuits is currently one of the most troublesome limitations to the large-scale use of MMICs in commercial applications.

10.3 CIRCUITS

10.3.1 Devices

Although bipolar-transistor technologies are sometimes used, monolithic circuits are usually based on FET technologies, both conventional MESFET and heterojunction devices. MMIC mixers are almost evenly divided between active and passive FET structures and diodes; many of the latter are "FET diodes," realized by the channel and gate junction of a FET. Because these devices are not optimized for use as diodes, they often have relatively low cutoff frequencies, often no more than a few hundred gigahertz; modifying the geometry (within the limitations of the FET technology) and selectively ion implanting the ohmic contacts can improve the cutoff frequency [10], [11] to 600 GHz or greater. Even so, this is much lower than the 2 to 2.5 THz or more commonly achieved with discrete diodes. Some sacrifice of performance caused by these low cutoff frequencies is often acceptable in view of the ease with which the mixer can be integrated with amplifiers or other circuits. FET diodes need not have the same geometry as MESFET gates; using a different geometry can often improve the diode's cutoff frequency. The most common modification is to increase the width of the gate stripe, thus reducing the metal resistance. FET diodes having wide rectangular, square, and even circular anodes have been realized.

Mesa diodes exhibit better electrical characteristics, sometimes having cutoff frequencies greater than 1 THz. Mesa diodes are more difficult than FET diodes to integrate with active FET circuits: the need for a mesa to isolate the diode from other devices requires extra processing steps. For high-performance millimeter-wave applications, the high cutoff frequencies of mesa diodes are virtually a necessity (Section 2.4.6).

Both single- and dual-gate FETs realized in both conventional epitaxial and heterojunction technologies are employed in both active and passive (resistive FET) MMIC mixers. An advantage of monolithic technology for active mixers is that the FET's gate width and geometry can be optimized to achieve the desired trade-off among gain, noise figure, and intermodulation; this degree of freedom is rarely available with discrete devices.

The current "workhorse" FET technology is the single 0.5-μm gate device fabricated on conventional GaAs. Although FETs having much shorter gates are sometimes produced, the reduction in yield that such devices entail is only acceptable when high performance is essential. Similarly, MMICs are sometimes fabricated on heterojunction materials in HEMT technologies; again, the use of such devices requires a trade-off between cost and performance.

10.3.2 FET Circuits

Although early MMIC circuits occasionally used single-device mixers [2], The logical way to realize a monolithic mixer is as a balanced structure. Balanced mixers are generally easier and less expensive to fabricate than single-device mixers: although the two, four, or eight devices needed for a balanced mixer are a little more difficult to fabricate than a single device, and the inherent isolation and spurious-response rejection of a balanced mixer reduces–often obviates–the need for filters or diplexers. Another option for the design of FET mixers is a distributed mixer.

Distributed mixers are not balanced and do not exhibit the IM and LO-noise rejection of balanced mixers; however, they are very broadband.

Singly Balanced FET Mixers

Singly balanced FET mixers were examined in Section 9.3. Many of the types of mixers described there have been used in monolithic circuits. In spite of the need for a relatively large number of components–two complete mixers plus two hybrids–both of the circuits in Figure 9.12 have been employed in MMICs. However, much more effort has been devoted to the development of circuits that are more compact and to circuits that need only baluns–not four-port hybrids–to provide the necessary phase splits. Baluns, unlike hybrids, are often more broadband and can be realized in active form.

Figure 10.1 shows two realizations of singly balanced FET mixers. The mixer in part (a) is a single section of the doubly balanced mixer described in Section 9.3.2 and operates in the manner described in Sections 9.2 and 9.3.2; part (b) is a single section of the much more practical doubly balanced mixer described below.

The individual FETs in the mixer of Figure 10.1(a) operate as conventional dual-gate mixers (Section 9.2.1). However, the circuit is a true balanced mixer, not simply two dual-gate mixers in parallel. Because the LO drives the gates out of phase and the RF is applied in phase, the RF terminal is a virtual ground for LO leakage. This gives the mixer considerable RF-to-LO isolation beyond that provided by the simple single-device dual-gate mixer. Of course, it exhibits the same spurious-response and LO-noise rejection properties of the other singly balanced FET mixer in Figure 9.12(a). These are described in Section 9.3.1.

Although superficially similar to the dual-gate mixer, the FETs in Figure 10.1(b) operate somewhat differently from the individual FETs in a single-device or a balanced dual-gate mixer: because the upper FETs in each pair are driven from opposite sides of the LO balun, the common source connection for these devices is a virtual ground for the LO. Thus, the upper FETs do not act as source-follower amplifiers, as they would in dual-gate mixers. Like the dual-gate FET, either the lower or the upper FETs can be in current saturation, but generally not both. The IF current in the drains of the upper FETs is proportional to the RF current in the drain of the lower FET; therefore, to maximize conversion gain, the lower FET must be in current saturation and the upper FET must be operated as an unsaturated switch (this contrasts with the dual-gate mixer, in which the upper FET was in saturation). Consequently, the lower FET must be biased in much the same way as an amplifier: the gate-bias voltage should be set to a value that provides about 50% of the FET's I_{dss}. If the gate dc bias is too great, the lower FET will leave its saturation region and the upper FET will enter current saturation.

This mixer operates fundamentally as an amplifier (the lower FET) followed by a switch operating at the LO frequency (the upper pair of FETs). The upper FETs—the switches—interrupt the lower FET's output current periodically at the LO frequency. Mixing occurs between the fundamental-frequency component of this switching function and the applied RF signal. The resulting IF currents in the drains of the upper FETs have a 180-deg phase difference and thus must be subtracted by an IF balun or hybrid. In this sense, this circuit is a polarity-switching mixer, operating like the circuit of Figure 7.27 with the RF and IF ports interchanged.

Design of such mixers is straightforward. The lower FET is simply an amplifier whose drain is terminated in the parallel input impedances of the upper FETs' sources; this impedance is fairly low. The RF input impedance of the lower FET's gate is given, to a good approximation, by (9.2) or (9.3). Any conventional matching circuit can be used to match the RF source to the gates (for simplicity, matching circuits are not shown in the figure; they should be used in practice). Similarly, because the sources of the upper FETs are virtual grounds for the LO, the LO input impedance is also given by (9.2) or (9.3).

It is difficult to form generalizations about the IF load impedance. The output of the upper devices is, essentially, the output of a resistive mixer. Thus, it is probably

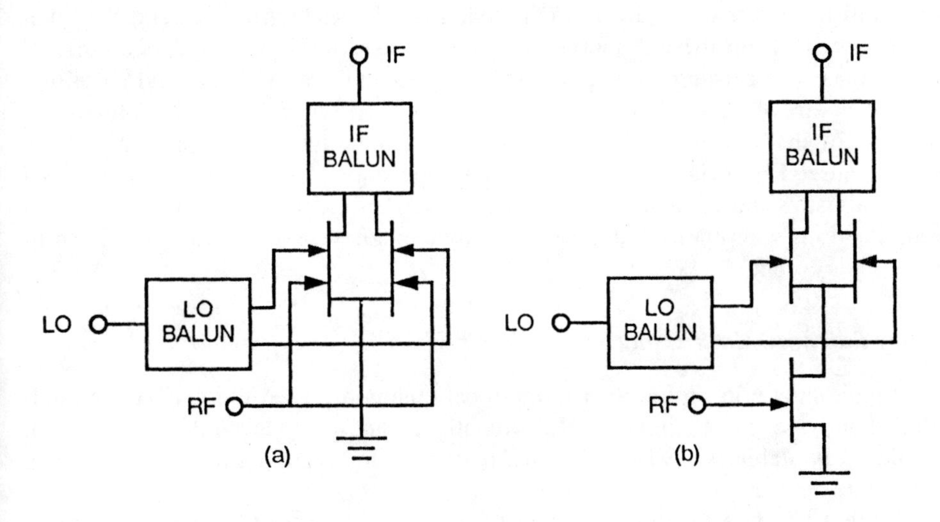

Figure 10.1 Singly balanced FET mixers: (a) dual-gate and (b) single-gate realizations. In spite of some superficial similarities, the operation of these mixers is very different.

little different from that of most diode or FET resistive mixers, on the order of 100Ω to 200Ω per device; more accurate values require the use of nonlinear analysis. In any case, this impedance is relatively high and may complicate the design of the IF balun.

Unlike the comparable doubly balanced mixers, the drains of the upper FETs in these singly balanced mixers are not virtual grounds for either the RF or LO signals. Therefore, some type of output filter must be used to reject the RF and LO and to provide an optimum short-circuit termination to the drains. In downconverting mixers, lumped L-C low-pass circuits are probably most practical for this purpose.

This filtering requirement may be difficult to meet when the mixers are used as upconverters. In an upconverter, the LO frequency normally is close to the output balun's passband, is not rejected by the balun, and cannot be filtered. This difficulty can be circumvented in the circuit of Figure 10.1(a) by placing the balun at the RF port instead of the LO (the lower gates should remain the RF gates); in this configuration the LO leakage excites the output balun in an even mode and will be rejected. (Care must be taken here: as described in Section 7.1.3 and Figure 7.17, most microwave baluns present an open circuit, not the optimum short-circuit termination, to even-mode excitation.) No similar remedy is available for the circuit of Figure 10.1(b).

In some cases, it is possible to eliminate the LO balun. Figure 10.2 shows a version of the mixer of Figure 10.1(b) having one LO gate grounded and the other driven directly from the LO source. This circuit uses the lower FET as a current-source transistor, causing the upper pair to operate much as a differential amplifier. In an ideal structure, the LO voltage is divided evenly between the gate-to-source junctions of the two upper FETs. Unfortunately, this division is equal only if the current-source FET is ideal; the finite output impedance of the lower FET prevents a perfect division and unbalances the mixer. Nevertheless, this circuit may be useful in applications where a high degree of balance is not necessary and elimination of the balun is essential.

Doubly Balanced Mixer Using Single-Gate FETs

The circuit shown in Figure 9.14 is the most commonly used topology for a dual-gate, doubly balanced mixer. Unfortunately, dual-gate technologies are not uniformly available, so it is sometimes necessary to create a similar mixer using single-gate devices.

Figure 10.3 shows a doubly balanced, single-gate mixer [4]–[13]. This mixer consists of two sets of FETs connected as in Figure 10.1(b). However, an RF balun is used to drive the lower FETs of the pairs out of phase; as a result, the drains of the

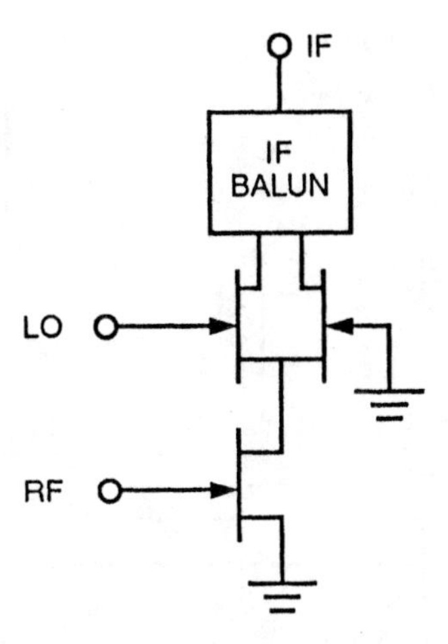

Figure 10.2 Singly balanced mixer that requires no LO balun. This mixer is similar to the circuit in Figure 10.1(b), but uses a differential configuration to eliminate the LO balun. Unfortunately, the balance of this circuit is not as good as that of Figure 10.1(b).

upper devices must be interconnected in the relatively complex manner shown. The points where the upper FETs' drains are connected are virtual grounds for both the RF and LO signals, so the optimum termination is provided to the devices without the need for any additional filtering. Another advantage of this circuit is that the parallel combination of the drains causes the IF output impedance to be half that of the singly balanced version, providing an impedance that is more practical to match to the IF load, usually 50Ω.

Distributed Mixers

Because they require a large number of devices, distributed mixers ([14]–[17] and Section 9.5) are especially suitable for fabrication as MMICs. Although such mixers can use either single- or dual-gate FETs, single-gate distributed mixers require some type of broadband diplexer or power combiner to couple both the LO and RF to the single gate line. Such combiners may be realized in either active or passive form. Both have disadvantages that make them undesirable for monolithic integration: passive combiners are large and may not have bandwidth as wide as the mixer, and

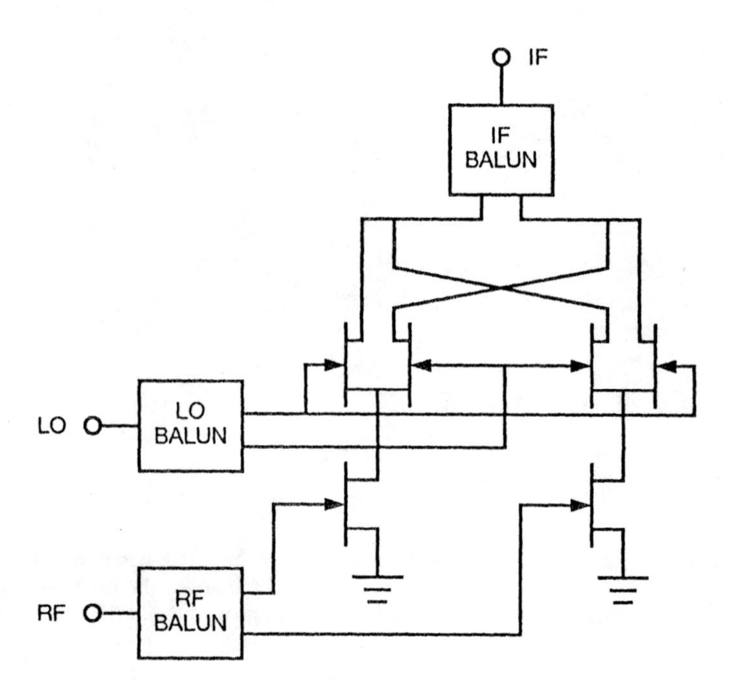

Figure 10.3 Doubly balanced FET mixer using single-gate FETs.

active combiners introduce noise and distortion. Consequently, we shall focus on dual-gate realizations.

The main advantage of distributed mixers over other types is their extraordinarily broad bandwidths. Such mixers are capable of achieving RF and LO bandwidths nearly from dc to many tens of gigahertz, with IF bandwidths on the order of several GHz. The 2- to 26-GHz RF-LO bandwidth reported in [15], for example, is not unusual; this mixer had an IF bandwidth of 0.4 to 2.5 GHz. Including an integral single-stage IF amplifier, this mixer's conversion gain varied from 5 to 10 dB, depending on the combination of RF, LO, and IF frequencies, and its LO and RF input return losses were better than 10 dB.

Such mixers need not have gain to be useful. The mixer in [14] exhibits 2 to 4 dB of conversion loss, as does the mixer reported in [17]. The single-gate mixer in [16] exhibits 2 to 4 dB conversion gain from 2 to 12 GHz, although much of this gain may come from its active LO-RF power combiner.

10.3.3 Baluns and Hybrids

Virtually all balanced mixers require hybrids or baluns. Unfortunately, in many circuits, transmission-line baluns are too large to be acceptable, and lumped-element or active baluns must suffice. The cost of these circuits, in terms of performance, is very high: lumped-element baluns (Section 7.1.2) are often narrowband, and active baluns introduce noise and distortion.

Distributed Baluns

Distributed (transmission-line) baluns and hybrids, described in Chapter 7, are not well suited to MMIC use. They are useful only at high frequencies, where their small size does not monopolize a large amount of expensive substrate area. Even so, attempts have been made to realize such circuits on substrates.

The most successful distributed hybrids (Section 7.1.2) are the simplest circuits. Single-section branch-line hybrids are frequently used in monolithic quadrature-hybrid diode mixers, and rat-race hybrids are used somewhat less often in 180-degree diode mixers. Lange hybrids are used in monolithic amplifiers, but not often in quadrature mixers. The primary reason for the disuse of Lange hybrids is the need for narrow microstrip lines and air-bridge interconnections between the transmission-line strips. These can be difficult to fabricate, have high loss, and reduce circuit yield.

Attempts to design parallel-line baluns (Section 7.1.3) have been somewhat more successful than Marchand baluns, despite the fact that Marchand baluns should be less sensitive to the inevitable even-mode parasitics of a planar structure. The reason for this may be that reported Marchand structures are fairly complex multilayer structures, which have large parasitics and often entail nonstandard processing techniques, while parallel-line baluns are relatively simple. The author has had some success designing planar Marchand baluns using parallel-line structures, often with multiple coupled lines and thick substrates, to obtain low even-mode and high odd-mode impedances.

Active Baluns and Combiners

Active power combiners and baluns are sometimes used to provide the phase split necessary for balanced mixers. Unfortunately, they suffer from a number of problems:

1. An active balun must often be designed primarily to achieve broad bandwidth in combination with good phase and amplitude balance. It is often not possible to optimize its noise figure within such constraints. Therefore, the balun introduces a substantial amount of noise.
2. Similarly, constraints on the design, introduced by the need for good balance and broad bandwidth, may make it impossible to design the balun to achieve low IM distortion.
3. Amplitude and phase balance of the balun are often poor.
4. The balun's impedances at frequency response are often different at its two outputs.

Figure 10.4 shows an active balun. It uses the well-known property of a transistor amplifier in which the signal at its drain and source have, ideally, a 180-deg phase difference. In practice, this property exists only at low frequencies, and the large number of low- and mid-frequency poles in its equivalent circuit introduce substantial phase shifts. Also, the voltage gain between the input and the two outputs is, in general, unequal, and the difference varies with frequency. For these reasons, active baluns using this principle have not been very successful.

Active baluns can also be realized in distributed form. Such baluns have broad bandwidth, but (for much the same reasons as with the single-device active balun) often do not have very good balance. Reference [17] reports an active, distributed balun; this balun has ±30-deg phase and ±0.5-dB amplitude balance over a 1- to 18-GHz bandwidth. This is not adequate to achieve reasonable balanced-mixer performance. The balun's noise figure and intermodulation intercept points are not stated.

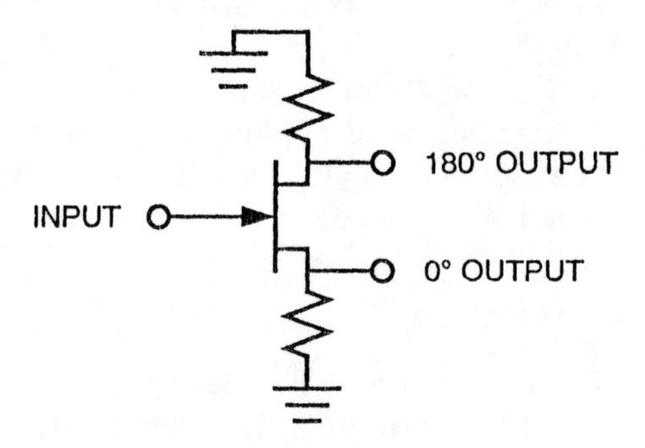

Figure 10.4 Active balun.

Figure 10.5 shows an active power combiner that can be used to diplex the RF and LO onto the single gate line of a distributed mixer. It could also be used to drive a single-device FET mixer. It consists, in effect, of two distributed amplifiers sharing a common drain line. Standard design techniques for distributed amplifiers can be used to design the active combiner; the only modification is that the drain-line capacitances in the combiner are twice those of a conventional distributed amplifier. An active balun similar to this is reported in [16].

Lumped-Element Hybrids

Two circuits for lumped-element hybrids and baluns were examined in Section 7.1. Although other circuits for lumped baluns exist, most are based on the same principles. Lumped-element baluns are quite practical; they were used in the image-

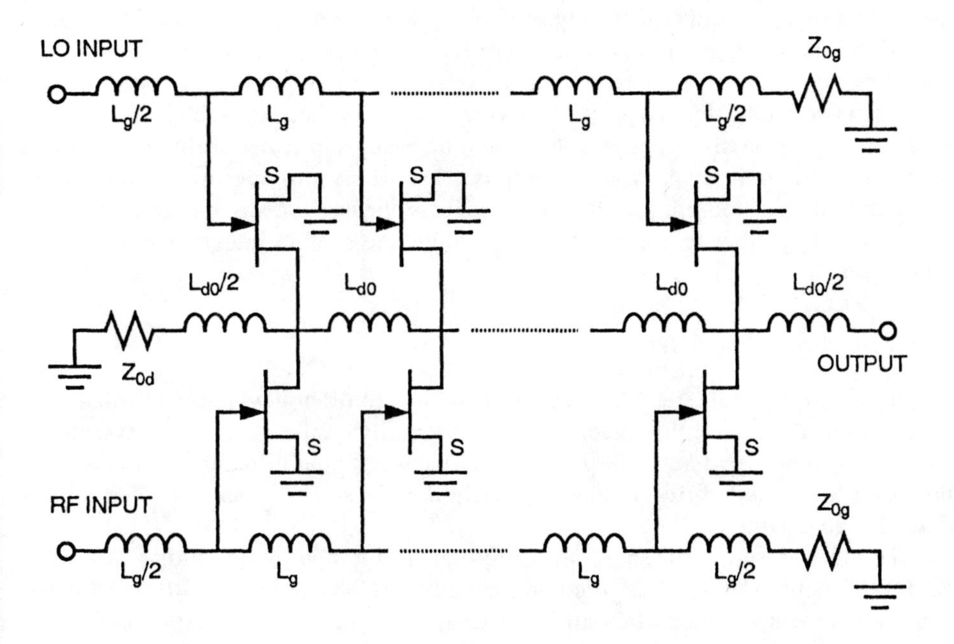

Figure 10.5 Active power combiner used to diplex the RF and LO inputs to a single-gate distributed mixer. Such a combiner could also be used for nondistributed mixers.

rejection mixers reported in [18] and [19]. A perennial difficulty in realizing such structures is that the reactive parasitics of the lumped elements destroy their performance above a certain frequency. The bandwidths of lumped-element hybrids are often very narrow; even those that are reasonably broadband are not nearly as broadband as distributed baluns such as the Marchand.

Some success has been achieved in designing transformer hybrids consisting of bifilar spiral inductors. Because of difficulties in achieving adequate coupling without the use of magnetic materials, such transformers are not as broadband as one might hope. Furthermore, as frequency increases, the inductance and size of the windings decreases, and adequate coupling becomes progressively more difficult to achieve. Nevertheless, such transformers have exhibited good performance over octave bands at frequencies up to X band, and occasionally into Ku band.

10.3.4 Diode Circuits

For much the same reasons as with FET mixers, diode mixers are usually realized in balanced structures, either as singly or doubly balanced mixers. The variety of such mixers is very great, and mixers operating well into the millimeter-wave region have been reported.

A fundamental dilemma in diode-mixer circuits is that the fraction of a chip's area covered by passive elements that could be easily fabricated as hybrids is often much greater than in other types of circuits. This brings into question the suitability of the monolithic medium for diode mixers. This apparent waste of substrate area is best justified by other requirements: for example, the need to integrate a mixer with other circuits.

Singly Balanced Diode Mixers

In contrast to FET mixers, where special circuits are often developed to satisfy the requisites of the monolithic medium, monolithic diode mixers are often essentially the same as their hybrid counterparts. Singly balanced monolithic diode mixers take the same forms as hybrid circuits, primarily the classical 90- and 180-deg mixers described in Chapter 7.

The quadrature-hybrid structure using branch-line hybrids (Section 7.2.3, [11], [20], [21]) is used quite often in monolithic circuits. This may be a little surprising, since we were not kind to this mixer in Chapter 7. The reason for the use of this circuit is probably not any misconceptions about its electrical performance, but more likely the simplicity and planarity of its layout: 180-deg mixers using rat-race hybrids often require some type of microstrip crossover. The stray coupling and parasitic inductance of such crossovers may be troublesome in high-frequency circuits. To avoid the crossover problems in rat-race mixers, 180-deg hybrid mixers have even been realized by branch-line hybrids with an extra quarter-wave transmission line in one output [22], [23].

Many different types of 180-deg singly balanced mixers have been reported, some operating at frequencies in the millimeter range; 94-GHz mixers are not uncommon [23], [24]. These mixers often use rat-race hybrids in either distributed [24] or lumped-element [19] form. Singly balanced crossbar mixer chips have also been reported; the diodes and filters are on a single substrate that fits in a cutout in a millimeter waveguide [25]. References [26] and [27] show other waveguide-mounted structures (the structure in [27] uses a single device).

Doubly Balanced Diode Mixers

Unlike singly balanced MMIC mixers, doubly balanced MMIC diode mixers often must be very different from hybrid realizations. One of the main reasons for the difference is that the best baluns used in hybrid mixers—forms of the Marchand and parallel-line balun—are best realized in nonplanar forms that are not adaptable to MMIC circuits. The paucity of planar balun structures is a significant impediment to the design of doubly balanced diode mixers.

One solution to this problem is to use lumped-element baluns. These are quite practical as long as the required bandwidth is not too great [10], [19]; an advantage of lumped baluns is that they occupy minimal substrate area. Ring-diode mixers using parallel-line baluns (Figure 10.6) are quite popular. In one design, coplanar-waveguide (CPW) RF and LO input lines are tapered to form parallel-line baluns [28]; an example of such a balun is shown in Figure 10.7. In another balun design, CPW inputs excite slotlines, and the edges of the slotline form balanced transmission lines (Figure 10.8). A circuit based on this design is illustrated in [29]; three such baluns are connected to eight diodes in two rings. The structure is a double doubly balanced mixer, equivalent to the circuit in Figure 7.30. A nice property of these parallel-line baluns is that they can be folded to minimize their use of substrate area.

Because of its widespread use, the ring-diode mixer having parallel-line baluns deserves some extra examination. As described in Section 7.1.3, the balun can be viewed as a pair of coupled transmission lines. The fundamental requirements for obtaining good balance and isolation from such a balun are that (1) the even-mode characteristic impedance be quite high, preferably ten times the odd-mode impedance, and (2) the even- and odd-mode phase velocities be the same, or at least closely matched. Although the latter requirement is substantially met, the former

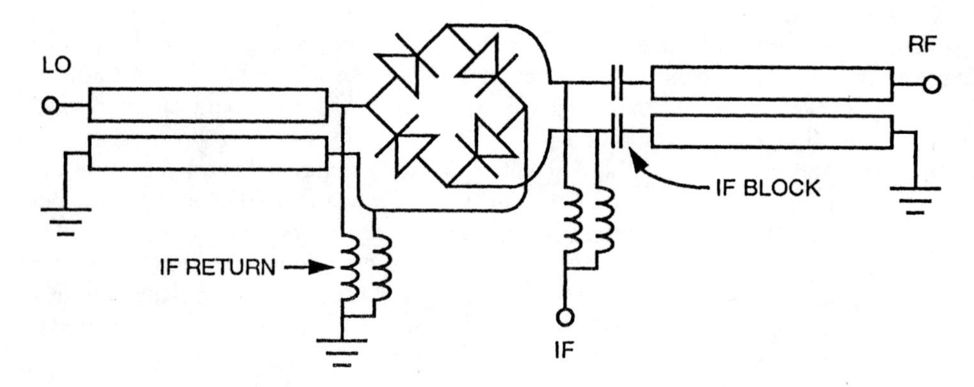

Figure 10.6 Ring-diode mixer using parallel-line baluns.

invariably is not. The effect of this low even-mode impedance on the mixer's performance is degradation of the RF-to-LO isolation. This is particularly severe on GaAs, which has a high dielectric constant of 12.9.

To increase the even-mode impedance of the parallel-line structure, the lines must be made very narrow and the substrate as thick as possible. Occasionally the substrate is mounted on a low-dielectric-constant spacer, often quartz ($\varepsilon_r = 3.8$);

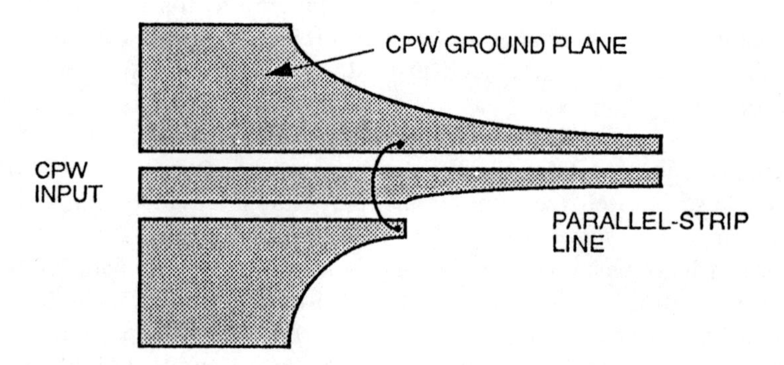

Figure 10.7 CPW transition to a parallel-line balun. The CPW input line and its ground plane are tapered directly to form the balun.

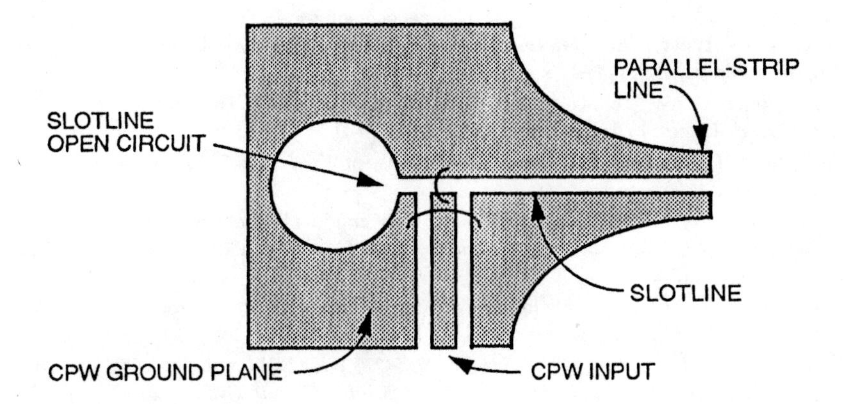

Figure 10.8 CPW transition to a parallel-line balun. The CPW line excites a slotline mode, and the slotline ground is tapered to form a parallel-strip line.

however, the added complexity of the mounting structure obviates many of the advantages of a monolithic circuit. Another possible response to the limitations of parallel-line baluns is the use of Marchand baluns, which tolerate low even-mode impedance much better. Unfortunately, however, many proposed Marchand-balun structures are not very practical; they require nonstandard processing and are difficult to arrange in a manner that uses substrate area economically. The use of edge-coupled lines, perhaps in multistrip form (Figure 7.12), may be a solution to this problem.

Another difficulty in the use of parallel-line baluns with doubly balanced ring mixers is the need for an IF return. The IF return is necessary because the baluns behave as open circuits for even-mode excitation, not short circuits, as would, for example, the transformer balun of Figure 7.26. The inductors or other structures needed for the IF return limit the IF bandwidth substantially and may occupy a large amount of substrate area.

10.4 REFERENCES

[1] Dessert, R., et al., "All-FET Front End for 12-GHz Satellite Broadcasting Reception," *Proc. 8th Annual European Microwave Conf.*, 1978, p. 638.

[2] Sugiura, T., K. Honjo, and T. Tsuji, "12 GHz-Band GaAs Dual-Gate MESFET Monolithic Mixers," *IEEE GaAs IC Symposium Digest*, 1983, p. 175.

[3] Phillipe, P., and M. Pertus, "A 2-GHz Enhancement Mode GaAs Down Converter IC for Satellite TV Tuner," *IEEE Microwave and Millimeter-Wave Monolithic Circuits Symposium Digest of Papers*, 1991, p. 61.

[4] Ablassmeier, U., W. Kellner, and H. Kniepkamp, "GaAs FET Upconverter for TV Tuner," *IEEE Trans. Electron Devices*, Vol. ED-27, 1980, p. 1156.

[5] Autriche, P. D., et al., "GaAs Monolithic Circuits for TV Tuners," *IEEE GaAs IC Symposium Digest*, 1985, p. 165.

[6] Ducourant, T., et al., "A Three-Chip Double-Conversion TV Tuner System with 70 dB Image Rejection," *IEEE Microwave and Millimeter-Wave Monolithic Circuits Symposium Digest of Papers*, 1988, p. 87.

[7] Rosenzweig, R., "Commercial GaAs MMIC Applications," *IEEE Microwave and Millimeter-Wave Monolithic Circuits Symposium Digest of Papers*, 1991, p. 59.

[8] Nakata, T., S. Miyazaki, and K. Shirotori, "0.5-2.6 GHz Si-Monolithic Wideband Amplifier IC," *IEEE Microwave and Millimeter-Wave Monolithic Circuits Symposium Digest of Papers*, 1985, p. 58.

[9] Buechler, J., et al., "Silicon Millimeter-Wave Circuits for Receivers and Transmitters," *IEEE Microwave and Millimeter-Wave Monolithic Circuits Symposium Digest of Papers*, 1988, p. 67.

[10] Ton, T. N., et al., "An X-Band Monolithic Double-Double-Balanced Mixer for High Dynamic Range Receiver Application," *IEEE Microwave and Millimeter-Wave Monolithic Circuits Symposium Digest of Papers*, 1990, p. 115.

[11] Colquhoun, A., and B. Adelseck, "A Monolithic Integrated 35 GHz Receiver Employing a Schottky Diode Mixer and a MESFET IF Amplifier," *IEEE GaAs IC Symposium Digest*, 1987, p. 151.

[12] Adelseck, B., et al., "A Monolithic 60 GHz Diode Mixer in FET Compatible Technology," *IEEE Microwave and Millimeter-Wave Monolithic Circuits Symposium Digest of Papers*, 1988, p. 91.

[13] Bharj, S., et al., "An L-Band Monolithic Microwave Receiver," *Microwave and Optical Technology Lett.*, Vol. 1, 1988, p. 95.

[14] La Con, M., K. Nakano, and G. S. Dow, "A Wide Band Distributed Dual-Gate HEMT Mixer," *IEEE GaAs IC Symposium Digest*, 1988, p. 173.

[15] Titus, W., and M. Miller, "2-26 GHz MMIC Frequency Converter," *IEEE GaAs IC Symposium Digest*, 1988, p. 181.

[16] Robertson, I. D., and A. H. Aghvami, "A Practical Distributed FET Mixer for MMIC Applications," *IEEE MTT-S International Microwave Symposium Digest*, 1989, p. 1031.

[17] Pavio, A. M., and R. M. Halladay, "A Distributed Double-Balanced Dual-Gate FET Mixer," *IEEE GaAs IC Symposium Digest*, 1988, p. 177.

[18] Hirota, T., and M. Muraguchi, "K-Band Frequency Up-Converters Using Reduced-Size Couplers and Dividers," *IEEE GaAs IC Symposium Digest*, 1991, p. 53.

[19] Putnam, J., and R. Puente, "A Monolithic Image-Rejection Mixer on GaAs Using Lumped Elements," *Microwave J.*, Vol. 30, No. 11, 1987, p. 107.

[20] Meier, P. J., et al., "IC Techniques Slice Cost of Integrated Receiver," *Microwaves & RF*, June 1985, p. 85.

[21] Chu, A., et al., "Dual Function Mixer Circuit for Millimeter Wave Transceiver Applications," *IEEE Microwave and Millimeter-Wave Monolithic Circuits Symposium Digest of Papers*, 1985, p. 78.

[22] Jacomb-Hood, A. W., et al., "30 GHz and 60 GHz GaAs MMIC Microstrip Mixers," *IEEE GaAs IC Symposium Digest*, 1986, p. 195.

[23] Trippe, M. W., et al., "mm-Wave MMIC Receiver Components," *IEEE Microwave and Millimeter-Wave Monolithic Circuits Symposium Digest of Papers*, 1991, p. 51.

[24] Chang, K. W., et al., "A W-Band Monolithic Pseudomorphic InGaAs HEMT Downconverter," *IEEE Microwave and Millimeter-Wave Monolithic Circuits Symposium Digest of Papers*, 1991, p. 55.

[25] Chao, C., A. Contolatis, S. A. Jamison, and P. E. Bauhahn, "Ka-Band Monolithic GaAs Balanced Mixers," *IEEE Trans. Microwave Theory Tech.*, MTT-31, 1983, p. 11.

[26] Nightingale, S. J., et al., "A 30-GHz Monolithic Single-Balanced Mixer with Integrated Dipole Receiving Element," *IEEE Trans. Microwave Theory Tech.*, MTT-33, 1985, p. 1603.

[27] Clifton, B. J., et al., "Cooled Low-Noise Monolithic Mixers at 110 GHz," *IEEE MTT-S International Microwave Symposium Digest*, 1981, p. 444.

[28] Maesel, M., et al., "A Double-Balanced 6-18 GHz GaAs MMIC Mixer," *IEEE GaAs IC Symposium Digest*, 1991, p. 49.

[29] Eisenberg, J., J. Panelli, and W. Ou, "A New Planar Double-Double Balanced Mixer Structure," *IEEE Microwave and Millimeter-Wave Monolithic Circuits Symposium Digest of Papers*, 1991, p. 69.

Index

The Artech House Microwave Library

Acoustic Charge Transport: Device Technology and Applications, R. Miller, C. Nothnick, and D. Bailey

Algorithms for Computer-Aided Design of Linear Microwave Circuits, Stanislaw Rosloniec

Analysis, Design, and Applications of Fin Lines, Bharathi Bhat and Shiban K. Koul

Analysis Methods for Electromagnetic Wave Problems, Eikichi Yamashita, ed.

Automated Smith Chart Software and User's Manual, Leonard M. Schwab

C/NL: Linear and Nonlinear Microwave Circuit Analysis and Optimization Software and User's Manual, Stephen A. Maas

Capacitance, Inductance, and Crosstalk Analysis, Charles S. Walker

Design of Impedance-Matching Networks for RF and Microwave Amplifiers, Pieter L.D. Abrie

Digital Microwave Receivers, James B. Tsui

Electric Filters, Martin Hasler and Jacques Neirynck

E-Plane Integrated Circuits, P. Bhartia and P. Pramanick, eds.

Feedback Maximization, Boris J. Lurie

Filters with Helical and Folded Helical Resonators, Peter Vizmuller

GaAs FET Principles and Technology, J.V. DiLorenzo and D.D. Khandelwal, eds.

GaAs MESFET Circuit Design, Robert A. Soares, ed.

GASMAP: Gallium Arsenide Model Analysis Program, J. Michael Golio, et al.

Handbook of Microwave Integrated Circuits, Reinmut K. Hoffmann

Handbook for the Mechanical Tolerancing of Waveguide Components, W.B.W. Alison

HEMTs and HBTs: Devices, Fabrication, and Circuits, Fazal Ali, Aditya Gupta, and Inder Bahl, eds.

High Power Microwave Sources, Victor Granatstein and Igor Alexeff, eds.

High-Power Microwaves, James Benford and John Swegle

Introduction to Microwaves, Fred E. Gardiol

Introduction to Computer Methods for Microwave Circuit Analysis and Design, Janusz A. Dobrowolski

Introduction to the Uniform Geometrical Theory of Diffraction, D.A. McNamara, C.W.I. Pistorius, and J.A.G. Malherbe

LOSLIN: Lossy Line Calculation Software and User's Manual, Fred E. Gardiol

Lossy Transmission Lines, Fred E. Gardiol

Low-Angle Microwave Propagation: Physics and Modeling, Adolf Giger

Low Phase Noise Microwave Oscillator Design, Robert G. Rogers

MATCHNET: Microwave Matching Networks Synthesis, Stephen V. Sussman-Fort

Materials Handbook for Hybrid Microelectronics, J.A. King, ed.

Matrix Parameters for Multiconductor Transmission Lines: Software and User's Manual, A.R. Djordjevic, et al.

MIC and MMIC Amplifier and Oscillator Circuit Design, Allen Sweet

Microelectronic Reliability, Volume I: Reliability, Test, and Diagnostics, Edward B. Hakim, ed.

Microelectronic Reliability, Volume II: Integrity Assessment and Assurance, Emiliano Pollino, ed.

Microstrip Lines and Slotlines, D.C. Gupta, R. Garg, and I.J. Bahl

Microwave and RF Component and Subsystem Manufacturing Technology, Heriot-Watt University

Microwave Circulator Design, Douglas K. Linkhart

Microwave Engineers' Handbook: 2 Volume Set, Theodore Saad, ed.

Microwave Materials and Fabrication Techniques, Second Edition, Thomas S. Laverghetta

Microwave MESFETs and HEMTs, J. Michael Golio, et al.

Microwave and Millimeter Wave Heterostructure Transistors and Applicatons, F. Ali, ed.

Microwave and Millimeter Wave Phase Shifters, Volume I: Dielectric and Ferrite Phase Shifters, S. Koul and B. Bhat

Microwave and Millimeter Wave Phase Shifters, Volume II: Semiconductor and Delay Line Phase Shifters, S. Koul and B. Bhat

Microwave Mixers, Second Edition, Stephen Maas

Microwave Transmission Design Data, Theodore Moreno

Microwave Transition Design, Jamal S. Izadian and Shahin M. Izadian

Microwave Transmission Line Couplers, J.A.G. Malherbe

Microwave Tubes, A.S. Gilmour, Jr.

Microwaves: Industrial, Scientific, and Medical Applications, J. Thuery

Microwaves Made Simple: Principles and Applicatons, Stephen W. Cheung, Frederick H. Levien, et al.

MMIC Design: GaAs FETs and HEMTs, Peter H. Ladbrooke

Modern GaAs Processing Techniques, Ralph Williams

Modern Microwave Measurements and Techniques, Thomas S. Laverghetta

Monolithic Microwave Integrated Circuits: Technology and Design, Ravender Goyal, et al.

Nonlinear Microwave Circuits, Stephen A. Maas

Optical Control of Microwave Devices, Rainee N. Simons

Practical Applications of Network Synthesis, Max Medley

PLL: Linear Phase-Locked Loop Control Systems Analysis Software and User's Manual, Eric L. Unruh

Scattering Parameters of Microwave Networks with Multiconductor Transmission Lines: Software and User's Manual, A.R. Djordjevic, et al.

Solid-State Microwave Power Oscillator Design, Eric Holzman and Ralston Robertson

Stripline Circuit Design, Harlan Howe, Jr.

Terrestrial Digital Microwave Communications, Ferdo Ivanek, et al.

Time-Domain Response of Multiconductor Transmission Lines: Software and User's Manual, A.R. Djordjevic, et al.

Transmission Line Design Handbook, Brian C. Waddell

Printed in the United States
43921LVS00004B/41

9 780890 066058